# PRAISE FOR *ELECTRIC BRAIN*

"*The Electric Brain* tells the story behind one of the most basic and important features of the nervous system—its electrical properties. Forget the hype you've heard about 'brainwaves.' Read Doug Fields's facile account of the role of brain electrical activity in mind and behavior."

—Joseph LeDoux, author of *The Deep History of Ourselves: The Four-Billion-Year-Story of How We Got Conscious Brains*

"Brainwaves, not heartbeat, determine the threshold between life and death. They can control drones and in turn be controlled by machines. Doug Fields has done a remarkable job of lifting the veil on these mysterious, fascinating emanations and their potential to change our behavior, influence our personality, and teach us about the core of consciousness itself."

—Florence Williams, author of *The Nature Fix*

"A lively personal account of the wild west of brain wave research, opening multiple windows to deep mysteries underlying our conscious and unconscious selves. Lucid discussions of dreams, mental disorders, mind control, and more provide the reader with a vibrant look into brain science and its impacts on our lives. Strongly recommended."

—Paul L. Nunez, PhD, neuroscientist and author of *The New Science of Consciousness: Exploring the Complexity of Brain, Mind, and Self*

"Douglas Fields explains how electricity is exploited in neuronal circuits for both simple and complex functions. He does it in accessible language, embedded in beautiful prose, where beyond discussing fascinating scientific topics of brain works, the scientists come to the fore. A great book to everyone interested in the most complex matter that exists."

—György Buzsáki, author of *Rhythms of the Brain* and *The Brain from Inside Out*

# ELECTRIC BRAIN

# ELECTRIC BRAIN

## How the New Science of Brainwaves Reads Minds, Tells Us How We Learn, and Helps Us Change for the Better

R. DOUGLAS FIELDS

BENBELLA
BenBella Books, Inc.
Dallas, TX

BenBella

BenBella Books, Inc.
10440 N. Central Expressway, Suite 800
Dallas, TX 75231
www.benbellabooks.com
Send feedback to feedback@benbellabooks.com

*BenBella* is a federally registered trademark.

Printed in the United States of America
10 9 8 7 6 5 4 3 2 1

Library of Congress Control Number: 2019952185.

ISBN 9781946885456 (trade cloth)
ISBN 9781948836296 (electronic)

Editing by Sheila Curry Oakes and Alexa Stevenson
Copyediting by Elizabeth Degenhard
Proofreading by Sarah Vostok and Cape Cod Compositors, Inc.
Indexing by WordCo Indexing Services, Inc.
Text design by Publishers' Design and Production Services, Inc
Cover design by Faceout Studio, Spencer Fuller
Cover image © Shutterstock / Sebestyen Balint;
image editing by Faceout Studio, Spencer Fuller
Printed by Lake Book Manufacturing

Distributed to the trade by Two Rivers Distribution, an Ingram brand
www.tworiversdistribution.com

*Dedicated to my father, Richard L. Fields*

# Contents

## PART III

## Harnessing Brainwave Power

# PREFACE

Blue panel lights blink ominously on rows of black computers sealed inside a glass room. I walk past the supercomputer nerve center and proceed down an empty, dark corridor. Arriving at a door, I open it and enter a cavernous room painted entirely black. Twenty-four digital cameras stationed around the room pinpoint my every move. Two men stand up from behind a bank of computer screens. I know that by using their sophisticated instruments, they can eavesdrop on electrical transmissions flashing through my brain. The most intimate details of my mind are theirs to see. They can read my thoughts before I have them. They can watch my brain learn. They can glean my intelligence, my propensity for adventurism, identify telltale signatures of mental illness and neurological disorders, and predict my ability to learn specific types of material. Am I a good reader? Bad at arithmetic? Prone to depression? Developing early stages of Parkinson's disease? These deeply personal insights into my mind are available to them and other brain scientists who are propelling a revolutionary leap in neuroscience that will transform our world.

I've come to this place to meet these two neuroscientists and experience for myself how they can watch my brain learn by tracking my brainwaves. There is no need to open my skull and stick electrodes into my brain to tap into my neural circuits. Electricity zipping through thousands of neurons deep inside my cerebral cortex creates electromagnetic waves of energy that penetrate my scalp. These electrical discharges can be picked up by touching a wire to my head and feeding the electrical signals from my brain through the wire into a computer. In

*Electric Brain* I will take you to this and many other laboratories around the world to see this exciting new brain science currently unfolding, and to trace back in time to find the roots of this discovery.

The detection of brainwaves in the early twentieth century is one of the most important developments in the history of neuroscience, yet brainwaves and their momentous implications are poorly understood by the general public. Brainwaves aren't taught in school. In fact, most college neuroscience textbooks lack a chapter on brainwaves. This situation has persisted for nearly a century. Yet, learning about brainwaves isn't complicated—anyone who is truly interested can easily understand this new science.

But brainwaves have always been surrounded in controversy. From the moment a reclusive German doctor in the 1920s discovered waves of electricity radiating out of the heads of his patients in a mental hospital, brainwaves sparked astonishment and deep intrigue. His secret experiments revealed that these cryptic electromagnetic emanations from the innermost workings of our mind change with our thoughts and mental state. Today we know that these bioelectric broadcasts expose the most intimate privacies: our conscious thoughts, our unconscious cognition, and the emotions stirring inside our brain.

From the moment scientists first glimpsed brainwaves at the turn of the twentieth century, they were seen as complex and mysterious emanations. The controversy continues, as brainwaves are hotly debated by neuroscientists today. Some dismiss brainwaves as the electrical noise of the brain at work, like the roar of an automobile engine is a by-product of its mechanism of operation. Others believe brainwaves are how the brain functions at its most sophisticated level. Brainwaves, these scientists believe, explain many complex aspects of the human mind that have long mystified philosophers and scientists alike.

While scientific debate rages over the origin and function of brainwaves, no one questions the extraordinary capabilities that can be achieved by monitoring and altering them. Neuroscientists can feed brainwaves into computers to control software, machines, and prosthetic limbs. This melding through brainwaves can fuse man and machine into unparalleled levels of perception, analysis, and problem-solving that far exceed the capability of a digital computer or the human brain working alone. Conscious and even preconscious thoughts can control everything from a motorized wheelchair to a fighter drone. Consciousness and brainwaves are so tightly interwoven that brainwaves, not the heartbeat, now define the threshold between life and death. Psychologists analyzing the pattern of brainwaves in a person's brain as they sit quietly letting their mind wander can see how that individual's brain is wired—normally or abnormally. If the brain's electromagnetic oscillations are atypical, the practitioner can program a computer to signal the patient when their brain's electrical activity shifts in the appropriate way, and given this feedback, the brain will correct its brainwave activity—healing itself without drugs.

Other scientists are using optical and magnetic methods that can read the brain's electrical activity through a person's head, and still others are implanting electrodes and computer chips into people's brains. Fusion of mind and machine means that the flow of information can go both ways. Human brain function can be controlled by computer-generated signals stimulating neurons by delivering electric current or electromagnetic pulses through a person's skull to manipulate brainwaves. Brainwaves of one person can be picked up and directly downloaded through a computer into another person's brain to transmit information from brain to brain as if by telepathy. This once obscure area of science has suddenly exploded into the mainstream as billionaires Elon Musk of Tesla and Mark Zuckerberg of Facebook pour millions of dollars into developing brain-computer interface technology for commercial purposes, which Musk claims will provide a third "digital superintelligence layer" to the human brain. Are such proclamations real or fantasy-fueled marketing? Here are the facts and open questions:

For the first time in human history, *Homo sapiens* has developed the ability to directly interrogate and manipulate the human brain. But with this new ability come many questions both practical and ethical: What are the risks and benefits posed by this noninvasive method of knowing what your particular brain can and cannot do well, and by being able to link your mind to a computer? How will such profound insights into our innate mental capabilities influence education and career choices? For some, the new technology raises Big Brother fears of mind control via radio-controlled beams or brain implants.

All of these profound developments, transforming our future and our understanding of the brain, are possible because the brain, unlike most bodily organs, operates by electricity. How was that remarkable discovery made? Who were the first people to discover brainwaves? What motivated them to pursue the bold idea that electromagnetic waves might propagate out of a person's head? What did they think they had found? How did other scientists react? Why was the discovery of human brainwaves kept secret for years? Why are most people unaware of the name of the person who discovered them? Why was there no Nobel Prize for this work? The answers are found by unraveling a fascinating tangle of science and society that surround the electric brain.

Whenever science crosses the threshold into a new frontier, the strange phenomena will confuse people and spark controversy, but I did not write this book to argue my viewpoint. I wrote *Electric Brain* to share the excitement of science in action, as seen through the eyes of a neuroscientist working in the field; to provide you with the latest scientific research (some of it not yet published) and empower you to form your own opinions. I will, however, separate fact from hyperbole—there is an enormous amount of hype and superficiality in the popular press about brainwaves. I will explain at a scientific level what brainwaves are, what they can do, what we know, and what we do not know—yet. I will take you into laboratories around the world, introduce you to my colleagues who are doing this research, allow you to see the data as I have, and let you learn and draw your own conclusions. That is, after all, what science is—a process of discovery. But first, we need to trace back through time to find the roots of the "electric brain."

## PART I

# Discovering the Electric Brain

# CHAPTER 1

# Broadcasts from the Mind

The fifteen-year-old boy was shot in the head—the circumstances lost to history, but the consequences made history. The bullet, lodged in the boy's brain from an accidental gunshot, could not be removed by surgery, leaving him paralyzed on one side of his body and suffering from vertigo. After surviving the bullet wound with a piece of his skull missing, the patient, sent to a mental hospital in Jena, Germany, was just what his doctor, Hans Berger, was looking for. It was not the now twenty-three-year-old's mental condition that interested Dr. Berger but rather the hole through the man's skull.

## SEEING THROUGH THE SKULL

In eager anticipation, in November 1902 Berger ordered the young patient brought to the laboratory on the ground floor of the psychiatric clinic where he conducted his solitary research. Whether the man believed it to be a treatment of some kind is not known. What is now known is that Dr. Hans Berger, who would become the director of the hospital in 1919, did not regard all human life as equally sacred.

The young man sat in the chair as instructed, his paralyzed arm dangling on one side. Behind him, he felt fingers pinch tufts of his hair and heard the snipping of scissors. He heard water and the hollow rattle of wood against the walls of a ceramic cup; there was a pause and then

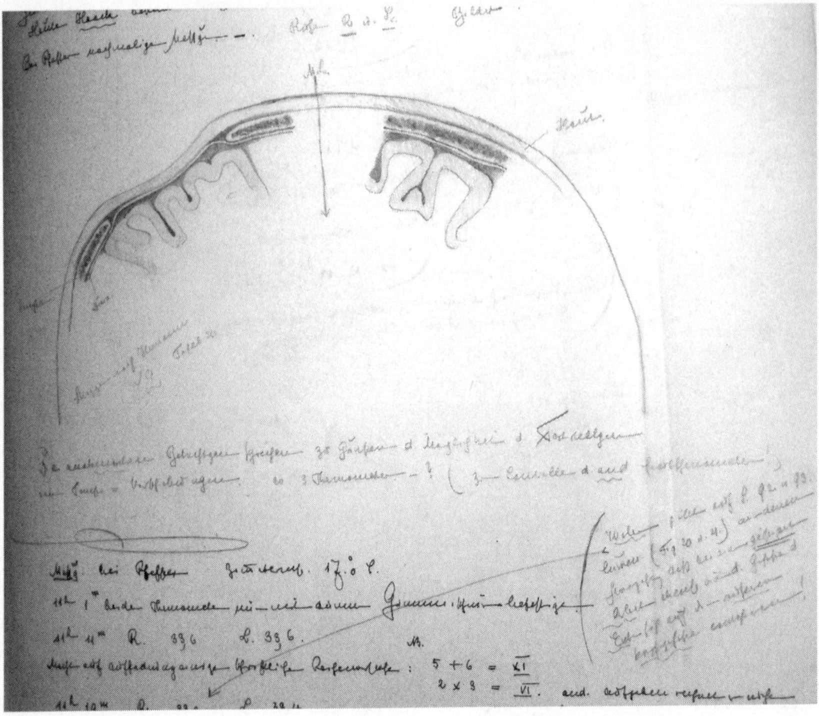


**FIGURE 1:** *From Hans Berger's laboratory notebook from the turn of the twentieth century, recording his experiments on the connection between psychic energy and physical energy. Cranial defects, covered only by a thin layer of skin, in his patients who had lost skull bone enabled Berger to monitor changes in brain volume pulsations in response to sensory stimulation, drugs, postural changes, and emotions.*

the luxurious warmth of a frothy shaving brush lathering his head and dispersing the sharp, spicy smell of soap. With long, slow strokes of a straight razor, the doctor shaved the young man's head and toweled it dry, taking extra care with the area around the head wound, where the missing skull bone had left an irregularly shaped hole covered only by a thin layer of skin and scar tissue.

Over this hole the doctor placed a strange-looking skullcap made of gutta percha—a natural latex rubber—and connected the cap to a fluid-filled rubber tube. The tube resembled a manometer, used for measuring changes in atmospheric pressure, and Dr. Berger was careful to seal it tightly to the fitting in the cap so that there were no leaks. The other end of the tube was connected to a device on the side table that held a sharp metal pen suspended on a pivot. Once the cap and tube were connected, the pen began to scratch out a thin white trace on a piece of soot-covered paper attached to a slowly rotating drum.

Blood, as it pumps through mushy brain tissue, causes the brain to swell and then retract slightly, like the chest of a panting dog, and these brain pulsations, accessible to Berger thanks to the window of missing skull, were what intrigued him. The fluid inside the tube rose and fell with minute pressure changes in the head wound, and the attached recording device traced jagged, regularly spaced undulations on the paper of the drum. Berger stared intensely at the wavy line as it formed in the wake of the shaking pen, tracking the swelling and shrinking of the young man's brain as it pulsed in time with his heartbeat.

Dr. Berger was not providing the young man with medical treatment, and his purpose was not to understand or alleviate any of his mental or neurological dysfunctions; Berger's interest was in testing his theory that mental states interact with physical processes inside the brain. At a time when the origins and mechanics of the human psyche, and mental functions in general, were still mysterious, this was a revolutionary concept.

While the brain's pulsations were regularly paced along with the young man's heart, their amplitude waxed and waned, and Berger believed that these fluctuations were associated with the mind's thoughts and emotions sweeping through brain tissue. He recorded in his notebooks, and later published in a monograph in 1904,[1] that the pulsations were affected by various drugs and changes in body position—sitting, standing, or tilting the head to one side—but more interesting by far to Berger was his observation that the waves also changed in association with conceptual tasks, emotions, and sensory input. He presented the young man, and other patients he subjected to similar experiments, with various stimuli and cognitive challenges: He instructed the patient to stick out his tongue and stimulated his taste buds with drops of sugar, lemon, or quinine. He passed cotton balls saturated with various pungent substances under the young man's nostrils, or had him sniff small vials containing pepper or other spices. Using a fine paintbrush,

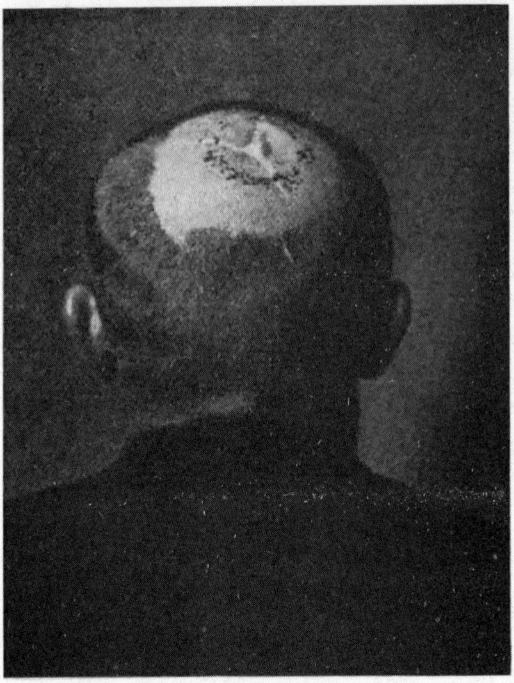

**FIGURE 2:** *One of the patients with a cranial defect who Hans Berger studied using a plethysmograph to monitor fluctuations in brain volume.*

he lightly stroked the back of the man's hand, arm, and cheek. All the while, he watched the line as it appeared on the rotating drum, tracking the swelling and shrinking of the young man's brain with the tide of its blood flow. Often, these stimuli perturbed the regular rhythm of the brain's throbbing, proof to Berger that the elusive psychic phenomena of the mind—such as the experiences of taste, smell, and touch—interacted with biological and physical processes.

Suspecting that human emotions, too, had a basis in the physical, Berger believed that his brain-volume monitoring device (called a plethysmograph) could record these manifestations of joy, sadness, or fear. Human emotion can only be studied in human beings, and Berger was about to use this young man's brain to put his theory to a test. He stepped quietly behind the chair where the man sat . . . and fired a pistol. The deafening report, a sound the man had not heard since the day he was shot in the head, plunged the young victim into a state of shock and absolute terror. Berger was thrilled to see the rapid decrease in brain circulation induced by the mental state of profound fear. He detailed his findings in a monograph published in two parts in 1904 and 1907, titled *Über die Körperlichen Äusserungen Psychischer Zustände* [*Physical Manifestations of Mental States*].[2]

Berger would move on from his research on changes in brain pressure to study the brain's electrical activity. He performed the first human electroencephalogram (EEG), becoming the first person to discover and document waves of electrical energy radiating out from the human skull and pioneering the technology that would give science its initial glimpse into the workings of the mind.

Given the significance of what he uncovered, you would think Berger's name would be a feature of textbooks. Instead, he remains a shadowy figure. Why? After all, the unfortunate truth is that the ethically unpalatable practice of experimenting on patients hardly makes him unique among his contemporaries in the world of psychiatry. So—who was Hans Berger? And how did he arrive at his shocking conclusions?

## A JOURNEY TO JENA

I traveled to the city of Jena, to see firsthand what might remain of Berger's work conducted there a century ago. Rocked by two world wars, subsumed by the Soviets into communist East Germany, and finally restored to a united and democratic German state after the fall of the Berlin Wall, Jena has been tumbled by the most monumental events of the last century. Berger is an enigma, and history, being a written construct, mutates along with political transformations, and the physical records and artifacts essential to reconstructing the true events of the past are all too easily scattered, lost, or destroyed with the passage of time. By seeing Hans Berger's world and work with my own eyes, I hoped to reach a better understanding of the discovery of a fundamental feature of the human brain: brainwaves.

Behind a medieval church of stone blackened with centuries of soot and encased like a Mayan temple by dead, ropy vines, I found a graveyard, little more than a field of green lumps overgrown with ivy. Large trees screened out the sun. There in a clearing, surrounded by ivy and bracken fern, sat a roughhewn granite headstone bearing Berger's name,

**FIGURE 3:** *Hans Berger's grave in Jena, Germany, where he is buried together with his wife, Ursula, and son, Klaus, near the psychiatric hospital where he worked and discovered human brainwaves.*

the two dates bracketing his life etched into the stone and painted black to contrast with the gray of the slab. In the silent stillness of the graveyard, Berger was transformed from a historical abstraction, and was now tangible and real. Here he was, buried beneath my feet, Hans Berger the human being, along with his wife and his son, who died in Russia as a soldier in World War II, just a few months after his father.

I left the grave and walked to the anatomy building of nearby Friedrich Schiller University to meet neuroanatomist Christoph Redies and his colleague, historian Susanne Zimmermann, who had kindly agreed to take me to the psychiatric hospital where Hans Berger worked. After a short walk from the university, we came upon the building, now named after Berger, three stories of yellow and red brick with dormer windows jutting from the steep roof. The building was vacant and undergoing renovation, one of the exterior double doors standing open, and out in front was a bust of the doctor himself. The cold stone suited Berger's formal, stern-looking countenance.

This is where Berger worked for thirty years, searching for the interface between the physical (brain) and the mental (psychic energy). His quest led him from studies of blood flow and brain volume to investigate the stunning new force of energy that was transforming technology and society in the early twentieth century: electricity. Berger began his scientific career at a time of horse-drawn transportation and gas lamps, and ended it in a world of electric light, automobiles, airplanes, radio, and the splitting of the atom. Over this same interval, the study of the human brain shifted from its roots in philosophy and psychology to become the germ of a new science—neuroscience, as it would come to be known—by applying microscopy for cellular analysis, biochemistry for analyzing chemical components, and electronics for investigating the electrical properties of nervous tissue.

In another building at the university, we located Berger's old notebooks. They are written in a peculiar stenographic notation, inscribed with precision, but the code was sometimes impenetrable even to my native German hosts. As I leafed through one of them, the pages yellow and the spine broken, I found evidence of a different approach Berger used to test his hypothesis that mental function interacted with physical and chemical processes. Berger reasoned from the fact that, according to the laws of thermodynamics, the work produced by energy is always accompanied by changes in temperature. From this law of physics, Berger concluded that changes in mental activity (which he called psychic energy) should be associated with changes in temperature in the brain as well.[3] A scene from another experiment is evoked by the sketches, graphed data, and notes recorded in fountain pen in Berger's tiny script.

His subjects were again patients at the hospital, this time two girls, a twelve-year-old suffering from epilepsy and an eleven-year-old with headaches. After a surgeon drilled holes through their skulls, Berger stabbed a rectal thermometer into the girls' brains—21 millimeters, according to a penciled anatomical sketch, or about an inch deep. He then proceeded to perform experiments similar to those he'd conducted with the plethysmograph, presenting the children with various sensory stimuli, this time looking for changes in temperature. He published his results in a monograph in 1910, titled *Untersuchungen über die Temperatur des Gehirns* [*Studies on the Temperature of the Brain*].[4]

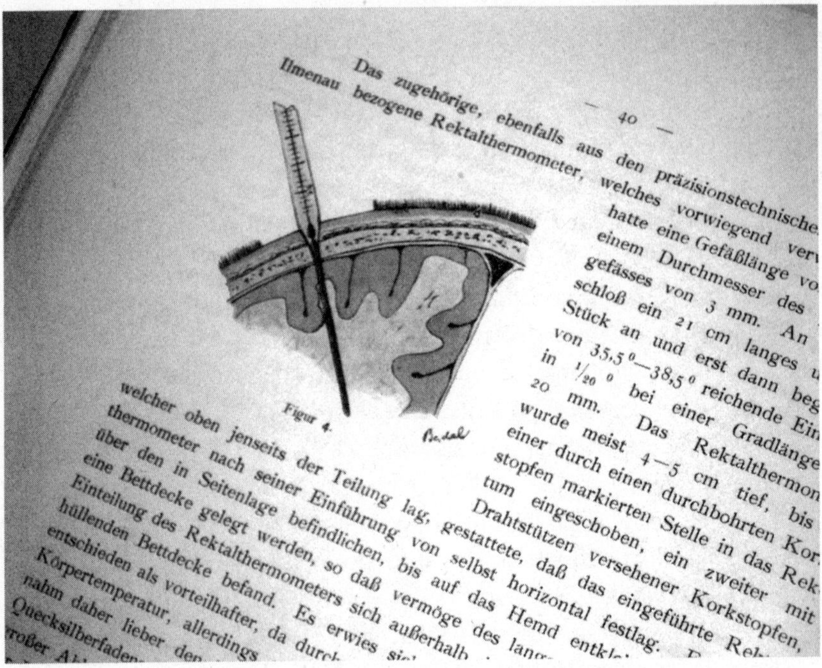
— 40 —
Das zugehörige, ebenfalls aus den präzisionstechnische
Ilmenau bezogene Rektalthermometer, welches vorwiegend ver
hatte eine Gefäßlänge vo
einem Durchmesser des
gefässes von 3 mm. An
schloß ein 21 cm langes u
Stück an und erst dann beg
von 35,5°—38,5° reichende Ein
in 1/20° bei einer Gradlänge
20 mm. Das Rektalthermon
wurde meist 4—5 cm tief, bis
einer durch einen durchbohrten Kor
stopfen markierten Stelle in das Rek
tum eingeschoben, ein zweiter mit
Drahtstützen versehener Korkstopfen,
welcher oben jenseits der Teilung lag, gestattete, daß das eingeführte Re
thermometer nach seiner Einführung von selbst horizontal festlag.
über den in Seitenlage befindlichen, bis auf das Hemd entk
eine Bettdecke gelegt werden, so daß vermöge des lan
Einteilung des Rektalthermometers sich außerhalb
hüllenden Bettdecke befand. Es erwies
entschieden als vorteilhafter, da dur
Körpertemperatur, allerdings
nahm daher lieber den
Quecksilberfaden
Figur 4.

**FIGURE 4:** *Reasoning from the laws of conservation of energy, Hans Berger believed that changes in psychic energy, associated with mental function and emotions, should be accompanied by changes in temperature. To test this theory, he inserted rectal thermometers into the brains of people and at least one monkey, and noted how the brain's temperature changed during mental effort and sensory stimulation.*

On the grounds of the old psychiatry clinic, Christoph Redies and I found the building that had once housed Berger's laboratory. In the early 1920s, working in secret in the basement of this outbuilding, Berger began to assemble the apparatus to record electricity from the human brain. He worked alone, after hours, using mental patients and his own son as experimental subjects. That son was Klaus, whose grave I had just left. Klaus would go on to become a doctor himself before being killed in battle at the age of twenty-nine in November 1941.

The historic laboratory building is now a small library; although arguably the most important discovery in electrophysiology of the last 100 years was made here, it bears no plaque commemorating the place where man first glimpsed brainwaves. But the librarian knew the building's history. Berger's lab had been in the basement, he told us, but no traces of it remain. The librarian showed me some faded photographs of the room filled with Berger's exotic electronic instruments, showing the first EEG subjects with wires attached to their heads, but none of Berger's equipment has been preserved.

In place of the plethysmograph device, which he still used to study brain pulsations, Berger stuck zinc-plated needles under the skin covering the missing skull bone in his patients, so that the tips of the needles touched the surface of the brain. He connected the needles to an instrument in an attempt to detect electricity, but the signals he extracted were weak and erratic. It was unclear to him whether he was seeing electricity flowing through the brain or instead electrical noise caused

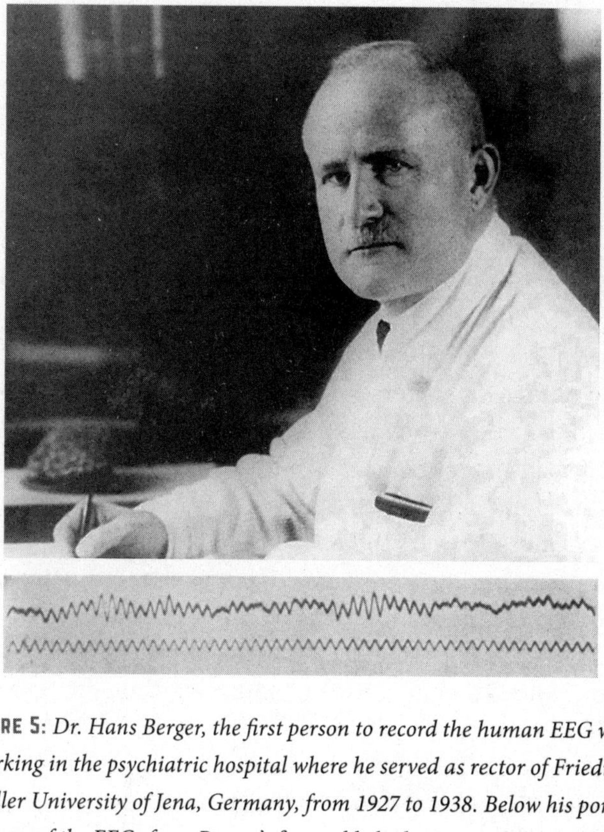

**FIGURE 5:** *Dr. Hans Berger, the first person to record the human EEG while working in the psychiatric hospital where he served as rector of Friedrich Schiller University of Jena, Germany, from 1927 to 1938. Below his portrait is one of the EEGs from Berger's first published paper on human EEGs.*

by the brain's pulsation, heartbeat, and subtle movement of muscle. He recorded signals that seemed to waver independently of the patient's respiration from several other patients' brains, including a forty-year-old man, but Berger was accessing the man's brain through the hole in his skull left from surgery to remove a brain tumor five months earlier. The man died a few weeks after Berger's experiment, raising doubts about whether this man's brain, swelled from disease, could be considered normal.

Over time his instruments and techniques improved. Inside this building one evening in 1924, Berger attached two large foil electrodes to twelve-year-old Klaus, one on the back of his scalp and another on his forehead, and fed the feeble electrical signals into a device, called a galvanometer, that recorded fluctuations in voltage by means of a light beam dancing on a strip of photographic paper scrolled through the device. As Klaus sat with his eyes closed, the beam plotted waves of electricity as they radiated from his brain, oscillating slowly at low frequency. When Berger asked his son to open his eyes, however, the brainwaves suddenly changed, vibrating erratically at high frequency. Berger's extensive experiments—on patients, his son, and even himself—showed that these oscillating waves of electricity changed with mental activity, arousal, attention, and sensory stimulation, and that they became perturbed in diseases such as epilepsy. It followed that these waves might not only give insight into how the brain operates and interfaces with the mind, but also enable doctors to identify brain disorders, mental aptitudes, and personality traits. But Hans Berger was hoping that brainwaves, like broadcasts of radio waves, might do even more.

The philosophy of Dualism regards the mind and body as separate and distinct from each other. Dualists believe that mental phenomena such as our thoughts and emotions are nonphysical in nature, but that the psyche, spirit, or mind interacts with the physical world. For most scientific Dualists of the time this simply meant that physical stimuli could produce mental events, while emotions or thoughts arising in the mind could in turn be located in effects on the brain. Berger, however, describes the mind as a force—a "psychic energy" that can interact with physical matter. He reasoned that by the laws of conservation of energy, changes in psychic energy producing mental work would necessarily require changes in other forms of energy in the brain, such as temperature and electricity. He believed that the waves of electricity he detected in the human brain were the rippling energetic reverberations transforming between psychic and physical energy. Stunningly, his discovery proved that the human brain's energy propagated through the skull and could be detected remotely.

In the building that had once housed Berger's laboratory, the librarian retrieved a monograph entitled *Psyche*, published by Berger in 1940.[5] The title page bears an inscription in his distinctive tiny script, nearly incomprehensible except for the ornate large capital letters initiating each sentence, then vanishing into a thin squiggle. In this paper, Berger relates his belief in mental telepathy and psychic energy. He describes an experience he had in the spring of 1893 while serving as a volunteer in the German army, in which he believed himself to have communicated telepathically with his sister. During a training exercise, Berger's horse suddenly reared and he was thrown into the path of a horse-drawn cannon. The driver of the artillery battery halted it just in time, leaving the young Berger shaken but with no serious injuries.[6] His sister, at home far away, had at that same moment a sudden strong feeling that Berger was in danger. She insisted that her father send him a telegram at once. The incident made such an impression on Berger that it changed the course of his career—after the war, he abandoned his plan to study astronomy for medical school as a result.[7] In this monograph years later, he wrote, "It was a case of spontaneous telepathy in which at a time of mortal danger, and as I contemplated certain death, I transmitted my thoughts, while my sister, who was particularly close to me, acted as the receiver."[8]

Even in his day, Berger's research at the fringe of paranormal psychology would have been viewed with skepticism by his scientific contemporaries, and it is almost certainly one of several reasons his work remained on the periphery. Nevertheless, his discovery that it is possible to receive the electrical waves of energy radiating out of the brain by placing electrodes on a person's head proved the interaction between the energy of the mind and the substance of the brain. Working alone in an isolated psychiatric clinic, Berger shattered the truism that one never really knows what is going on inside another person's mind. For the first time in history, it was possible to directly monitor cognitive

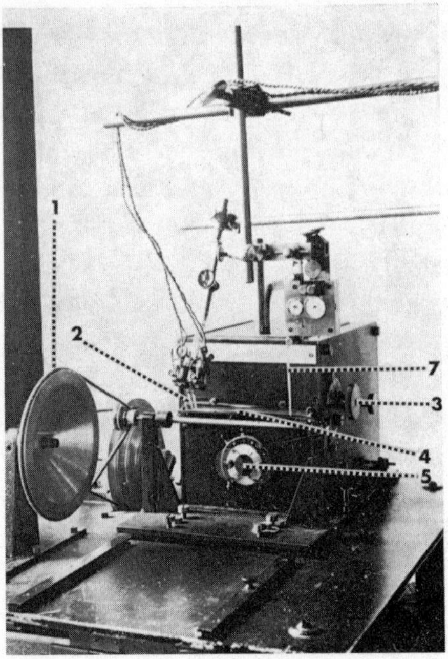


**FIGURE 6**: *The Edelmann string galvanometer first used by Hans Berger to record brainwaves from patients with trepanations (holes bored through the skull for medical purposes) and in people who had suffered injuries to their skull, leaving a portion of the brain covered only by a thin scar of skin. The oscillating string in the galvanometer was focused through a lens (2) onto a roll of silver bromide paper (4) that was rolled through the device by the belt drive motor (1) and developed as a photograph. Timing marks were made by a tuning fork (7) vibrating at 10 Hz and sustained by a clockwork mechanism driven by a windup spring. Later instruments had a tiny mirror attached to the string to bounce a beam of light onto the photosensitive paper that streamed past.*

and emotional states in the human brain. With this achievement, the feasibility of transmitting thoughts over a distance via brainwaves was inescapable.

Berger first recorded human brainwaves in his basement laboratory in Jena in 1924, but he told no one what he was doing; indeed, he shared his stunning results with none of his colleagues until 1929, when he published his first report on brainwaves in the *Archiv für Psychiatrie*, titling it "Über das Elektrenkephalogramm des Menschen" ["On the Electroencephalogram of Man"].[9] Few scientists took note. Researchers were making great strides in understanding the cellular and physiological operation of the brain by taking a reductionist approach—examining brain cells under a microscope, mapping in detail parts of the brain involved in different functions, determining how individual nerve cells generated and transmitted electrical impulses; Berger's approach of recording activity from the whole head at once made little sense to leading scientists of the time. His work seemed bizarre; most dismissed the noisy voltage fluctuations picked up by the scalp electrodes as originating in electrical disturbances from muscle, not the brain. (When muscles contract they generate electrical discharges that are much greater voltage than that generated by neurons; muscles are more massive than nerves and are directly under the skin where the sensors are attached.)

Berger's publications established the priority of his discovery, but he shared his work as narrowly as possible, guarding rather than promoting the fruits of his research. "During the several semesters I had been his lecture assistant I never knew him to speak during his course on his own field of research," said Rudolf Lemke, who joined the clinic as an assistant in 1931.[10] "Only in 1931 did I hear him give a paper at a meeting of the Medical Society at Jena on his discovery of the EEG," recalled Lemke. "I can well remember that the interest amongst his colleagues who listened to him was not very great."[11]

Berger's sensational findings did not escape the popular press, however, as indicated in this newspaper article published in the *Baltimore Sun* on January 4, 1931:

> [Hans Berger has measured] by means of a machine which he has invented to register the electrical energy set free in the brain in the course of mental activity. Similar attempts were made some years ago by an Italian physicist, Dr. Ferdinando Cezzemelli [*sic*], using a radio receiver of special design to pick up what were believed to be electric waves from the brain; a real example of a "brain wave." Difficulty was encountered, however, in sorting out from these waves of "brain radio" other impulses generated by the flow of the blood, the beating of the

> heart, the action of muscles and other vital activities. These difficulties Professor Berger believes that he has overcome by a special apparatus using electrodes attached to the body to pick up the supposed brain impulses instead of depending on accompanying radio waves.[12]

According to an article by Cyril Burtt [*sic*] in the book *Science and ESP* by J. R. Smythies, an Italian neurologist, M. F. Cazzamali [*sic*], recorded "electro-magnetic waves" conveying information directly from one brain to another. He states that Berger believed the brainwaves he had discovered could be the mode of telepathic communication, and he performed experiments on brain-to-brain transmission of brainwaves, but he failed to obtain experimental proof that such transmission by electromagnetic radiation occurred. He therefore concluded that telepathic communication took place by the transmission of psychic energy.[13] (Note that Cyril Burt's name is misspelled in print throughout Smythies's book as Burtt, and Cazzamalli's name is misspelled as Cazzamali.)

Interest in mental telepathy, and other forms of ESP such as clairvoyance and premonition, peaked in the 1920s and 1930s. Upton Sinclair, renowned Pulitzer Prize–winning author of *The Jungle*, was a firm believer. In 1930 he self-published a book, *Mental Radio*, detailing his extensive experiments on mental telepathy, most of them carried out with his wife. Radio was the rage of the Roaring Twenties, and the emerging science of radio waves and brainwaves converged in a flash of excitement that crashed as abruptly as the soaring stock market in 1929.

Ferdinando Cazzamalli was a fellow at the neuropsychiatric clinic at the University of Rome working in the 1920s and 1930s at the same time Berger was studying brainwaves in Jena, but Cazzamalli's studies were focused on metaphysics and telepathy. Instead of electrodes attached to the scalp, Cazzamalli positioned a radio antenna in the vicinity of a person's head to monitor the feeble brainwave broadcasts.[14] Enlisting the assistance of noted radio engineer Eugenio Gresetta to build the sophisticated electronic apparatus, Cazzamalli pursued his "mental radio" research for ten years, first publishing his findings in 1925. The bulk of his research was published in Italian,[15] and one paper in 1935 was published in French,[16] so Cazzamalli's research remained obscure to mainstream scientists who did not read those languages or have any interest in the paranormal.

Cazzamalli selected subjects whom he believed would be especially strong in mental telepathy: artists and musicians. One such subject was a painter who was also an Alpine mountaineer. The man was escorted into an unfamiliar room that had walls lined with sheets of lead and was illuminated only by the eerie dim glow of a red light bulb. The lead walls were necessary to screen out electromagnetic interference from man-made radio waves and alternating currents, so that the feeble brainwave broadcasts might be intercepted.

In a corner of the room was a metal chassis containing an electronic amplifier, a rectifier, and an oscillator tuned to a frequency of 300,000 kilocycles—all the components of a radio receiver. There was a couch next to one wall, and the painter was asked to lie down and relax. Cazzamalli positioned a dipole antenna 70 centimeters (about 2.5 feet) above the man's head and fed the signal through a wire into the radio receiver. The output of the oscillator was recorded on film.

Cazzamalli asked the painter to close his eyes and clear his mind. He recorded the signals for several minutes as the man remained in a passive state. The painter's mountaineering experiences had extended well beyond Italy and into the Andes where tragedy had struck. Exploiting this, Cazzamalli asked the man to recall the traumatic experience the mountaineer had suffered on Mount Tronador in Patagonia searching for the bodies of fellow climbers who perished on the treacherous mountain. Suddenly, the beam of light tracing out a steady streak on the streaming photographic film was abruptly interrupted. After a moment, the trace resumed. Cazzamalli concluded that all of these emotionally charged thoughts radiated out of the subject's brain into the radio receiver and interrupted the signals—clear proof of a physical basis for mental telepathy.

You will not find a single citation for Cazzamalli's research in the PubMed record of scientific publication, but while Cazzamalli is essentially unknown to neuroscientists, in Italy his reputation is sustained by paranormal researchers. In 1937, together with Giovanni Schepis and Emilio Servadio, Ferdinando Cazzamalli founded the Italian Society of

Metaphysics to support experimental research on paranormal phenomena.[17] The society continues today, and it publishes the journal *Metapsychic, The Italian Journal of Parapsychology*, which was founded by Cazzamalli in 1946. The publication is not sold, but it is freely distributed among members of the society.[18]

Lest one dismiss this esoteric episode as an historical anomaly, it is important to recognize that science, in exploring the unknown, is always caught up in a torrent of clear and murky currents, pulled into eddies along the way, and at times mixing science and pseudoscience. At the Society's annual meeting held in Bologna in December 2018, one of the speakers, William Giroldini, presented research on mental telepathy and EEG.[19]

Alas, then as it is now, the popularization and sensationalizing of scientific research can have a corrosive effect on how the work itself is perceived within the scientific community. In a German newspaper article published in July 1930, the potential of Berger's discovery was touted to the public with still more enthusiastic hyperbole: "Today, the brain still writes secret signs. Tomorrow, we will probably be able to read neurologic and psychiatric diseases in it. And the day following tomorrow, we will start to write our first honest letters in brain script."[20]

Such an achievement is a persistent dream. As I write this in 2019, billionaires Mark Zuckerberg, founder of Facebook, and Elon Musk, founder of the electric car company Tesla, are reportedly investing in methods of extracting thoughts and emotions from a person's brainwaves and transmitting them over the internet, directly downloading them into other people's brains, and using them to operate electronic devices and computers by thought alone.[21] In part III of this book we will visit researchers around the world pursuing this type of "mind reading" and "telepathic" communication through brainwaves, to investigate the reality behind the reports you may have read in the popular press, which is no less prone to sensationalism now than it was in Berger's time.

Over nearly a decade, from 1929 until 1938, Berger published one or more papers a year on his research, all fourteen with the same title as the first, "Über das Elektrenkephalogram des Menschen" ["On the Electroencephalogram of Man"], and all in the same journal: *Archiv. für Psychiat. Nervenkr.* This peculiar practice served to further shroud the results of his experiments. The titles scientists give to their publications are typically very descriptive of the specific new finding being reported, but no clue of what Berger had discovered in any of his experiments can be gleaned from the title of this series of publications. Any scientist who might be interested in what Berger had discovered in a particular experiment would have to dig through fourteen identically titled papers, as if to find a prize hidden behind one of not three, but fourteen, identical doors. It is also difficult to reference any of Berger's specific findings, as the citations for these fourteen papers differ only by the year of publication. Cloaking his findings in this way hid them from the larger scientific community and diminished their impact.

Having made the first recording of a human EEG in 1924, Berger's work remained almost entirely unknown until 1934, when Nobel Prize winner Edgar Douglas Adrian drew attention to the phenomenon of electric brain activity by repeating Berger's experiments in a paper coauthored with Bryan H. C. Matthews and published in a prominent scientific journal, *Brain*.[22] In that paper, Adrian and Matthews implicitly mocked the significance of what they called "Berger waves" by including a figure comparing changes in Adrian's own brain EEG when he opened and closed his eyes with the EEG of a water beetle doing the same—evoking identical brainwave responses in the insect and the Nobel laureate.[23] Adrian did not pursue studies of EEG afterward, and interest in Berger's EEG recordings within the medical field remained tepid.

"The lectures which Berger gave each year on his researches [*sic*] to the Medical Society of Jena found no greater appreciation as time went on," noted Rudolf Lemke. "In 1934, at the meeting of German Neurologists and Psychiatrists he reported on the EEG at Munster, and there too he did not arouse the interest he had expected."[24] In the final line of her excellent scholarly book published in 1961, *A History of the Electrical Activity of the Brain: The First Half-Century*, neuroscientist and historian Mary Brazier wrote, "Berger's contribution to the first fifty years [of research on electrical activity in the human brain] was known only to himself: the successful recording of the electroencephalogram of man."[25]

Leaving the place where Berger made his monumental discovery, I followed Christoph Redies to the basement of another library on campus that had just received an uncataloged collection of Hans Berger's letters, photos, and notes that had been saved by one of his former colleagues. I picked up a photograph and held the actual recordings of one of the first human EEGs—the same recording that had been held in the hand of Hans Berger nearly a century ago.

Berger worked for almost thirty years in nearly complete isolation in this psychiatric clinic a short walk from the cemetery where his body now rests; he hardly seems the most likely candidate for such a scientific breakthrough. Why was no one else at the time apparently pursuing this line of research?[26] What led Berger to perform his strange experiments and set him on the path to discovering the electric brain?

Tracing the source of Berger's inspiration would take me to the mountains of Turin, Italy, where in the 1880s, nearly two decades before Berger's experiments, Angelo Mosso was the first person to record human brain activity. But the activity Mosso recorded was not electrical; he developed the plethysmographic method later used by Berger, using it to monitor how blood flow in the brain varied in response to exertion, changes in respiration, and, among other things, adaptation to high altitude in the mountains surrounding Turin. Berger's early studies on brain volume changes were directly adapted from Mosso's methods and research, and Berger credits Mosso extensively in his first publication in 1901.[27] Today, functional MRI, imaging of the brain at work, uses changes in blood flow in localized regions of the brain to pinpoint where specific neural processes are taking place. It too can trace its origins directly to Angelo Mosso, who pursued similar investigations using only the primitive methods available to nineteenth-century scientists.

My route to Mosso's laboratory in Turin was a circuitous one.

## SWELLED BRAINS

Dodging a frigid rain in the coastal village of Alassio, Italy, I escape to the Caffé Roma, where Ernest Hemingway once scrawled his name on the wall and started an infectious custom I could not resist. The previous day I'd been in a hospital in Zaragoza, Spain, watching the ceiling slip away as I slid into the chamber of an MRI machine, my mother's scolding whine from weeks earlier echoing in my head: "Why would you want to damage your brain? You have such a good brain."

**FIGURE 7**: *Angelo Mosso, 1846–1910, an Italian scientist and one of the first people to study brain function quantitatively with instruments. He used a plethysmographic method to measure fluctuations in brain volume in patients who had cranial defects, which inspired Hans Berger's research. Of particular research interest to Mosso was how the brain is affected by high altitude, which he studied in his laboratory built on Monte Rossa, Turin, Italy.*

A good question.

It all started when I read a paper by a pair of Spanish researchers, radiologist Dr. Nicolás Fayed and neurologist Dr. Pedro Modrego.[28] In a study of thirteen mountain climbers on an expedition to Mount Everest, they found that all but one returned from the summit with physical brain damage, as clearly visible on an MRI as a bone fracture on an X-ray. The damage was permanent.

Mountaineers are well aware that the oxygen-depleted air at high altitude can cause serious sickness—and in the extreme, brain damage and death—from what is known as high-altitude cerebral edema (HACE). The body's response to the lack of oxygen increases the pressure inside blood capillaries, causing fluid to seep out of them into surrounding brain tissue, bloating the brain and crushing the life out of gray matter as it is compressed against the skull. This phenomenon is precisely what Angelo Mosso was investigating in the work that inspired Berger. Attaching his manometer device to the heads of people with cranial defects, Mosso monitored the swelling of their brains as they climbed the 15,000 feet of Monte Rosa, where he had built a mountaintop laboratory to study altitude sickness and the physiological mechanisms of adapting to high-altitude conditions.

The lasting cognitive effects of altitude on some veteran alpinists are well known among climbers—"spacey" is the usual description—but most high-altitude mountaineers believe that if one acclimatizes oneself properly, severe altitude sickness and brain injury can be avoided. In fact, the first reports of altitude illness are from monks accompanying the Spanish conquistadors advancing high into the Andes in the 1500s. The astute monks noticed that while the generals were stricken with altitude sickness, the soldiers under their command were spared. The reason, they soon realized, is that the generals, who were riding on horseback, ascended the mountains rapidly, while the soldiers climbed on foot at a much slower pace. (Monks, in addition to serving God, were both historians and brewmasters; all of their activities have left lasting impacts.)

The process of acclimatization requires ascending slowly to give the body sufficient time to adapt to the drop in oxygen at higher altitudes, which involves a series of fascinating and complicated physiological changes. Acclimatization takes time, and weekend warriors paying guides to help them bag a summit often pay a price for their impatience by becoming very sick. While the subjects of Fayed's and Modrego's paper had experienced the milder reactions to high altitude that virtually all climbers endure—headache, nausea, fatigue—none had exhibited the more severe symptoms that are the hallmark of HACE. In short, they weren't aware that they had injured their brains at all, much less permanently. The Spanish researchers then studied climbers scaling other popular mountains of lesser height, including Aconcagua, Kilimanjaro, and Mont Blanc, and found the same damage visible on an MRI, although it appeared with less frequency in mountaineers who had ascended these lesser peaks. Again, none of the climbers had suffered symptoms of HACE; nevertheless, their brains had swelled enough at high altitude to cause injury.

Being both a neuroscientist and a climbing enthusiast, I found Fayed's and Modrego's paper especially fascinating, and it occurred to me that the subject would make an excellent article for the general reader interested in mountaineering's effects on the brain. I approached an editor at *Outside* magazine and proposed that I climb Mount Rainier, "America's Mont Blanc," taking extra care to acclimatize myself, and have an MRI before and after to prove that mountaineering could be done safely. They loved the pitch and commissioned the article.[29]

That's when I encountered a roadblock: I could find no one who would agree to examine my brain by MRI for this project. It didn't matter whether I paid cash, or if the venue was a medical clinic or an experimental facility at a university: Apparently, there is some ethical problem with putting a person in a position where they might injure themselves just because it would be interesting to study the damage.

I was stuck—until I thought to contact Dr. Fayed, who had done the original study on climbers' brains. "Sure!" he responded agreeably. "Come on over; we'll give you an MRI after the clinic closes and all go out for dinner."

So that's what I decided to do. I wouldn't have a "before" picture, but at the very least I'd see whether my years of mountaineering had damaged my brain. And en route to Zaragoza, I would go to Turin, Italy, to see the laboratory where Angelo Mosso had done the research on brain swelling that had been the inspiration for Hans Berger. But first, I had a mountain to climb.

## CLIMBING MOUNT RAINIER

I set off for a rendezvous with my climbing partner and son, Dylan. Attempted by hundreds of amateur climbers in a season, Mount Rainier is a perfect setup for altitude illness. The glacier-shrouded active volcano rises steeply to an elevation of 14,410 feet from sea level in Seattle, Washington, where most climbers begin their ascent. Like Mont Blanc, Mount Rainier is often attempted in a weekend push that is far too short to allow the body to acclimatize properly to the thin air at the peak. Dylan and I, however, would take our time.

It was the end of climbing season, and the guided groups and most other climbers on Mount Rainier had bailed in advance of a ferocious approaching stormfront. Dylan and I decided to head up into the weather despite whiteout conditions, hoping to wait out the worst of it in our tent and make for the summit after the storm had passed. I carried a pulse oximeter to monitor the oxygen saturation of my blood and my heart rate. On one particularly strenuous part of the climb, my blood oxygen plummeted to 75 percent, while my heart rate rocketed to 165 beats per minute. (High heart rate is one of the mechanisms by which the body tries to make up for reduced oxygen supply.) I've seen my blood oxygen drop lower, while climbing the 19,347-foot Cotopaxi in Ecuador, and climbers on Everest have registered numbers in the 50s. At normal elevations, people with that low of a level of oxygen saturation are likely found only in the intensive care unit, and, from a medical viewpoint, anything below 90 percent would be cause for some concern. But at high altitude, changes in how tightly the blood's hemoglobin holds on to oxygen (among other things) lets us get by on less, demonstrating what a nimble problem solver the human body can be if given time.

Of course, all the physiological calibration in the world can't eliminate some of the risks of climbing. On the way back down, I snagged a crampon blade on the icy slope and took an invigorating header. When you hear about mountaineers falling to their deaths, this is usually how it happens: Unable to stop their screaming slide down the icy slope, the climber is rocketed over a cliff into space, shot into the abyss of a crevasse (I *hate* crevasses), or stopped, bug on a windshield style, by a rock outcropping. Careening down the peak headfirst, I knew I would only get one chance to sink my ice ax into the surface and break my fall as my acceleration built exponentially (32 feet per second squared, as every scientist knows). Once you get going fast enough, there is nothing you can do—your ax will skitter off the icy surface, made as impenetrable as cold concrete at 60 miles per hour, and your partner can only watch you sail by. That is, unless you are roped together, as Dylan and I were at that moment, in which case your buddy can try to use his own ax to brace his end of the rope and yank you to a gut-wrenching halt . . . or, if that fails, join you in tumbling toward death. So, appreciating the situation as I accelerated rapidly, I jackknifed quickly into the proper position (head upslope) and swung once, hard, with deliberation and purpose. My ice ax bit. Grasping the head of the ax with my left hand, I rammed my right shoulder down onto it, driving the blade into the snow and gouging a furrow down the mountain, slowing and finally arresting my death slide just before bowling over Dylan, who was poised like Paul Bunyan with his ax at the ready.

It was a good reminder. Hypoxia is bad, but gravity sucks.

After returning to sea level at the Seattle waterfront, the oxygen-rich air felt thick as cream, as if I could swim through it frogman-style; supercharged to operate in the thin air of the mountain, my strength and energy felt superhuman. This feeling will only last a couple of days, but it is a profound sensation. It's easy to see why, living among the mountains of Turin, Mosso might have been moved to study altitude's effects.

Climb completed, I was off to Italy to see where Mosso worked . . . and then to Spain, to get my head examined.

## THE CLIMBER'S BRAIN

Dr. Gianni Losano is waiting for me outside the Instituto de Fisiologia Umana, a complex of nineteenth-century stone buildings at the University of Turin. Losano, a gentle lion of a man, now works at the same desk Angelo Mosso once did, a massive oak and leather throne at the center of a spacious office with twenty-foot-high vaulted ceilings and a parquet wood floor. It's Saturday, and the buildings are empty as he takes me to meet his colleague, Dario Cantino, an expert on Mosso who is engaged in translating his work from Italian to English and making it available on the internet.

A scientist and mountain climber himself, Cantino is a lean sprite of a man with an exuberantly cheerful disposition. He is also an eclectic collector of scientific and mechanical apparatus. He takes me through room after room: In one, he has collected electron microscopes—from the earliest devices of the 1950s, which resemble something out of an old Boris Karloff movie, to recent models—and in another, printing presses of every design are stashed to the ceiling, with trays of lead letters spilling their contents. Cantino's collection of printing presses has grown so large that he has no place to store them all; some are kept outside covered with tarps to protect them from the elements.

"Why printing presses?" I ask him. Cantino thinks a minute before offering that perhaps it is because printing presses translate and record information and ideas, which is what he does himself, both as a scientist translating data into insight and in translating Mosso's work into

English. Cantino walks over to an enormous press and selects a single letter the size of my hand to give me as a souvenir: a question mark—a most appropriate selection for one scientist to give another. (Equally appropriate as a punctuation mark to my present quest, my mom would have thought.)

Next, Losano and Cantino take me to the library, where Angelo Mosso's notebooks and instruments are kept neatly organized and cataloged in a room open only to scholars. Smoked recording drums that turned by brass clockwork gears and devices to measure changes in blood pressure and brain volume were displayed on tables or housed in small cubbies. There are photographs of subjects having their brain swelling monitored and of Mosso's various other high-altitude experiments. My fingers become lightly soiled with soot as I hold in my hands a strip of blackened paper, covered with the delicate oscillating tracings that were the first recordings of human brain activity. I am mindful

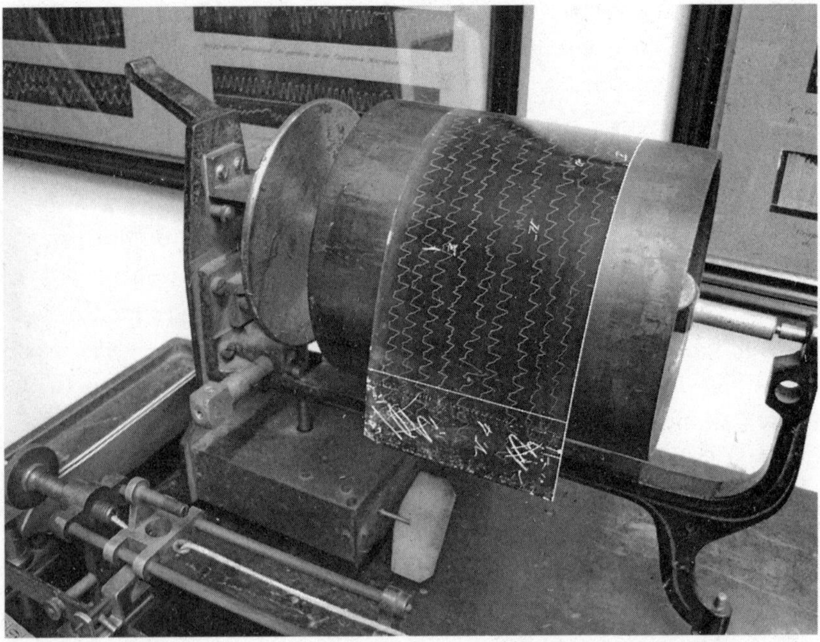

**FIGURE 8:** *Apparatus used by Angelo Mosso to study the effects of high altitude on the brain and body. Fluctuations in brain volume, detected by a plethysmograph, were recorded by a pen scratching soot off paper on a slowly rotating drum.*

that I have now had the privilege of holding both of the first two such records ever made—one by monitoring blood flow and the other electrical signals—each bearing the fingerprints of the scientist who made the recording a century ago.

From Turin, I drive all day and most of the night to reach Zaragoza, Spain, after taking a detour through Switzerland (I think) when I make a wrong turn somewhere in France. Dr. Nicolás Fayed, radiologist at the Clínica Quirón de Zaragoza, is tall and pewter-haired, with a casual manner and friendly sense of humor. He asks about my climb as his colleague, Dr. Jaime Medrano, translates our exchange. "You have to be *poco loco* [a little crazy] to be a climber," Fayed says with a smile, implying that this is not entirely a bad thing.

We enter Dr. Fayed's office adjacent to the MRI control room, and he begins pulling out files; the desk is soon covered with scans of climbers' brains.

"Atrophy of the frontal lobes," he says, pointing to the black-and-white

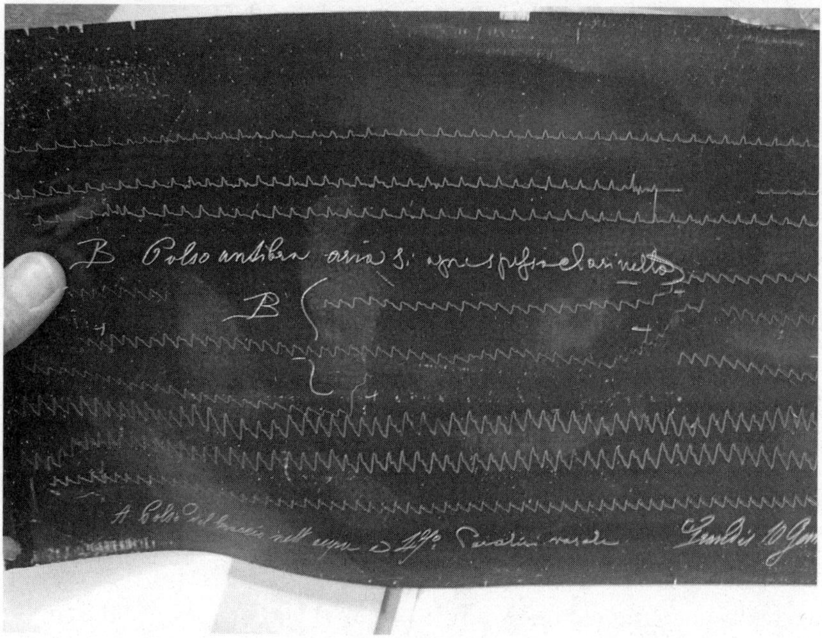

**FIGURE 9**: *The author holds a recording of physiological responses recorded by Angelo Mosso using a plethysmograph.*

slice of brain on the film. The forebrain is located behind the forehead; in this scan it is shriveled like a dried fruit. This is the brain region severed in a prefrontal lobotomy, a surgery that leaves patients intellectually unimpaired but with deficits in higher-level executive functions like focus and planning—in other words, "spacey." I have scrutinized similar images with clinical interest in published papers, but suddenly this is different. Each of the images is labeled with a climber's name.

"José suffered the most serious damage," says Fayed. He hands me a picture of a robust young climber in red Gore-Tex standing on the snowy slopes of Aconcagua, with windblown dark hair and a whiskered beard, bronzed by the high-altitude sun. He looks fit and determined, someone I would enjoy roping up with. "When José came back, he couldn't remember his own phone number. His wife would send him to the store for a loaf of bread and he would forget why he was there and come home without it."

"See the lesions in the forebrain?" he asks, pointing at another scan. Lesions like these are caused by small strokes or hemorrhages. They show up as bright spots on an MRI throughout the brains of climbers but are especially common in the white matter regions. White matter is like the core of a baseball, composed of millions of tightly bundled cables that connect neurons in the surface layers of the brain together into networks. Damage to white matter anywhere in the brain can have severe and wide-ranging consequences, just as when a backhoe digs up power cables in one yard and causes extensive blackouts throughout the city. White matter seems to be particularly vulnerable to hypoxia because it is not as richly supplied with blood capillaries.

Eleven of the climbers scanned had enlarged Virchow-Robin (VR) spaces, caused by swelling and loss of brain tissue. You don't need a medical degree to see these: They look like white shotgun pellets lodged deep in the brain. VR spaces are the areas that surround blood vessels in the brain; we all have them, and as our brains age, the spaces widen. However, while multiple or much-enlarged VR spaces are common in the brains of the elderly and those with Alzheimer's disease, they are not normally encountered in healthy people in their twenties and thirties—the ages of these climbers.

Next I meet Coral Roya, at the controls of the MRI machine. To my surprise she reveals that she was one of the test subjects on the Kilimanjaro climb (5,895 meters, or 19,340 feet, above sea level). As is typical, all seven climbers on that climb experienced mild symptoms of mountain sickness, but no one suffered HACE or other serious altitude illness. They all had MRIs before the climb, which showed no abnormalities, but one of them returned with dilated VR spaces in his brain. On Mont Blanc (4,810 meters, or 15,781 feet, above sea level), one of the seven amateur climbers returned with a subcortical lesion, and two had multiple dilated VR spaces.

I ask how many of the climbers in the studies, most of them physicians and engineers, stopped climbing after seeing the damage to their brains. "They are all still climbing," Fayed tells me.

"Our purpose is not to stop climbing," Dr. Modrego emphasizes. "It is to make people aware of the dangers and the need to acclimatize properly."

Now it is my turn—the moment of truth. I strip to my underwear and don green paper booties and a powder-blue hospital robe. The costume makes me feel instantly infirm.

A smiling assistant, Elisabeth Pérez, hands me a pair of sticky pink wax plugs to stuff into my ears to blunt the noise of the MRI machine as it takes pictures of my brain, slice by virtual slice. I lie down rigid as a mummified pharaoh on the narrow plank sticking out of a monstrous machine that looks like an industrial clothes dryer on steroids. She adjusts the position of my head with reference to a luminous grid projected onto my skull. Then she lowers what looks like a catcher's mask over my face and activates the machine, which feeds me headfirst into a tunnel. If this were a James Bond movie, now would be the time to break out one of Q's inventions, grab Ms. Pérez by the waist, and make my escape, but I brave the claustrophobic confines and try not to move a muscle so as not to blur the photos.

Then the racket starts. Even with the earplugs, it is deafening. It sounds like a massive construction job is being conducted inside my skull, with hammering and buzz saws punctuated by bizarre video-game sound effects. Suddenly it occurs to me that maybe this isn't such a great idea. What are they seeing as they virtually slice my brain like thin prosciutto? To hell with climbing scars—what if they find a freaking brain tumor? My mind taunts me with the unforgettable image of Homer Simpson's head X-ray, a walnut-size brain lost inside his cavernous cranium.

I can see the doctors and technicians framed between my green booties through a mirror suspended above my eyes. The mirror is intended to relieve the stress of patients reluctant to bury their head in a machine that from my current perspective suddenly resembles a guillotine. They are gathered behind a plate glass window watching the scan on computer screens, but without my glasses on, I can't tell if they are disturbed by what they see or just looking up my skirt and making jokes.

When it is over, I anxiously approach the group standing around computer screens displaying a part of me far more intimate than my jockeys.

"Perfectly normal," Dr. Fayed says, a bit too quickly for my comfort. Dr. Fayed and Dr. Modrego leaf through the slices on the screen, magically peeling away layers of undulating cerebral cortex, penetrating into the dark core of my imaged brain and emerging out the other side, as my living brain tries frantically to take it all in. Dr. Fayed stops the shuffling and points: "A small VR space." Flipping through a bit more: "Another one."

"Perfectly normal," Dr. Modrego reassures me. "For your age."

Shit! What does that mean?

Dr. Fayed burns the stack of 3-D images of my brain onto a CD and, with a smile, hands it to me like a deck of tarot cards.

## NAZIS AND BRAINWAVES

Hans Berger committed suicide. Perhaps ironically, for a scientist intimately familiar with the effects of restricted blood flow on the brain, he chose to do it by hanging.

Berger served as director of the Psychiatry Clinic at the University of Jena from 1919 until 1938, when he was made a professor emeritus and retired. He continued his work under the new director after his retirement until his death in 1941. Rudolf Lemke, an associate who'd assisted Berger by taking photographs of the EEG traces, and who was himself a professor at the University of Jena for the rest of his career, shared his personal remembrances of Hans Berger in 1956 in an article published in a scientific journal devoted to research on EEG.[30] "Berger was no adherent of Hitler and so he had to relinquish the service of his University," wrote Lemke. "Not having expected this, he was gravely hurt."

In 1930s Germany, leaders in academia and elsewhere who were not allied with the Nazis were purged. Berger came to be seen as yet another tragic victim of the brutal Nazi regime. His suicide inside the hospital in Jena, using an electrical cord to hang himself (yet another irony), was viewed as either an act of protest against the Nazis or an act of desperation driven by Nazi reprisals against him. This belief was echoed and reinforced in the neurology community for the next fifty years. In a 2001 book, epilepsy and EEG expert Dr. Eli Goldensohn wrote that "Berger showed his dislike of the regime and they retaliated. In 1938, he was humiliated by Nazi functionaries who abruptly removed him."[31] In 2005, Ernst Niedermeyer, another pioneer in the field, wrote that "his relationship to the Nazi regime was not good and Berger was most unceremoniously made a professor emeritus at earliest convenience."[32]

It is well known that the horrors of the Nazi "Final Solution" originated in mental hospitals. The concept of "racial hygiene" involved ridding society of people perceived to be inferior and therefore detrimental to the ideal of genetic purity, either by preventing them from reproducing or exterminating them altogether, and psychiatric institutions were ground zero for the eugenics craze of the early twentieth century. The forced sterilization of the mentally or physically handicapped was common practice in mental institutions long before the rise of Nazism—in fact, the United States was one of the first countries to implement such practices—and Hitler's first extermination programs involved the euthanasia of the mentally ill. The methods and instruments of killing used by the Nazis, including the gas chamber, were developed in German mental institutions for use on psychiatric patients, and later transported and expanded into the death camps where they were applied to mass genocide. As director of the Psychiatric Clinic at this time, Berger would have been at the epicenter of the eugenics movement and the racial hygiene policies that marked the rise of Nazism.

After the war, Jena came under Soviet control, but many of the people who held positions of authority at the university and hospital during the war remained in charge afterward, a situation that promoted cover-ups and dissuaded the Soviets from investigating former Nazis or preserving documentation. And with many records lost or destroyed, the history written by those left standing tended to reflect a narrative that meshed with the sentiments of the day. A narrative that may or may not have been accurate, as I found out when I visited Jena and met historian Susanne Zimmermann.

Rudolf Lemke, the colleague who placed Berger in opposition to the Nazis in his postwar remembrances, was in fact a Nazi himself. A member of the NSDAP (the Nazi party), Lemke worked at the *Erbgesundheitsgericht* (Hereditary Health Court) from the time it was established and onward. The *Erbgesundheitsgericht* was introduced by the German Reich on January 1, 1934, to carry out the forced sterilization of the mentally and physically unfit, a broadly defined category that included the disabled, psychiatric patients, and alcoholics, among others.[33] By May 1945, about 350,000 such people had been subjected to compulsory sterilization via the genetic health authorities.[34] After the war, Lemke stayed on at Jena, and his activities, as well as his virulently anti-Semitic and anti-homosexual views, were covered up by authorities. He became director of the Psychiatric Clinic from 1945 until 1948, and his distinguished academic career lasted until his death in 1957—one year after he published his whitewashed recollections of Hans Berger.

What were Berger's feelings about racial purity and the atrocities of euthanasia and forced sterilization taking place around him? Zimmermann showed me comments in Berger's notebooks that complained that there were too many Jews on Hiddensee, a popular island destination in the Baltic, making it no longer fit as a vacation spot. These were his personal notes, and thus presumably reflect his true sentiments, but what would provoke him to make such anti-Semitic comments in his laboratory notebook? Was he perhaps performing experiments on someone of Jewish faith? We do not know, but the racism and vitriol are there, written in his own hand.

"Berger was a supporting member of the SS in 1934," Zimmermann told me. "Supporting members" were not SS members per se, but they had usually made a financial contribution in exchange for membership.

"Many people became members to enhance their career," explained Zimmermann. "NSDAP membership is not known, which probably means that he was not a member of this hard-core Nazi organization, but at the time there were many small Nazi organizations. There was not just one big party."

Zimmermann pulled out stacks of records she obtained from the Stasi police files of former communist Berlin, handing me official paperwork from court proceedings in which Berger participated in the hearings of people appealing their sterilization orders. There was a mentally disabled eighteen-year-old, a schizophrenic, someone with epilepsy, a sixty-one-year-old alcoholic, and a woman with lower intelligence and memory deficits accompanied by her husband, who pleaded with the court not to sterilize his wife. Berger denied every appeal, condemning them all to forced sterilization.

I asked, "Was he perhaps coerced to comply with the Nazis out of necessity to maintain his position?"

"No," Zimmermann replied. Berger served on the upper court of appeals for health and hereditary disease (forced sterilizations) from 1938 to 1939. "This was after his retirement," she observed. "At this time, he no longer had a professional motive to cooperate with Nazis or sterilizations."

Another close associate of Berger's was Karl Astel, a member of the NSDAP and notorious for his activities pursuing "racial purity." Astel became rector of the University of Jena in 1939 and worked to reshape it into a model SS institution. He was also the Nazi head of Health Affairs and director of the Institute for Racial Hygiene. On March 4, 1941, Berger replied to Astel's invitation to continue his service overseeing forced sterilizations at the hospital, writing, "I am willing to serve again as a consultant in Jena and I thank you for this." However, he became ill, preventing him from serving, and he committed suicide only two months later. Berger suffered from chronic depression, and he committed suicide while in the hospital in search of a cause for his physical discomfort. "An autopsy found no physical indications of a medical disorder," Zimmermann said. On April 3, 1945, Karl Astel turned a gun on himself, ending his life, as did many who were complicit in Nazi atrocities and found themselves losers at the end of the war.

After my conversation with Zimmermann, Christoph Redies took me to the university's anatomy department, which has a renowned collection. He unlocked the door and we walked into a room stuffed with neatly arranged jars of human anatomical specimens and skeletons in glass cabinets. I saw an amazing assortment of dissected human anatomy: female reproductive organs, a skinned erect penis, testicles, and human skeletons ranging from those of babies and children to adults; there were many fetuses in jars, several with horribly monstrous birth defects. There was a two-headed child. There were heads sliced in various sections and limbs, hands, arms, and legs, all stripped of skin to reveal muscle and bone.

These specimens are priceless and essential aids for teaching human anatomy. Training scientists and doctors in human anatomy is what brings us medical treatments and relief from suffering, not to mention all the understanding gleaned about the human body. Still, it is difficult to imagine the generosity of those who give their bodies to science, knowing that students of medicine will pick them apart, willingly suffering in death the indignity of their naked bodies being sliced open, bit by bit, finger by finger, organ by organ, their face peeled away, their head carved into slices as thin as baloney to expose their brain, their genitals handled and inspected, then cut apart to see what is inside.

The vast majority of us choose the conceit of having our dead bodies embalmed to preserve some likeness of what they were when alive—pointless for a body entombed in darkness six feet deep in the soil. Yet those who give their bodies to anatomists and students know that their dead corpse is otherwise utterly useless, except as food for worms and bugs, and respect it for what it was—a miraculous vessel for the soul. They give their bodies to save lives, to train physicians and allow scientists to uncover new insights into this wondrous human machine. They are the courageous ordinary people to whom every single one of us owes our lives and our health.

At the time of my visit, Redies and Zimmermann had just discovered and reported in a scientific paper that approximately 203 bodies in the anatomical collections at the University of Jena were from prisoners executed by the Nazis between 1933 and 1945; 200 others were patients in mental hospitals and retirement homes who were euthanized; and several dozen others had died at forced labor camps.[35]

"Some people supported euthanasia in the hospital because anatomists needed specimens and they welcomed the supply," said Zimmermann. "This also solved the Nazi problem of disposing of corpses."

Having held human brains in my hands and dissected them, I can say that one always thinks of the person when doing so. It is a sacred, solemn, and grateful feeling of deep respect for the individual who selflessly gave his or her body for science. But the anatomical display of a body of a murder victim is an abomination that goes beyond disrespect to depravity; no scientist or doctor of any decency would willingly be a party to it. After Redies's and Zimmermann's work, any specimens that could be identified as having come from those who were euthanized or died in labor camps were removed from the anatomical display, and a plaque was hung outside the building to honor them. By nature of the anatomical work, though, the origin of many body parts cannot be traced. As a result, this valuable collection exists under a pall that can never quite be dispelled.

Berger's legacy is colored by a similar uncertainty. Was Hans Berger a man a century ahead of his time, struggling with primitive instruments to achieve intellectual greatness, or was he a quack using simple devices—manometers, thermometers, and galvanometers—in bizarre and sometimes cruel ways in an attempt to find evidence for the paranormal and an answer to the eternal mystery of what separates the spirit from the body?

"Was he a genius, a mad genius, or just mad?" I asked a small group of scholars at the end of my day in Jena. "He was crazy," one in the group answered. No one else advanced an opinion.

How could he shove a rectal thermometer into a young girl's brain?[36] How could he condemn his patients and other unfortunates to forced sterilization, even when they pleaded with him to reconsider? Disregard for the humanity of those with mental ailments is not unique to Berger or Germany, but it is an atrocity wherever it occurs. Berger was a Nazi at heart, whatever the records say. He also made a momentous discovery that will intrigue and benefit mankind for generations. Both are true.

## BERGER AND BECK

The story thus far would give the impression that Berger's decision to search for electrical activity in the brain was the result of a brilliant leap of genius, but that is not the case. It had been known since the eighteenth century that signals are transmitted in the nervous system by bioelectricity, although how the nervous system worked, like the nature of electricity itself, were great mysteries at the time. The idea that mental functions like thoughts might generate waves of electrical activity that emanated out from the human skull—much less uniquely identifiable, readable waves—was nowhere on the horizon. However, Berger was not the first to investigate the possibility of an electric brain that one might tap into using electrodes attached to electronic instruments. In fact, just as with Mosso's methods for monitoring blood flow, he was simply copying what someone else had done—someone you have likely never heard of.

## HITLER'S BLACK BOOK

In many ways, Polish scientist Adolf Beck is Berger's mirror opposite, the two men driven by the same curiosity and passion, engulfed by the same toxic politics, but their lives polarized—one allied with evil, the other evil's victim. The fates of these two scientists are entangled by the atrocities of the Nazi frenzy that set the world at war, and connected by inspiration: It was Beck's work in Poland that inspired Berger's secretive research in Jena, exploring the possibility that the human brain functioned by generating electromagnetic energy.

In the 1880s, brain scientists were preoccupied with attempting to determine which parts of the brain controlled specific functions. The first experimental approach, which was extensively used into the early twentieth century, was to remove bits of brain tissue from animal subjects and observe the resulting effects. In many cases, this produced useful insights—for example, in localizing motor function by finding that damage to certain regions of the cerebral cortex in either hemisphere caused paralysis on opposite sides of the body—but other times the approach failed to provide any clues. Beginning in the 1920s, for example, American psychologist Karl Lashley trained rats to run a maze, then systematically removed portions of their brains before reintroducing them to the task, in an attempt to pinpoint where memories are stored. Lashley found that the location in the cortex where the bits were removed didn't seem to matter, and after decades of research he gave up in defeat, concluding in 1950 that it was not possible to localize a "memory trace" anywhere in the brain.[37] Today, we've come a long way toward unraveling the enigma of how mental functions map onto the physical brain, and we understand that the assumption underlying the search for a specific locus of a particular mental function is false. Although specific locations in the brain do indeed correspond to hubs of neurons with a particular purpose—for example, vision at the back of the brain and speech in Broca's area of the cerebral cortex—many functions require sending information broadly over networks that encompass large territories of the brain. Still, pinpointing brain regions and brain circuits that contribute to specific mental, motor, or sensory tasks is essential, not only for understanding how the brain works, but for practical necessity in treating traumatic brain injury and performing neurosurgery.

In the 1880s, while experiments to locate the regions of the brain

**FIGURE 10:** *Adolf Beck (1863–1942), Polish scientist who recorded brainwaves from electrodes placed on the exposed brain and spinal cord of experimental animals, beginning with his doctoral dissertation published in 1891. Beck was imprisoned by the Russians during WWI for two years. Just shy of his eightieth birthday, Beck swallowed a potassium cyanide capsule to end his life as the Nazis came to take him to the Janowska extermination camp during WWII.*

that controlled motor function by simply damaging parts of the cerebral cortex were fruitful, the process of determining where sensory information was processed was considerably more complicated. It had been known for some time that nerves transmit their messages via electrical impulses, and the development in the nineteenth century of instruments to record electrical signals provided a new technique to track down the neural circuits carrying sensory information: placing a recording electrode on a spot in the brain to test for a response to sensory stimulation, much as an electrician would probe an electronic device with a voltmeter to trace an active circuit.

However, an electrician with a voltmeter has a far easier task than did the neuroscientists of the late nineteenth century; it is important to appreciate that all of this advanced study of electrical events in the nervous system was taking place long before the advent of electronic amplifiers, or in fact any other electrical device that we now take for granted. These experiments were conducted in rooms lit by gas lamps and candles, reached by horse and buggy, and in a time prior to an understanding of the fundamental nature of electricity. Imagine that you are attempting to record electrical responses in the brain of a test animal using nothing more than nonelectrical devices and whatever other materials you might find in your kitchen. How would you do it?

At the time, the most sensitive instrument available for detecting electrical phenomena in the brain and nervous system was the string galvanometer. In a string galvanometer, a magnetized needle is suspended by a fiber between two coils of wire. When a voltage passes through the circuit, the electrical field surrounding the needle is disturbed, deflecting the needle slightly, much like a compass needle when brought close to a metal object. The deflection of the indicator needle by the weak electrical events of the nervous system is so miniscule that viewing it required using a magnifying lens; however—just as a laser pointer greatly magnifies the subtle shaking of a person's hand during a PowerPoint presentation—the response of the needle could be amplified by bouncing a beam of light off a tiny mirror attached to the fiber. The mirror bounced the beam onto a ruled scale where the deviations of the needle would be more easily measured, like plotting the height of a child against a doorframe. Thus, instead of the intensity of the response

**FIGURE 11:** *Adolf Beck used a Wiedemann galvanometer to record electrical brain activity in his early experiments. Various versions of these instruments were used during this period. Shown here is a D'Arsonval galvanometer. Note the small circular mirror suspended on the string, which reflected a beam of light onto a scale in proportion to the electrical signal applied.*

being recorded in millivolts, electrical events in the brain and nerves were measured by millimeters of deflection of the light beam. Mind you, this beam was not a laser pointer; the light that bounced off the miniature mirror where it hung inside the elegant polished brass and wood device was generated by a flame.

When it was developed in 1874, the Wiedemann galvanometer, designed by Gustav H. Wiedemann, Hermann von Helmholtz, and John Tyndall, became the most advanced instrument of its kind to measure voltage changes. It was used by a young Adolf Beck in his dissertation research, after he entered the University of Jagiellonski in Kraków (now called Jagiellonian University) in 1886, where he performed his studies under the supervision of distinguished Polish physiologist Napoleon Cybulski.[38] The other equipment necessary to investigate the electrical activity of the nervous system was electrodes, and these too presented a challenge in the nineteenth century. Metal electrodes have the serious problem of creating a battery when placed in contact with salty (and thus conductive) bodily fluids; the electrochemical voltages created by these electrodes themselves compromise attempts to record the subtle electrical events generated by nerve tissue. Indeed, controversy raged for decades between Italian scientists Alessandro Volta (1745–1827) and Luigi Galvani (1737–1798) as to whether the body generated any bioelectricity at all or was instead simply responding to the artificial battery created when metal probes touched bare nerve, shocking it like tinfoil on a metal filing. To overcome this problem, nonmetal electrodes or strips of paper or fabric soaked in salt solution were used to pick up the weak bioelectric signals from muscle, nerve, and brain. Adolf Beck described in his dissertation the electrodes he used: "These [electrodes] were of clay soaked in 1 per cent sodium chloride solution protruding from a glass tube filled with zinc sulphate [*sic*], from which zinc wires made the connexion [*sic*], to a Wiedemann galvanometer."[39] (Beck's thesis research clearly met the challenge posed earlier; sodium chloride is of course table salt, and zinc sulfate is a dietary supplement—both might be found in the cabinets of a modern kitchen.)

Adolf Beck began his research with frogs, carefully dissecting them to expose the brain and spinal cord down to the hind legs. Like a variation of the recipe pheasant under glass, Beck placed the frog with its filleted legs under a Bell jar containing wet cotton to moisten the atmosphere and delay desiccation. His clay electrodes, connected to the string galvanometer, were carefully placed on the frog's brain, ready to pick up the response from the sciatic nerve in the leg, which he would stimulate with a pulse of electricity. The pulse of electricity was generated by an induction coil—a coil of wire wrapped around a cylinder, surrounded by a second wire-wrapped cylinder that has many more wraps of wire than the inner coil. The inner coil was energized by a homemade battery, and when the connection from the battery to the coil was broken, the electrical field surrounding it collapsed, and the force of the collapsing electric field induced a sudden jolt of electricity through the outer coil that was of much higher voltage than that generated by the battery but concentrated into a brief flash. The same mechanism is what generates the spark in a modern automobile engine: The 12 volts from a car's battery is transformed into 25,000 volts when the current running between the battery and the inner winding of an induction coil is repeatedly interrupted in time with the piston, generating instantaneous short bursts of high-voltage electricity in the outer winding to provide a powerful spark across the engine's spark plugs.

Frog and equipment in place, Beck commenced his experiment, staring expectantly at the quavering spot of light from the galvanometer. He fired the induction coil to shock the sciatic nerve—the light beam bounced! The tiny deflection indicated that the point in the frog's brain in contact with Beck's probing clay electrode was generating electrical energy triggered by impulses sent from the sciatic nerve when he shocked it. By moving the electrode to various regions of the brain and spinal cord and stimulating different sensory nerves, Beck was able to locate where in the brain different senses are processed.

In the course of this experiment, Beck made an even more fundamental discovery: He noticed that the beam of light quivered energetically even when he was not stimulating a nerve. Many people would have dismissed these subtle vibrations as noise inherent to the delicate measurement system, but Beck astutely concluded that he was seeing

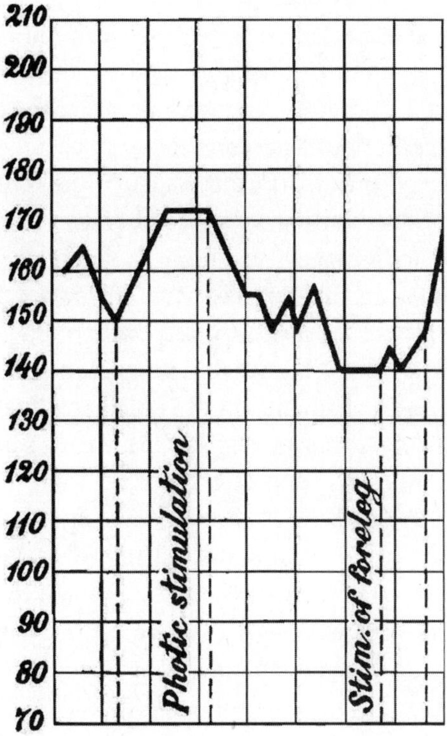


**FIGURE 12:** *A plot of electrical responses in a dog's brain responding to light stimulation of the retina and to stimulation of the foreleg in experiments performed by Adolf Beck for his doctoral thesis published in 1891. Electrical signals cause small movements of a thin filament inside the galvanometer measuring device. Reflections of a beam of light off a mirror, suspended on a filament that twisted when electricity passed through the device, were projected onto a ruled scale and read through a telescope.*

spontaneous oscillations of electricity surging through the brain. "We are dealing with spontaneous excitation of the nervous centres," he wrote in his doctoral thesis, published in 1891.[40] In short—brainwaves.

Beck moved from frogs to rabbits and dogs, removing the skull from one side of the animal's head to expose the creature's brain. He found that, when he activated the animal's various senses by presenting them with sensory stimuli, specific parts of the cerebral cortex exhibited abrupt changes in their electrical activity:

> The optic nerve was stimulated with light, the auditory with sound, and the different sensory nerves of the skin with an induction current. For the light stimulation of the eye, I used a burning magnesium ribbon, which was moved by a special clock mechanism so that the flame reflected by a mirror into the eye remained constant.[41]

Curiously, stimulation of the senses did not increase the ongoing electrical activity; it *decreased* it. In each case, he saw that sensory input quelled the ongoing electrical fluctuations in the cortex of the animal's brain. What we now understand is that, following sensory stimulation, the ongoing activity of brainwaves within the cerebral cortex becomes desynchronized. Just as a child rhythmically pushing against the walls of a wading pool will build large waves that splash exuberantly over the sides, while erratic or "desynchronized" pushing by two children generates smaller irregular ripples on the surface, new sensory input somehow disrupted the large-amplitude, synchronized oscillations of electrical activity seen in the brain before the stimulus was presented.

With only the most primitive instruments and his brilliant intellect, Beck had made two fundamental discoveries: In addition to detecting the electrical events caused by signals entering the brain from the senses, Beck correctly perceived that there is spontaneous rhythmic electrical activity taking place in the cerebral cortex all the time, and that this ongoing activity is perturbed by sensory input. He found it reasonable that "stimulation of some centres causes a blocking of activity in others . . . hence some sort of active state must have existed in those centres and it was only the intrinsic activity that was suppressed."[42] This intriguing brain response later led him to propose that pain might be blocked by manipulating these waves of electric activity, something that is at the forefront of research today. Transcranial direct electric current stimulation and deep brain stimulation (DBS) with implanted electrodes have both been used to dampen brainwaves in appropriate spots to regulate, among other things, the sensation of pain. What's more, as we'll see in a later chapter, the sudden desynchronization of brainwaves that Beck discovered coinciding with sensory and cognitive activity has been used in developing brain-computer interfaces to enable a person to operate prosthetic devices through thought alone.

These pioneering experiments performed in Poland by Adolf Beck (and others we will soon meet), revealing electromagnetic waves coursing through the brains of animals, are what would inspire Hans Berger to attempt to record electrical activity through the scalp of the human head. Adolf Beck extended his work from frogs, rabbits, and dogs to the study of monkeys, all of them showing the same brainwave responses. Human beings were the likely next step for Beck, but unfortunately, he was pursuing his research at the epicenter of international conflict during the outbreak of World War I. On the cusp of making monumental strides in brain science, Beck's research was squashed by political events, war, and bigotry.

In 1895, Beck had accepted the chair of the newly established department of physiology at the University of Lwów, which was then in the Austrian section of partitioned Poland. The city of Lwów has changed hands many times in the upheavals of the past two centuries, and at the turn of the twentieth it was a diverse center of Polish, Ukrainian, and Jewish culture, the home of many Polish national societies as well as the first Yiddish daily newspaper. But in 1914, at the start of the First World War, the city was taken by the Russians. Poles struggled to retain their cultural identity and language, making Polish academics de facto resisters of, and threats to, Russian occupation and domination. The year after the invasion, Beck—along with nine other prominent citizens—was abducted by the Russians in their retreat from the combined German and Austrian armies, and he was held hostage in Kiev.

While a prisoner, Beck wrote to Professor Ivan Pavlov, the prominent Russian physiologist best known for his studies conditioning dogs to salivate at the sound of a bell, appealing to Pavlov to plead his case to Russian authorities. In 1916, after enduring two years of captivity, an exchange of hostages was arranged through the International Red Cross, and Adolf Beck returned to Lwów, eternally grateful to Pavlov for securing his release. Lwów in 1916 was still very much on the battle front, however, now surrounded by Ukrainian forces. (Ukraine was not at that time part of Russia.) World War I was followed for Lwów by the Polish-Ukrainian War and the Polish-Soviet War. In 1919, Poland won its independence for the first time since 1795, but Lwów continued to be contested territory until 1923, when it was officially recognized as part of the Polish state.

Beck did not take his research into the electric brain further after his imprisonment, focusing instead more generally on the science of physiology. With his mentor, Napoleon Cybulski, he published a textbook on the subject, and he went on to publish a textbook specifically on neurophysiology as well. He continued to teach at the university and was a devoted father and a pillar of the academic and cultural communities. He retired in 1930, at the age of sixty-seven, but his retirement was not to be a peaceful one. Turmoil erupted once more when the Germans invaded Poland on September 1, 1939. The Soviets invaded in response on September 17, and on September 22 the city of Lwów fell to the Red Army. The USSR annexed the eastern half of the Polish Republic and renamed the city Lviv. During this period, the university was renamed after a Ukrainian writer and Ukrainian was made the official language of instruction. Many Polish academics were dismissed from their positions, and a dozen were killed by the Soviets over the next year and a half.

Then on June 22, 1941, Nazi Germany invaded the USSR, and Lviv was taken by the Germans. What followed was a series of massacres directed at the city's Jewish population and orchestrated by the *Einsatzgruppen*, the Nazi death squads under the direction of Heinrich Himmler and Reinhard Heydrich. The first of these, in which some 7,000 Polish Jews were murdered, was carried out with the help of Ukrainian national groups; the second targeted a list of some 2,500 citizens believed to be hostile to the Nazi regime. In addition to orchestrating mass killings of civilians, the *Einsatzgruppen* targeted specific undesirables for extermination after invading a country, guided by a carefully cataloged hit list. The version drawn up for Britain was known as "Hitler's Black Book," and 20,000 copies of the 144-page book were printed, listing the names and identifying details of 2,820 people to be tracked down and murdered.[43] A similar list was used by the *Einsatzgruppen* after the invasion of Poland. The list included not only Jews, but also members of the clergy, politicians, writers, artists—and scientists. The intelligentsia, free thinkers, academics, and scientists were as dangerous to Hitler's domination as politicians and partisans. "The Polish lands are to be changed into an intellectual desert," wrote the German general Hans Frank, while serving as governor of Poland after the Germans invaded.[44] As an intellectual, a Pole, and a Jew, Beck was very much in danger.

On July 3, 1941, at the direction of the *Einsatzgruppen,* Nazi occupation forces arrested twenty-five Polish academics, along with their families.[45] The detainees were taken to a repurposed dormitory, where they were tortured and interrogated, and then killed.

At the time of the Nazi-Soviet invasion of Poland at the beginning of World War II, an estimated 150,000 Jews resided in the city of Lwów, including 90,000 children.[46] Almost none of these 150,000 Jews of Lwów were alive at the end of the war.[47] A former factory on 134 Janowska Street in the northwestern suburbs of Lwów was converted into a concentration camp where more than 200,000 people were killed, and among its victims was Professor Adolf Beck.

Beck remained at his home near the university for a year after the massacre of Lwów professors, refusing to flee.[48] Just shy of his eightieth birthday, he became ill, and while in the hospital the Germans came to take him to the Janowska extermination camp. Beck's son, a physician, had supplied all members of the family with capsules of potassium cyanide. Beck swallowed his, choosing to end his own life rather than have it taken by Nazis in the gas chamber.[49]

History in some parts of the world is not linear; it is a vortex, repeatedly sucking the inhabitants back into the same whirlpool of disaster, destruction, and death. Today, Lviv is a part of Ukraine. After the fall of the USSR in the 1980s, Ukraine became independent, but in 2014 the Russian military took control of the Ukrainian territory of Crimea, annexing it into the Russian Federation, and the area is once again awash in conflict. Reviewing these stories from the dawn of neuroscience reminds us that we are always living against the backdrop of history, and, one hopes, learning from it. The distance between Beck's past and our present is not as great as one might think: One of the Gestapo officers who orchestrated the massacre of Lwów professors was found decades later living in Argentina under a false name, and was only arrested in 1985. He died before he could be extradited.

## RICHARD CATON

The world is a small place. A few miles from my home, near Baltimore Washington International Airport, is the town of Catonsville, Maryland. It was originally settled by Europeans in 1720, and the Caton family came from Great Britain to establish their home there in 1787.[50] In 1887, Richard Caton, an MD from Liverpool, England, visited the town founded by his ancestors while on a trip to present his research at a meeting held at Columbia University (now Georgetown University) in

**FIGURE 13**: *Richard Caton (1842–1926), an English physician, was the first person to discover waves of electrical activity in the brain, in his research on rabbits and a monkey. He is shown here in his thirties, at about the time he was conducting his research on brainwaves.*

Washington, DC. Caton's presentation was titled "Researches on Electrical Phenomena of Cerebral Grey Matter."[51]

"Read my paper on the electrical currents of the brain," he wrote in his diary. "It was well received but not understood by most of the audience."[52]

A consequence of the relative isolation of scientists in different countries speaking and writing in different languages is that important discoveries can go unseen by the larger scientific community. This is still true now, but it was even more so in the days before the internet, or in fact any fast and reliable method of international communication. Indeed, when Adolf Beck published his dissertation in 1891, he had no idea that someone working sixteen years earlier in the United Kingdom had already discovered brainwaves in experiments on rabbits and monkeys. Beck had just begun his own study of electrical responses to sensory stimulation when Richard Caton presented his research in Washington, DC, in 1887, and, remarkably, Caton had initiated this research a full twelve years earlier. He had published a brief preliminary report of his results in 1875, in the proceedings of the meeting of the British Medical Association, describing that by using a galvanometer he had succeeded in recording spontaneous electrical currents from the brains of rabbits and a monkey: "Impressions through the senses were found to influence the currents of certain areas; e.g., the currents of that part of the rabbit's brain which Dr. Ferrier has shown to be related to movements of the eyelids, were found to be markedly influenced by stimulation of the opposite retina by light."[53] This brief published account of the electric brain of a rabbit detailed the discovery of the EEG, although almost no one—including Adolf Beck—knew of it.

Caton expanded his studies to forty rabbits, several cats, and monkeys, describing the results in a more extensive report in the *British Medical Journal* on May 5, 1877, concluding that "all brains examined have shown evidence of the existence of electric currents . . . The current is usually in constant fluctuation; the oscillation of the index being generally small, about twenty to fifty degrees of the scale. At other times great fluctuations are observed, which in some instances coincide with some muscular movements or change in the animal's mental condition."[54] However, these publications remained obscure.

In general, biological science in that era was conducted primarily by medical practitioners, such as Caton, who earned their income from their medical practice while their scientific work afforded no income and little notoriety. Physiology (never mind neurophysiology) was not yet recognized as a scientific discipline and lacked its own academic departments, scientific societies, and scholarly journals, and Caton's work was of little practical use or relevance to medical doctors at the time. After twelve years of research, the audience at the Medical Congress Caton traveled to in 1887 failed to grasp the significance of his findings. Although research did not resonate with the medical community, Caton was nevertheless well respected as a physician and professor,

FORTY-THIRD ANNUAL MEETING

OF THE

BRITISH MEDICAL ASSOCIATION.

*Held in EDINBURGH, August 3rd, 4th, 5th, and 6th, 1875.*

*The Electric Currents of the Brain.* By RICHARD CATON, M.D., Liverpool.—After a brief *résumé* of previous investigations, the author gave an account of his own experiments on the brains of the rabbit and the monkey. The following is a brief summary of the principal results. In every brain hitherto examined, the galvanometer has indicated the existence of electric currents. The external surface of the grey matter is usually positive in relation to the surface of a section through it. Feeble currents of varying direction pass through the multiplier when the electrodes are placed on two points of the external surface, or one electrode on the grey matter, and one on the surface of the skull. The electric currents of the grey matter appear to have a relation to its function. When any part of the grey matter is in a state of functional activity, its electric current usually exhibits negative variation. For example, on the areas shown by Dr. Ferrier to be

**FIGURE 14**: *The first report of brainwaves was by London physician Richard Caton. Caton announced his discovery at the annual meeting of the British Medical Association in August 1875 held in Edinburgh, but the breakthrough went unnoticed. Caton traveled to the United States to deliver a paper on his research at what is now Georgetown University in Washington, DC, in 1887, but his research was not understood and the discovery was lost to history until half a century later when Adolf Beck began his own research on brainwaves without knowledge of Caton's earlier work.*

and he became the dean of the medical faculty at Liverpool College in 1904.

In 1890, Beck published a summary of his dissertation research detailing his recordings of brainwaves, which kicked off a flurry of controversy over who had first discovered brainwaves. Russian physiologist Vasili Danilevsky claimed to have recorded brainwaves from animals during his dissertation work in 1877, prior to Beck's experiments. He had not published his findings, depositing them instead in a sealed letter in a vault at the Imperial Academy of Sciences in Vienna. This manner of recording and preserving scientific discoveries was not unusual in an era before the establishment of scientific societies and journals to record and disseminate findings in various fields. But in February 1891, Caton wrote a letter to the journal *Centralblatt für Physiologie*, pointing out that he had beat them all with his 1875 paper.[55] Quoting from his letter, "In the year 1875 I have a presentation before the Physiological Section of the British Medical association in which the electrical currents of the brain in warm-blooded animals was demonstrated and, in addition, their undoubted relationship with regard to function was established. May I be permitted to draw your attention to the following publication: *Brit. Med. J.* 1875, 2: 278."[56]

This settled the matter. Richard Caton, a British MD, was indeed the first person to observe brainwaves. He found them in the brains of experimental animals using the unsophisticated equipment available in the 1870s, long before the world was prepared to recognize the magnitude of this achievement. In 1891 he resigned his professorship at the age of forty-nine and abandoned research on the electrical activity of the brain, turning his attention to medical issues related to fever, poisoning, and acromegaly (a form of gigantism). Interestingly, in 1897 Caton attended the International Medical College Congress held in Moscow, where Napoleon Cybulski—Adolf Beck's mentor—was also in attendance.[57]

Richard Caton died in 1926. Unbeknownst to him, Hans Berger in Jena, Germany, was then two years into his studies on the electroencephalogram of man, the findings of which he would first publish in 1929—and which, like Caton's first animal EEG more than fifty years earlier, would be initially ignored and dismissed.[58]

◇◇◇◇◇◇◇◇◇◇◇

From the perspective of modernity, it is easy to judge those who failed to see the significance of Caton's presentation, or to wonder why it took so long for someone to make the leap from animal brainwaves to those of humans. But Caton's and Beck's pursuit of the idea that electricity flowed through the brain was particularly bold when you consider that earlier in the 1800s, electricity flowed through nothing in the man-made world. Except in isolated laboratories where a few eccentric scientists either captured electricity from the heavens or made it themselves to ponder this force of nature at a time before anyone fully understood it, electricity was as rare and mysterious then as dark energy is today.

# CHAPTER 2

# It's Alive! The Spark of Life

*The busy morning of half the London population had begun . . . It was market-morning. The ground was covered, nearly ankle-deep, with filth and mire; a thick steam, perpetually rising from the reeking bodies of the cattle, and mingling with the fog, which seemed to rest upon the chimney-tops, hung heavily above. All the pens in the centre of the large area, and as many temporary pens as could be crowded into the vacant space, were filled with sheep; tied up to posts by the gutter side were long lines of beasts and oxen, three or four deep. Countrymen, butchers, drovers, hawkers, boys, thieves, idlers, and vagabonds of every low grade, were mingled together in a mass; the whistling of drovers, the barking of dogs, the bellowing and plunging of oxen, the bleating of sheep, the grunting and squeaking of pigs, the cries of hawkers, the shouts, oaths, and quarrelling on all sides; the ringing of bells and roar of voices, that issued from every public-house; the crowding, pushing, driving, beating, whooping, and yelling; the hideous and discordant din that resounded from every corner of the market; and the unwashed, unshaven, squalid, and dirty figures constantly running to and fro, and bursting in and out of the throng; rendered it a stunning and bewildering scene, which quite confounded the senses.*

*—Charles Dickens,* Oliver Twist, *chapter 21*

The center of the universe, London, was the largest city in the world in the early to mid-1800s. Swollen by a deluge of humanity seeking work, its 1.3 million inhabitants toiled in dirt, mud, manure, and filth. Workers traveled to punishing jobs at factories and markets on foot or by jolting carts and buggies that filled the streets with the sound of muddy hooves thumping over dirt or clopping along cobblestone. Medicine mired in ignorance and superstition operated amid the slums spanning the gray city—accretions of decrepit one-room dwellings heated by smoldering coal, lacking running water and sanitation. Women died miserably in childbirth or shortly afterward from infection; others were taken abruptly by cholera, tuberculosis, and diseases of unknown origin.

Yet the upper crust enjoyed lives of leisure and luxury, living in opulent multiroom homes and estates graced with exquisite furnishings and elaborate gardens; the house and grounds were kept (and the lavish meals prepared) by servants. Never having cause to venture into the less-desirable parts of town, the wealthy met the poor only in their role as master of those in their employ. Wealthy women were expected to occupy themselves with running their homes and maintaining intricate social lives, while gentlemen of means sought comradery, entertainment, and commerce among other men of means. Many such men satisfied their curiosity by dabbling in natural philosophy, becoming dilettante experimentalists. This particular hobby was available only to men who had the means to obtain the equipment and resources necessary for scientific experimentation—as well as the time to devote to scholarly but unprofitable pursuits. Essential to the advancement of this amateur science was membership in the many social clubs and societies where gentlemen assembled to share and discuss their observations, launching, for example, the Geological Society in 1807 and the Astronomical Society in 1820. At the same time, public lectures by professors at universities were a source of entertainment and enlightenment for interested and educated citizens. The wealth of these gentlemen patrons helped support the research of university doctors and scholars, but such public displays also required that academics be part scientist and part showman.

On May 16, 1837, scientific instrument maker Edward Clarke met with several other London gentlemen, including a chemist and a wealthy wine merchant, to launch a new society devoted to the scientific exploration of electricity. Whereas other scientific societies were defined by specialization, the study of electricity could be applied across diverse swaths of the biological and physical sciences, "giving us insight into the operations of nature as connected with the animal, vegetable, and mineral kingdoms . . ." said H. M. Noad in closing remarks at a London Electrical Society meeting.[1] This broad scope of inquiry was in addition to the obvious technological potential of electricity to revolutionize industry, bring light to the night, replace steam power, and—although this was beyond imagination at the time—lead to the development of microelectronics that would usher in the digital age.

The technological and industrial applications of electricity would be enduringly transformative, but the electrically inspired medical remedies that blossomed throughout the nineteenth century largely died in the next. Applying electricity to a spectrum of maladies ranging from gout to deficient female orgasm, the majority of "cures" were dubious or outright quackery. Still, the appeal of electricity in the 1800s as nature's most powerful and mysterious force drew many to explore the intersection between electrical energy and biology, with the suspicion that the enigmatic force of electricity might also be the spark and essence of the life force itself. Electricity might explain the operation of the living body, especially the mechanism controlling muscles and the transmission of sensation to our brain; it might power the beating heart, and rule the operation of the mind. Understanding and harnessing this force might even give mankind the God-like power to apply it to restore life. Mary Shelley captured this shared ambition of the time in her classic work *Frankenstein*. Published in 1818, her novel was born of this age of nineteenth-century gentlemen scientists seeking to harness the power of lightning and use this electrical power to decode the living and raise the dead.

## LIGHTNING BUG GENESIS

In the Sedgemoor district of Somerset, England, stands Fyne Court, the country estate of the Crosse family, which was constructed in 1780 and set amid trim English landscaping surrounded by sixty-five acres of pristine woodland.[2] Following the deaths of his parents, Andrew Crosse assumed management of the family estate in 1805. Abandoning the practice of law, he established a laboratory at the isolated Fyne Court and devoted himself instead to the study of electricity. Crosse suspended wires from poles and trees to span over one-and-a-quarter miles in length to capture electrical discharges in the atmosphere, storing the frightful power of lightning in Leyden jars for use in his experiments.[3] Leyden jars were large capacitors capable of collecting and holding large amounts of electric charge, which could then be discharged at the experimenter's will. In addition to the electricity captured in Leyden jars, Crosse used electricity he created himself, building enormous batteries to generate powerful voltages. When discharged, the bolts of electricity reverberated with a noise as if a cannon had been shot, and Crosse became locally known as "the thunder and lightning man."[4] Mary Shelley reportedly knew Crosse.[5] She recorded in her diary that on Wednesday evening, December 28, 1814, she attended the lecture "Electricity and the Elements," by the "Electrician from Somerset," Andrew Crosse. This was the first time Crosse had spoken in public, and Mary, then going by her maiden name of Godwin, attended together with her paramour, Percy Bysshe Shelley. In 1816 the former mistress became the poet's second wife.[6] Though there is no proof that Mary Shelley ever visited Fyne Court, some believe Crosse's laboratory inspired the setting of *Frankenstein*.[7] Whether or not Crosse influenced Shelley's creation, there is no doubt that some of his later experiments, conducted almost twenty years after *Frankenstein*'s publication, evoke the immortal words: "It's alive!"

Crosse began studying the effects of passing electricity through various salt solutions and marveled at the formation of crystals around the electrodes that "exactly resemble nature." Crosse was moved to conclude that electricity was "the secondary cause of every change in the animal, mineral, vegetable, and gaseous systems."[8]

Working in his laboratory in 1836, he concocted a solution of silicate of potash and muriatic acid that he allowed to drip slowly over a chunk of porous red oxide stone—volcanic iron from Mount Vesuvius. Two platinum wires ran from the rock to the poles of a battery Crosse had constructed called a voltaic pile, that could provide a continuous current. Over the next month he carefully recorded his observations of the moist electrified rock, inspecting it every day through a magnifying lens. His hope was that, by subjecting the fluid to a steady stream of electricity with the aid of the porous rock, he would cause silica crystals to form. What he saw instead was an astonishing creation of nature at the hand of man. Reporting in 1837 to the London Electrical Society, of which he was a member, he wrote:

> At the end of fourteen days, two or three very minute white specs or nipples were visible on the surface of the stone, between the two wires . . . On the eighteenth day these nipples elongated and were covered with fine filaments. On the twenty-second day their size and elongation increased, and on the twenty-sixth day each figure assumed the form of a perfect insect, standing on a few bristles which formed its tail. On the twenty-eighth day these insects moved their legs, and in the course of a few days more, detached themselves from the stone, and moved over its surface at pleasure, although in general they appeared averse to motion, more particularly when first borne.[9]

Crosse repeated the experiment many times with the same result. Swarms of these minute insects proliferated in his experimental apparatus, which delivered the power of electricity to inert iron stone cast out from a volcano. Crosse had no idea what the bugs were, so he took a sample and sent it to an expert, Monsieur Turpin, at the Académie des Sciences in Paris, who identified the creature as being from the genus *Acarus*—a type of mite. However, Monsieur Turpin reported that it was not a mite he had ever seen before and suggested that it may be a new species. Crosse then sent specimens of the strange insects to several eminent authorities in London, all of whom agreed with Turpin's assessment. One of these experts was comparative anatomist Richard Owen at the Royal College of Surgeons in London, remembered today as the biologist who coined the immortal word "Dinosauria" (meaning "terrible reptile"). Without endorsing their apparent miraculous origin, Owen identified the minute creatures as indeed some mite of the genus *Acarus*. Crosse christened this form of life that he had seemingly created with his own name: "*Acarus crossii*."

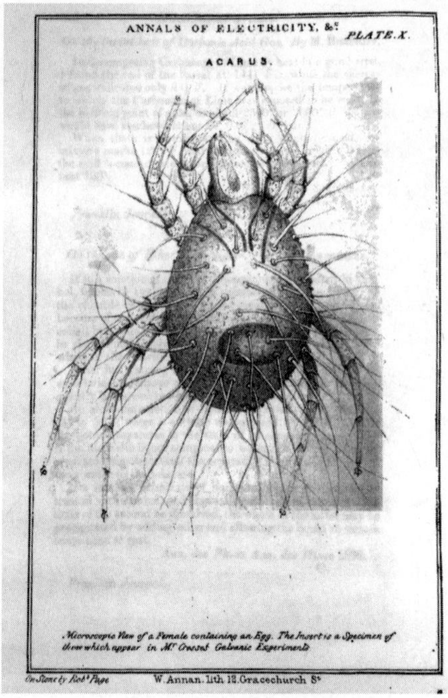


**FIGURE 15**: "Acarus crossii," *the species of mite Andrew Crosse believed he had created by passing electricity through a porous stone from the Mount Vesuvius volcano, which he reported to the London Electrical Society in 1838. This specimen, identified as a new species and a female containing an egg, is from a description by M. Turpin, published in 1838 in* The Annals of Electricity, Magnetism, and Chemistry and Guardian of Experimental Science *2: 355–360. That egg raised doubts in Turpin's mind that the mite had been created by electricity: ". . . there are distinct sexes; among which there is a coupling and fecundation necessary for the reproduction of individuals of the species, and consequently we must admit there are genital organs; and, finally, the females make and lay eggs, whence the young ones are hatched . . ."*

The popular press grasped the momentous importance of Crosse, an amateur electrician, having recreated genesis in his laboratory: "And the Lord God formed man of the dust of the ground," Genesis 2:7, and from Psalm 29:7, "The voice of the Lord strikes with flashes of lightning." Some condemned Crosse for his blasphemous usurpation of God's domain, and he reportedly received death threats. Farmers blamed their failed crops on the mite Crosse had foolishly created in his home laboratory.[10]

Crosse was understandably peeved, and he took pains to note that he had not offered an opinion about *how* these mites had come to be, only that they had, and remarked that he was "sorry that the faith of his neighbours could be overturned by the claw of a mite."[11] Many, notably Michael Faraday, whose experiments with electricity were already famous, favored the conclusion along with most scientists that the moist environment of Crosse's apparatus had simply provided suitable conditions for dust mites to flourish.[12] Some scientists believed that electricity had not created the mites from scratch, as it were, but somehow "revivified" eggs that had been present in the porous rock.[13]

After he was invited to present his findings at a meeting of the Electrical Society in June 1837, the Society commissioned its members to repeat Crosse's experiments in a sealed environment to exclude the possibility of contamination by common dust mites. Electrical Society member W. H. Weekes, a surgeon, successfully replicated the results of Crosse's experiments within a bell jar in 1841,[14] and grasping this momentous achievement, Francis Maceroni championed the findings by writing in *Mechanics' Magazine* that Crosse's experiments were "clear experimental proof that . . . the identity of all life, with the electric, galvanic, magnetic, or solar substance."[15] Others, however, failed to repeat Crosse's results, thus sustaining the controversy. A prominent Utopian Socialist, Thomas Simmons Mackintosh, delivered a series of lectures on the "Electrical Theory of the Universe,"[16] mentioning Crosse's experiments, and in *Mechanics' Magazine*[17] he argued against critics protesting that scientists should not play God, saying, "It is better to view man as an organized machine, and to search for the seat of those impulses in the functions of his physical nature, where assuredly they are to be found, than to trace them to sources beyond our knowledge and above our control."[18]

## THE ORIGIN OF LIFE?

Jumping through time one century into the future, we find a parallel approach with the same motivation—to uncover the origin of life and reproduce it in the laboratory with the aid of electricity—driving a classic twentieth-century experiment. In the early 1950s, working in the laboratory of Harold C. Urey at the University of Chicago, Stanley L. Miller sealed purified water inside a glass globe filled with a mixture of gases that replicated the atmosphere on Earth when life first arose 3.5 billion years ago. Miller delivered electrical sparks through platinum electrodes, creating lightning inside a "snow globe" of the primordial world in miniature.

After two days of electricity arcing within the glass, the water began to turn pale yellow and a tarry residue collected around the electrodes on the inside of the flask. Stanley collected the ooze and used paper chromatography to determine whether any organic molecules had formed from the simple elements in the atmosphere. In particular, he was looking for amino acids, the building blocks of proteins and thus of life. His analysis revealed conclusively that the amino acid glycine had been produced. After prolonging the electrical stimulation, several other types of amino acids were found. From these amino acids, themselves formed from inorganic molecules energized by electricity, large proteins would grow as the amino acids automatically self-assembled into strings. With these proteins and other molecules that came together, in time a living cell could form. Simple cells might then assemble into primitive organisms—all initiated by a spark—and through eons of evolution the tree of life would expand and ultimately yield *Homo sapiens*, "the wise one," with the cognitive power to recreate life one day in a Chicago laboratory, inside a globe of glass.

"If God did not do it this way, then he missed a good bet," said Harold Urey in reply to a question by Enrico Fermi at a seminar describing Miller's results in 1953.[19] Many renowned scientists of the Manhattan Project were in attendance when these results were announced; in fact, Edward Teller (father of the hydrogen bomb) had worked with Stanley Miller on his early efforts to understand the genesis of life, but he gave up and moved on to investigate the God-like power of the atom.

From the nineteenth century to the twentieth—indeed, throughout the span of history—we see mankind grappling with this great mystery of science: What is life and how did it arise? What does it mean to be human, and how do body and spirit derive from one another?

Of course, to apply the scientific experimental method of analysis to the question of how life infuses and departs the body and whether the force of electricity is indeed the life force, it is necessary to go beyond experiments on water and rocks and work instead with the bodies of subjects who were, or at least had once been, alive.

## ARTIFICIAL BRAINS

Today's news is filled with accounts of scientific progress in artificial intelligence, in creating artificial cells,[20] and in making "brains in a dish" from stem cells for experimentation,[21] but the quest to build an artificial brain stretches back to the earliest days of neuroscience.

Another supposed source of inspiration for Mary Shelley's *Frankenstein* is the chilling research by Karl August Weinhold,[22] which predated Crosse's mite-producing experiments by twenty years. Weinhold's bold goal was precisely the premise behind Shelley's book—to bring life to a dead body by replacing its brain with an artificial one made in his laboratory. Seeing how electricity from Leyden jars could stimulate the nervous system even after death (as will be described below), commanding the body to respond under the control of a scientist, Weinhold sought to make the leap of directly inserting the source of electrical power directly inside the empty skull cavity. All that was required to make an artificial brain, he reasoned, was to mix the proper types of metals together to create a battery, replacing the animal's bioelectricity with this man-made electrochemical energy source.[23]

The first step was to work out the details of his battery-brain. Weinhold cut the head off a three-week-old kitten and applied electricity to stimulate the animal's muscles. The kitten twitched in response to this electrical probing for about fifteen minutes before it became unresponsive. Then Weinhold destroyed the headless kitten's spinal cord, using a sponge attached to a screw-probe to auger out the spinal canal, which he then filled with an amalgam of silver and zinc—the components of a battery. The results, as he described them in his 1817 paper,[24] were astonishing. According to Weinhold, as soon as the metal filled the spinal canal, the kitten's heart resumed beating and the muscles of the body began to contract strongly. "There is no noticeable difference between the natural and the artificial spinal column," he wrote. The headless kitten was restored to "life," and "hopping around was once again stimulated after the opening in the spinal column was closed. The animal jumped strongly before it wore down."[25]

But the ghoulish spectacle of a blood-soaked headless kitten hopping around a laboratory bench was only the beginning. Next Weinhold would attempt to revive a dead kitten by replacing its brain with an electric one, made from his bimetallic battery. He took another kitten, cut open the back of its head, and scooped out its brain with a small spoon. The kitten's cerebrum, cerebellum, and spinal cord were all removed, and the cranial and spinal cavities filled with Weinhold's silver-zinc amalgam. Weinhold wrote:

> For almost twenty minutes the animal got into such a life-tension that it raised its head, opened its eyes, stared for a time, tried to get into a crawling position, sank down again several times, yet finally got up with obvious effort, hopped around, and then sank down exhausted. The heartbeat and the pulse, as well as the circulation, were quite active during these observations, and continued after I opened the chest and abdominal cavities fifteen minutes later . . . Body temperature was also completely restored.[26]

Weinhold was indeed a "Frankenstein," a mad scientist on the fringe, working in isolation—and therein lies the problem. There were no witnesses to these miraculous results, which were clearly highly exaggerated. Whatever responses Weinhold did observe were not a result of electrical resurrection but rather simple reflexes to the battery's voltage, which induced the muscles to twitch while the kitten remained quite dead. Even in his own time, other scientists rejected Weinhold's fantastic claims. "Time need not be wasted on criticizing this experiment and the bold conclusions drawn by its extravagantly gifted creator," said Austrian physician Max Neuburger.[27]

Weinhold had been a respected professor and member of the German scientific community, but after 1820, his reputation began to deteriorate. Students reported that his lectures were "scandals," repeating obsolete information. He was criticized as being "rough and awkward" with patients in the surgical clinic, "not a surgeon, but a slaughterer." A formal investigation by the clinic director concluded that Weinhold was a "fantastic writer," with "a great inclination to charlatanism and lies."[28]

According to medical historian Hans-Theodor Koch, Weinhold felt personally persecuted and threatened; Koch suspects the underlying cause of his professional decay was chronic depression. Weinhold never married, and autopsy revealed that his genitals were deformed.[29] This finding is ironic, considering Weinhold's other pet project: In 1829, almost a century before fellow electric brain researcher Hans Berger presided over forced sterilizations, Weinhold published a paper advocating that beggars and other impoverished and unmarried people—along with those unable to work, those suffering from protracted diseases, and so on—have their genitals yoked (infibulated), pierced through with a metal ring to prevent procreation:

> The operation itself is easy . . . as is the soldering and metallic seal, which is my invention. The foreskin is advanced, and gently clamped between a pair of perforated metal plates, so that the piercing of a hollow needle in which a four to five inch long lead wire is present can hardly be felt.[30]

Weinhold's experiments were appalling and heartless, without question, but in the fundamental principles behind his research, he was correct. The brain and nervous system do indeed operate by generating their own electricity. His ideas were handicapped by both primitive technology and the neuroscientific naivete of the time; if he were doing his experiments now, it is interesting to speculate on what his approach might be . . . microelectronic brain implants and brain-computer interfaces seem likely.

There is no better way to test one's scientific knowledge of any natural phenomenon than to recreate it in the lab, whether that be crystallizing a pure diamond from carbon dust, replacing the heart, lung, or kidney with machines that replicate the organ's functioning, or by creating life itself inside a snow globe version of Earth's earliest days.

The research of the early nineteenth century led to the conclusion that the energy powering the nervous system, and indeed the vital energy of the human body as a whole, was electricity. The only way to prove this, however, was to experiment on human beings themselves.

## WE NEED A HUMAN BODY

Scientific investigation of the body electric was made possible by an ageless and enduring human activity: crime and punishment. London had no police force before 1829, and the stark disparity in wealth between the elite class and the masses fueled rampant crime. Criminals convicted of a range of offenses, from burglary to murder, were executed by hanging—a public spectacle attended by thousands of morbid, rowdy onlookers until the practice was moved indoors in 1868.

Execution by hanging has a long history in England, spanning from the 1700s until capital punishment was abolished there in 1965.[31] By the 1960s, the technique of hanging had been refined to render death quickly, by using a length of rope calibrated according to the condemned's body weight so as to snap the neck from the fall without either decapitating the victim or leaving him kicking and slowly strangling to death. In the past, however, death by hanging had been intended to be gruesome and prolonged. In the eighteenth and nineteenth centuries, a short length of rope was used to leave the victim struggling above the ground with their hands and feet tied; "mercifully," friends and family were frequently permitted to hasten death by pulling down on the person's legs as they hung by their neck. Drawing thousands of onlookers from all walks of life, hangings were carried out in public not only to dissuade criminal activity, but because public humiliation was considered a justified part of the punishment.

Sometimes part and parcel of that humiliation was postmortem anatomization. Doctors and anatomists were frequently permitted to dissect the bodies of hanging victims, furthering scientific knowledge while having the added benefit of increasing the severity of the punishment by debasing the condemned's body, often before the public.

In 1779, James Brodie, twenty-three, was convicted of murdering an eight-year-old boy and sentenced to death by hanging. Brodie implored the court that his body not be anatomized, but the judges rejected the appeal because Brodie had murdered a child. His body was given to surgeons for dissection and then displayed in a public hall. "Crowds of men, women and children indulged their morbid curiosity by thronging to the scene of this most repulsive spectacle," wrote E. T. Hurren in his account. He quoted from the county records:[32]

> The mutilated and frightfully excoriated corpse lay extended on a long table . . . Words can scarcely give an idea of the coarse and brutal jests, the obscene remarks, and the horrid blasphemous oaths to which the display gave voice . . . In the height of the exhibition, when amongst the crowd were a number of women, a disgruntled fellow treated the body in a manner which most completely violated every feeling of decency and which his conduct excited the disapprobation of a few, and the screams and exclamations of the females, but to the great majority it served to elicit great peals of laughter.

Explained Hurren, "One man allegedly grabbed a body-part in a sexual way."[33]

It was an ignoble end. Parts of the criminal's body not kept for study or display in anatomical museums were disposed of down drains and discarded as trash. Many of the condemned feared the postmortem disposition of their dead body as much as death itself.

But these freshly killed bodies did indeed advance scientific understanding. Not only serving as subjects for general anatomists, they enabled scientists to test out various theories, including the theory that bioelectricity (called galvanism at the time) controls the human body. In 1818 in Glasgow, Scotland, Matthew Clydesdale was sentenced to death by hanging. The same court had also sentenced Simon Ross, James Boyd, and Margaret Kennedy to the same fate. Ross and Boyd had been convicted of burglary, and Margaret Kennedy of passing forged banknotes. Clydesdale was accused of killing a man with a coal pick in a drunken fit of rage. For such a hideous crime, Clydesdale was denied burial in the prison graveyard after he was hanged; instead, he was sentenced to being publicly dissected and anatomized by the professor of anatomy at the University of Glasgow, James Jeffray. The death sentences of Margaret

Kennedy and James Boyd were commuted—hanging of women was exceedingly rare, with only four women having been hanged in Glasgow since public executions began in 1765, and Boyd was only fourteen years old—leaving one burglar, Ross, and the murderer, Clydesdale, to be executed on November 4, 1818.

Inside his cell, Clydesdale could hear crowds beginning to gather on the Green. Throngs of spectators had traveled from around the city and the surrounding rural areas to participate in the special occasion of a public hanging and anatomization. Dressed in a white gown, like a nightshirt with a hood, Clydesdale was escorted through the crowd by guards, with his hands bound.

After ascending the stairs of the gallows, built only about shoulder height to permit better viewing by the crowd, Clydesdale and his fellow condemned prisoner Simon Ross stood and faced the executioner, Tammas Young, and the four assistants who had constructed the scaffold the night before. The prisoners' legs were tied to prevent them from resisting the drop through the trapdoor. Now in a state of shock, both men began to pray, and Clydesdale fell to his knees. The executioners positioned the men and slipped nooses around their necks. Clydesdale could not see because of the hood over his head, but he could hear the massive jeering crowd that had come to see him hanged and then dissected.

The town clock struck three and Tammas Young released the trapdoor from beneath Clydesdale's feet. He fell through the air, only to halt when the rope jerked taut around his neck. Ross and Clydesdale were left to hang in public for an hour to ensure that they were dead. Then the executioner cut the men down; Ross's body was released to his father for burial, while Clydesdale's body was taken by horse and cart, with a military escort, from the gallows to the nearby University of Glasgow School of Anatomy as the crowd, eager to witness the spectacle, followed in a vast parade. All through the town along the route people were packed together to watch from rooftops and other vantage points, cheering as the body passed. At the university they were met with another crowd; the assembled masses roared as the body was taken inside the building.

Clydesdale's body had been given by the court to Professor Jeffray for dissection, but knowing the unique opportunity that a freshly deceased human specimen would provide his colleague, Dr. Andrew

Ure, Jeffray generously permitted Ure to perform experiments on the convict. Andrew Ure, who gave popular lectures on physics and physiology, had a special interest in electricity. He was keen to perform an experiment using a voltage pile of 270 discs, generating a powerful electric jolt to test his theory that electricity was the force that directed bodily movement, propelled the beating heart, and energized mental function in the brain—and that it thus had the potential to resuscitate the dead. Ure's intention was not to resurrect this hanged criminal, but rather to understand the electrical basis of nervous system function to develop the means to apply electricity to revive drowning victims and others who died suddenly.

The lecture theater was packed with spectators. Newspapers reported that Clydesdale's body, still clothed in the white gown he wore for the hanging, was placed on the table in front of the professor. The hood covering his ghastly face and the ropes binding his hands and feet were removed, and the murderer was lifted into a seated position in a chair directly facing the rapt audience.

A few minutes before Clydesdale arrived, Ure charged his battery with nitrosulfuric acid to bring it to full power. Because Ure was not a surgeon, Jeffray's assistant, Thomas Marshall, was appointed to conduct the dissection under Ure's direction. Ure directed Marshall to begin, making a large incision into the back of Clydesdale's neck to expose the cervical vertebrae and the "spinal marrow" (spinal cord). Next he made a large incision into the left hip through the buttock to expose the sciatic nerve and another incision into the heel of the left foot. Grasping the long conducting rods from the battery, Ure touched one metal probe to Clydesdale's spinal cord and the other to his sciatic nerve, which instantly evoked a dramatic reaction that stunned and appalled the crowd. "Every muscle of the body was immediately agitated with convulsive movements, resembling a violent shuddering from cold," Ure wrote, describing the proceedings in an address to the Glasgow Literary Society on December 10, 1818.[34] Ure's descriptions, which were objective and clinical, contrasted with the sensational and alarming accounts that would appear in the press of Ure raising the dead criminal monster back to life. Ure's address continued: "The left side was most powerfully convulsed at each renewal of the electric contact. On moving the second rod from the hip to the heel, the knee being previously bent, the leg was thrown out with such violence, as nearly to overturn one of the assistants, who in vain attempted to prevent its extension." The crowd was in an uproar.[35]

Ure instructed Marshall to make an incision to expose the phrenic nerve, which controls breathing, and which Ure believed also controlled the heart. He hoped that electricity would restore breathing and set the stilled heart pulsing again:

> The success of it was truly wonderful. Full, nay, laborious breathing instantly commenced. The chest heaved and fell; the belly was protruded, and again collapsed with the relaxation and retiring diaphragm. This process was continued without interruption, as long as I continued the electric discharges.[36]

The experiment took a still more dramatic turn, as far as the audience was concerned, when a cut was made in Clydesdale's eyebrow to expose the supraorbital nerve. When Ure applied electric current to the cut, the dead man's face contorted in a series of animated grimaces; Ure manipulated the expressions by adjusting the stimulus strength. He recounted: "Every muscle in his countenance was simultaneously thrown into fearful action: rage, horror, despair, anguish, and ghastly smiles, united their hideous expression in the murderer's face surpassing far the wildest representations of Fuseli or a Kean [famous actors]." Several of those watching fled the room in terror. Others were overcome with nausea, and at least one spectator fainted.[37]

But the public display of Ure's horrible "resurrection" of the dead murderer was not yet complete; indeed, Ure was working up to the climax. Ure touched the electrode to the dead man's ulnar nerve while he placed the opposite electrode in contact with the spinal cord—Clydesdale's fingers instantly pitched into a frantic but nimble pantomime of a violinist. Ure broke the electric circuit and the fingers ceased their twitching and clenched into a fist. Then he touched the electrode to the tip of Clydesdale's forefinger and the dead man's arm shot out forcefully, seemingly pointing directly at the spectators.

Many in the crowd, as the newspapers reported, were convinced that Clydesdale had come back to life. Ure held no such illusions, but the experiments furthered his determination to find a means of using electricity to someday revive victims of asphyxia, stroke, or heart attack: "The experiments I did on the hanged criminal did not aim at reanimating the cadaver, but only to acquire a practical knowledge as to whether galvanism can be used as an auxiliary, and up to which it can override other means of reanimating a man under such circumstances."[38] He concluded, "It is possible, indeed, that two small brass knobs, covered with cloth moistened with a solution of sal ammoniac [ammonium chloride], pressed above and below, on the place of the [phrenic] nerve and diaphragmatic region, may suffice, without any surgical operation [to start the heart beating again]."[39] Ure had conceived the modern automated external defibrillator (AED) now common in many public buildings such as schools, businesses, airports, and where social and athletic events take place. They are found in every ambulance and hospital.

Most argue that it is chronologically impossible for Ure's experiment on Matthew Clydesdale in November 1818 to have inspired Mary Shelley's writing of *Frankenstein*, because this public display took place ten months after publication of the novel in January of that year. Others argue against galvanism as the inspiration for her book not on the basis of chronology, but on the basis of methodology.[40] In contrast to Hollywood portrayals of the monster jolted to life by dramatic bolts of lightning, Shelley is quite vague about how the monster is brought to life, and chemistry, not electricity, seems more strongly implicated:

> It was already one in the morning; the rain pattered dismally against the pane, and my candle was nearly burnt out, when, by the glimmer of the half-extinguished light, I saw the dull yellow eye of the creature open, it breathed hard, and a convulsive motion agitated its limbs.[41]

Yet chemistry and electricity were intimately entangled at this time, when Crosse and other electricians generated electricity with voltaic jars (homemade batteries). Crosse's estate was littered with 2,500 such jars, most of them formed by breaking the necks off empty wine bottles. Moreover, galvanism and lightning are threaded into Shelley's tale, which she relates through a frame narrative from letters and notes written by Robert Walton to his sister. After Walton, captain of a ship bound for the North Pole, rescues Victor Frankenstein, exhausted and near death from his grueling search to chase down the monster he had created, he records the tragic tale told to him by Frankenstein with his last breaths of life.

> Before this I was not unacquainted with the more obvious laws of electricity. On this occasion a man of great research in natural philosophy was with us, and excited by this catastrophe, he entered on the explanation of a theory which he had formed on the subject of electricity and galvanism, which was at once new and astonishing to me.[42]

It is likely that such research and the intention to use electricity to revive the dead were widely known in the lead-up to Ure's experiments on the hanged murderer; indeed, such demonstrations of the power of electricity to restore movement to a dead body have a longer history. Most famously, this history includes experiments performed by Italian scientist Giovanni Aldini on January 17, 1803, at Newgate Prison in London. Furthermore, Mary's eventual husband, Percy Shelley, had a long fascination with electricity, which is why the illicit lovers had attended Crosse's lecture together in 1814, two years before Mary began to write her book in 1816. She commenced writing it in response to a challenge by Lord Byron to his summer guests on the shores of Lake Geneva, to pen ghost stories.

George Forster was found guilty of drowning his wife and child in a canal and sentenced to death by hanging, after which his body was delivered to Aldini for public dissection. Aldini was the nephew of Luigi Galvani, the discoverer of "galvanism," or biological electricity. As reported in *The Times*, "On the first application of the process to the face, the jaw of the deceased criminal began to quiver, and the adjoining muscles were horribly contorted, and one eye actually opened. In the subsequent part of the process the right hand was raised and clenched and the legs and thighs were set in motion."[43] Aldini then applied electrodes to Forster's ear and the rectum, exciting violent contractions of the body. This display of the effects of electricity on a human subject was the culmination of numerous experiments on animals that Aldini had performed in

Bologna; the Italian scientist Antonio Maria Vassalli-Eandi (1761–1825) had previously conducted a series of experiments much like those performed by Aldini on several newly decapitated criminals in Turin, provoking the same shock and sensationalism in the public.[44]

## FROM ELECTRIC BRAIN TO ELECTRIC CHAIR

In 1881, a Buffalo, New York, dentist, Alfred P. Southwick, conceived of applying the life-and-death power of electricity to devise a more humane method of execution than hanging—the electric chair. Thomas Edison was an enthusiastic backer of this quest, bankrolling the dentist in a diabolical maneuver to defeat his rivals, Nikola Tesla and George Westinghouse, with whom he was locked in a battle for the nation's electrical grid. Edison promoted the direct current (DC) model he had developed, while Tesla and Westinghouse advocated for their alternating current (AC). In a campaign to convince the public that AC current was dangerous, Edison demonstrated its lethality by electrocuting dogs, calves, and horses in public demonstrations. Making sure his rival was firmly associated with the deadly dangers of AC current, Edison advocated that criminals be "Westinghoused" in an electric chair.

The first person to be executed by electricity while strapped to the dentist's invention was William Kemmler, convicted of murdering his wife with a hatchet. Kemmler was executed on August 6, 1890, by electric chair in Auburn Prison in New York, but the process did not go smoothly. Seventeen seconds of 1,000 volts of AC current rendered him unconscious and smoldering . . . but still alive. The generator was recharged and boosted to deliver twice the voltage. The resultant bolt of electricity ruptured blood vessels and burned the flesh in contact with the electrodes. The horror, smoke, and stench of burning flesh filling the air caused several in attendance to be overcome with nausea, and two witnesses fainted. The back of Kemmler's coat caught fire; minutes passed before Kemmler finally went rigid.

"They would have done better using an axe," said George Westinghouse.[45]

Electrocution is still used for executions in the United States in

Alabama, Arkansas, Florida, Kentucky, Oklahoma, South Carolina, Tennessee, and Virginia; execution by hanging is currently authorized only in Delaware, New Hampshire, and Washington State.[46]

This is the bizarre story of the discovery of the electric brain, including animating the dead, struggling through war and atrocities, and venturing into the paranormal terrain of mental telepathy. We now know that electrical brainwaves are generated by all complex brains, even those of octopuses and other invertebrates, but what are they, really? What can they tell us, and how can we discover their function? In the next chapters, we will investigate these questions, in part by exposing my own electric brain.

## PART II

# Understanding Brainwaves

# CHAPTER 3

# Brainwaves Explained

The tiny room is dark. Inside, the blue glow from a computer monitor, mounted in a steel tower of electronic instruments bristling with knobs and switches, illuminates my profile. In my white lab coat, bent over a microscope suspended on a boom, I peer through the binocular eyepieces. Beneath the lenses of the microscope, a brilliant white spotlight is focused on a rat's furry white head. The top of its skull has been removed to expose its fleshy brain; wet and glistening, the surface is laced with pulsating blood vessels. In the center of the spotlight, a thin, four-inch-long needle, resembling something an acupuncturist might use, sticks up out of the putty-colored flesh. Deeply anesthetized, the rat breathes peacefully, and the room is quiet except for the hum of electronic equipment, a gently hissing speaker, and the sound of gurgling fluids—artificial cerebral spinal fluid and solutions of drugs for the experiments, bubbled with 95 percent oxygen and 5 percent carbon dioxide.

Neurons communicate by generating electrical impulses of about one-tenth of a volt in flashes lasting one one-thousandth of a second, as brief as the flash of a camera strobe. To detect these weak and fleeting signals it is necessary to tap into individual neurons by probing them with a sharp microelectrode. The needle stuck in the rat's brain is such a microelectrode, so named because its tip is sharpened to pierce inside a nerve cell. It is clamped into a device called a micromanipulator, and connected by a pair of slender wires to a high-gain electronic amplifier that sits within the tower of instruments next to the stainless-steel table where the unconscious rat rests. The steel table floats on four air pistons that eliminate any vibration. By twisting three knobs that control the micromanipulator, I can position the electrode in 3-D space with minute precision, and advance or retract it by microscopic distances.

While I slowly turn the control knobs of the micromanipulator, I listen via a speaker to the signal fed through the amplifier from the microelectrode and watch a voltage trace as it sweeps across a computer screen. As I penetrate slowly through brain tissue, sounds like radio static flood the room; when I begin to hear faint popping sounds emerging from the background static, I know my probing has encountered a neuron and is pushing up against it. This popping is the sound of a neuron firing not far from the tip of my microelectrode. Now I slowly and carefully advance and retract the electrode like a dowser searching for water, listening to the popping grow fainter or stronger as I probe.

Suddenly the popping cracks loudly and furiously, like a burst of hail on the hood of a car in a thunderstorm. My electrode has made decisive contact—I am directly tapping into the bioelectrical signals of an individual neuron. With each pop of the neuron firing, I see the voltage trace on my computer screen spike and fall, looking something like a heartbeat monitor in a hospital.

I must be very careful not to bump anything that might move the electrode even microscopically, and I feverishly begin to collect data. I know that I will only be able to record this neuron's firing for a brief period of time. How much time? I never know because at any time the cell can become damaged or lose contact with the electrode so it can be a few seconds, minutes, or tens of minutes.

This method of recording the activity of neurons is very instructive. By studying neuronal firing in response to stimulation, for example, scientists called electrophysiologists have learned a tremendous amount about the electrical properties of neurons. Observing the action of individual neurons has taught us how they function together in neural circuits. As you can see, however, recording the impulses of a single neuron is a slow, difficult, and tedious process. It requires delicate brain surgery, carefully maintaining the respiration and body temperature in the animal under anesthesia, as well as the ability to extract information and efficiently perform an experiment in the heat of the moment.

In contrast to the painstaking process of this "single neuron" recording, the combined activity of large groups of neurons firing together causes fluctuations in the surrounding electric field that can be detected by electrodes placed on the scalp. Yes—we have returned to brainwaves. The activity recorded from outside the skull by an electroencephalogram (EEG) can be thought of as a bit like the roar of a stadium heard from the parking lot. You can't hear the conversations of individual spectators (i.e., the activity of individual neurons); instead, you hear collective disturbances—the roar changes quality when a goal is scored, for example. However, anywhere from a dozen to over a hundred electrodes can be placed on the scalp in precise locations, each acting something like an applause meter tracking the activity of everyone cheering in a specific area of the stadium and each one feeds the signals it picks up through the skull into an amplifier and then into a computer to be recorded for playback and analysis. Thus, a hundred different traces of voltage fluctuations at different points on a person's head can be collected simultaneously. So rather than spikes of electricity from single neurons, EEG recordings trace out wobbly lines and waves that look something like the tremors traced out by pens on a lie detector machine.

But are brainwaves simply the electrical noise of the brain at work or are they *how* the brain works? Can changing our brainwaves change our brains? If we are able to identify brainwave patterns associated with neurological or psychological dysfunctions, can altering those patterns correct those dysfunctions? This is precisely what the two women you are about to meet are attempting to do.

## FROM ELECTROSHOCK TO NEUROFEEDBACK

Joey was a beautiful baby, with chubby cheeks and bright blue eyes, but at six months of age his expression remained doll-like; he wasn't reacting joyfully to his mother's smiles, and while his eyes moved about, he did not connect with the gazes of others. Joey's parents had eagerly anticipated happy, playful time with their new baby, but instead these months were suffused with worry over their son's lack of social responses. By the time Joey was a year old, their concern had swelled into agonizing fear. Joey did not show any reaction to hearing his own name, something even the family's floppy-eared puppy had quickly learned to do. Joey did not engage in the babble so common to babies—to communicate, he would thrust out his tiny arms or wave them, but he had no words, and remained seemingly indifferent to interaction with those caring for him. Something was wrong.

The doctors tested his hearing, finding it normal, and performed a battery of other medical tests. All of the results pronounced Joey a perfectly healthy toddler. Then, one unforgettable day, just down the hall from the examination room, they sat across from Joey's doctor in his private office. From behind his polished oak desk before a wall of framed diplomas, the doctor delivered the news: Joey was autistic.

Autism spectrum disorder (ASD) is a psychological condition that arises when the brain develops atypically, resulting in differences in the way information is processed—especially sensory stimuli, which is often experienced as overwhelming. The neurological causes are being intensely studied by scientists, and while there are many intriguing clues, there is no clear understanding of what causes people to be autistic, although there appears to be a strong genetic component. According to the Centers for Disease Control and Prevention,[1] autism spectrum disorder is alarmingly prevalent, afflicting one in fifty-nine children in the United States. Statistically, boys are roughly four times more likely than girls to be diagnosed as autistic.[2]

As the name suggests, autistic individuals can experience a wide range of levels of difficulty as a result of their differences. In Joey's case, even minor changes in routine or in his surroundings overwhelmed and upset him. He repeated words and phrases over and over as he rocked and fidgeted in isolation. By the time he reached school age, he still had no close friends.

But after a practitioner named Jessica Eure worked with Joey, who was eight years old at the time, things changed dramatically. Joey's mother was so elated, she eagerly agreed to share Joey's experience. "Normally she can only complete one or two chores in a given day because if she does more he decompensates and is overwhelmed," Eure told me, reading from her notes on the changes reported by Joey's mother. "Today they went all over the place. He got a haircut at a new salon. He tried new foods at a restaurant. She popped into a new store and bought some clothes. They went to the post office and the grocery store. Her husband recently took Joey to a movie, and to the mall for lunch."

"This has never happened before," Joey's mother told Eure. "He's doing great and he's so happy . . . there are more changes that I'm forgetting—there have been so many firsts."[3]

Joey was not treated with a new drug. His transformation did not come from a novel form of psychotherapy or behavioral modification. Instead, Jessica Eure and her colleague Robin Bernhard helped Joey with a new and revolutionary approach. They recorded his brainwaves as he sat quietly in a chair doing nothing but letting his mind wander for a few minutes. Then they sent the EEG recordings to their colleague, Jay Gunkelman, in California, who analyzed Joey's brainwaves in an effort to pinpoint any patterns of abnormal electrical activity sweeping through the neural circuits of Joey's brain. After receiving Gunkelman's report, Eure and Bernhard invited Joey back and used a noninvasive and drug-free method to change his EEG patterns: neurofeedback.

The neurofeedback method they used involved monitoring Joey's brainwaves in real time as a computer gave him a positive indication (a tone) whenever his EEG patterns were responding to the training protocol based on previous assessments. There was no conscious effort involved in this process. The brain can correct itself, automatically and unconsciously, as long as it is given the guidance needed to do so via neurofeedback.

This approach, although well supported by the latest neuroscience research, is not currently covered by most health insurance. Some dismiss neurofeedback as a fad from the 1960s, when an explosion of research in the area resulted in a glut of practitioners offering biofeedback treatments—with varying levels of skill and success. But the science has progressed quite a bit from the lava lamp days. Still, it takes years before new findings from scientific research are ready to be applied by doctors treating patients. Eure and Bernhard may be on the cutting edge of a future where psychiatrists replace their couches with EEG machines, or they may instead be nibbling at the fringe of alternative medicine that in fact only engages the brain's powerful placebo effect. (The placebo effect is when people find benefit from a treatment that they believe will help them, as when given, unknown to the person, a sugar pill rather than a real drug.) To find out more for myself, I contacted Jessica Eure and asked if she would agree to monitor my brainwaves and permit me to experience the neurofeedback method she has used to help people with autism and a wide range of psychological conditions, from addiction to OCD. She enthusiastically agreed.

Jessica Eure's interest in EEG began in 2001, when she was doing an undergraduate honor's thesis on REM sleep. The mystery of sleep and the changes in the brain as we slumber were unlocked by EEG (as will be explored in detail in chapter seven), and Eure studied EEG recordings of the sleeping brain as part of her research. After college, in 2002, Eure began working in a psychiatric hospital for patients suffering from severe and persistent mental illnesses, where she soon became discouraged by the experience of seeing patients suffer while feeling powerless to really help them. "It was my first job in the mental health field, and I wasn't sure I wanted to stay in mental health, seeing what I was seeing there," she told me. She was especially disturbed by ECT—electroconvulsive or "shock" therapy—which many of the patients were receiving routinely.[4]

## ECT THERAPY

The idea of curing mental illness by shooting jolts of electricity through the brain originated in the 1930s by a Hungarian doctor named Ladislas von Meduna, working at a mental hospital. He was not a psychiatrist but a neuropathologist. Meduna examined human brain tissue at autopsy, or taken by biopsy during brain surgery, and in the course of doing so he made an interesting observation about a type of brain cell called glia. Unlike neurons, glia cannot fire electrical impulses, so they were (and still are) largely ignored by most neuroscientists.[5] But Meduna noticed that brain tissue taken from people who suffered from epileptic seizures had a greater number of glia, and their glia cells were unusually bloated. On the other hand, brain tissue from patients with schizophrenia and depression had far fewer glia than normal.

Meduna put his observations together with another curious set of facts: In 1926, Dr. Robert Gaupp, a psychiatry professor in Tübingen,

Germany, had reported that while schizophrenia and epilepsy are two very common disorders, very few people suffer both illnesses. In 1929, Drs. Albin Jablonszky and Julius Nyiro, at the state asylum in Budapest, reported that on the rare occasions that epileptics developed schizophrenia, their epilepsy was often cured. The recovery rate from epilepsy was sixteen times greater than normal after the first psychotic episode. Meduna had also observed from his time in the hospital that patients with schizophrenia who happened to suffer a spontaneous seizure often experienced a reduction in psychotic symptoms afterward.[6]

Piecing together these clues, Meduna concluded that schizophrenia and epilepsy were both diseases resulting from an imbalance in glia: too few glia in schizophrenia, and too many in epilepsy. He theorized that inducing a seizure in people with schizophrenia would increase the glia in their brain and cure them. Meduna's strange ideas about mental illness were viewed by his colleagues as aberrant, and—like Hans Berger—he carried out his preliminary research in secret.

On January 23, 1934, Ladislas von Meduna induced a violent seizure in a schizophrenic patient in an attempt to cure his mental illness. The man, suffering from schizophrenia, was hopelessly catatonic. He had lain in bed for four years, immobile and fed through a tube. Meduna injected a chemical, camphor, into the catatonic patient's bloodstream. Instantly, the injection induced a violent seizure that wracked the man's body for a full minute. Meduna's legs gave out in shock upon witnessing what he had done, and he collapsed. Two nurses had to help the doctor back to his apartment to recover.

But four days later, Meduna bravely continued his experiment, inducing another seizure in his patient. On February 10, two days after the patient's fifth seizure induced by injection—less than three weeks after the first—the man awoke, dressed himself, requested breakfast, and cheerfully greeted Dr. Meduna by name. "I hear them talking that you were going to make some crazy experiment? Did you do it?" he asked.[7]

The treatment Meduna had devised seemed barbaric—but it worked. That man was only the first of several mental patients to become Meduna's experimental subjects. Eventually, chemical injections to induce seizure were replaced by electrical stimulation. In Italy, Ugo Cerletti and Lucio Bini adapted electrodes that were used to stun pigs at the slaughterhouse for use in delivering electrical current to the heads of their mental patients. Working first with dogs, they determined how much voltage they could deliver and how long it could be applied without killing the animal. Then they set out to shock the mental illness out of a patient's brain. Bini asked Meduna if he thought convulsions induced by the pig–stun gun electrodes would have the same effect as chemically induced seizures. Meduna predicted that they would.

On April 11, 1938, electrodes were pressed against the temples of Enrico X, a schizophrenic patient, as Cerletti, Bini, and several other colleagues witnessed the first use of an electrically induced seizure to treat mental illness. After seven more shock treatments, Enrico X was discharged—cured of schizophrenia.

None of these experiments would have been permitted under the ethical standards for human experimentation today. Convulsions are frightening for observer and patient alike, and seizures can be dangerous. In the early years of ECT, many patients suffered fractures from the violent contractions—today these are prevented by using muscle relaxants and more controlled application of electricity. Yet Meduna's experiments not only led to a new effective treatment for a life-threatening disorder, they transformed thinking about mental illness.

Previously, psychotherapy was considered the only treatment for disorders of the mind, like depression. More severe mental illnesses, such as schizophrenia, were understood to be genetic defects beyond remedy, and patients suffering from these conditions were often locked up and condemned to a life without hope. Meduna's conclusion that mental illness might have a biological basis that is treatable through physical intervention was a turning point. The model of mental illness as a physical dysfunction in the brain is what ultimately led to the development of the psychoactive drugs that are the mainstay of modern treatments for depression, schizophrenia, and many other psychological disorders.

Meduna had no biological understanding of how glia might be involved in epilepsy or the mechanism by which these same cells might affect mental illness, but today we know that glia are in a position to control the balance between mental health and madness. A type of glia cell called astrocytes, for example, wraps around synapses and removes the neurotransmitter that is released from neurons in communicating across synapses. Generally, electrical signals are not transmitted between neurons by direct electrical connections; instead, a chemical messenger (neurotransmitter) is relayed across a microscopic gulf of separation between neurons called the synaptic cleft. When an electrical impulse reaches the tip of a nerve fiber, a neurotransmitter is released, which stimulates receptors on the recipient neuron. To terminate the signal, the neurotransmitter must be removed, and astrocytes perform this important function. An imbalance in neurotransmitters disrupts normal brain function, affecting our moods and emotions, and is responsible for the symptoms of many mental illnesses. All hallucinogenic drugs have their mind-bending effects by disrupting the normal balance of neurotransmitters in brain circuits involved in emotion and higher-level cognitive function. All drugs for treating depression and psychotic disorders work by regulating the level of specific neurotransmitters in the brain, such as SSRI (selective serotonin reuptake inhibitors) medicines used to treat depression and other mental illnesses, which regulate the amount of the neurotransmitter serotonin available in the brain. And regulating neurotransmitters is precisely what astrocytes do. While our understanding of how astrocytes influence synaptic transmission has only come about in the last twenty-five years, Meduna's hunch does not seem so wild now. Today, ECT remains the most effective therapy for major depression in cases that resist treatment by drugs, yet scientists are still unsure how it works.

ECT has been transformative for many, restoring countless men and women to healthy, productive lives, but it doesn't always work. Moreover, sending high-voltage jolts of electricity through a person's brain to induce a seizure can have some serious consequences—memory loss is the most common, but other cognitive impairments or personality changes can occur as well. Working in the psych ward, Jessica Eure saw many patients with severe depression being repeatedly treated with multiple rounds of ECT; even when it was effective, the improvements were often only temporary. Eure was seeing these discouraging failures firsthand, every day. It bothered her to see these people suffering, and she became disgusted by the cynicism she saw in others working in the mental hospital. Too many seemed to view the mental patients as incurable, and they applied ECT with apparent indifference, while at the same time regarding the treatment as pointless.

Having studied EEGs in college, Eure could well imagine what was happening inside the brains of patients as their brainwaves were subjected to a violent electrical storm via shock therapy. She could visualize the healthy waves and oscillations traced out on the EEG machine suddenly erupting into chaos, like the pen on a seismic monitor scribbling madly during an earthquake. Every circuit in the brain would be blasted, and the explosions of electrical activity reverberating through the cerebral cortex destroy the normal flow of information and operation of the brain, causing loss of consciousness, memory loss, and seizures. In the aftermath of this firestorm that devastated normal brainwave function, Eure could imagine how the brain might reset itself differently in recovery, but this approach seemed far too imprecise to target repair of the particular brain circuits that were not behaving normally. It was too blunt, too brutal, and too fickle. The intersection between the ECT treatments she encountered in the hospital and what she'd learned about EEG in college moved her to explore the possibility that EEG might be able to detect specific abnormalities or weaknesses in the brain, and then, by way of neurofeedback, be used to correct them.

## MY OWN ELECTRIC BRAIN

I drove to Charlottesville, Virginia, on a sweltering summer afternoon when the air was thick with humidity. The next morning I would meet Eure and her colleague Robin Bernhard, who together run the Virginia Center for Neurofeedback where I'd be having my brainwaves recorded, not far from the University of Virginia Medical School where I had given a lecture on my research on brain plasticity only a couple of months earlier.[8] Charlottesville, Virginia, is a college town, grown up around the stately red brick and white trim of the university founded by Thomas Jefferson, but when I arrived at 8 AM I found the Neurofeedback Center isolated from the buzz and urgency of the hospital and medical school, situated on a shady, quiet side street in a tidy white-and-green trimmed house with a wide porch spanning the front. I imagined the patients who are treated here, many of them children, might find this a relaxing respite from the stress that permeates hospitals and clinics. As I approached the house, a young woman wearing a bright blue sleeveless blouse patterned with cheerful blue and white hibiscus flowers rose from a wicker chair on the porch to greet me. Extending her hand, she introduced herself as Jessica Eure.

In preparation for my EEG exam, I had completed a survey of several hundred questions about any problems I might have and asking me to rank each of a series of statements on a numbered scale from "never" to "virtually always." These included "I have a special phrase that I repeat over and over" and "I have an inability to remember faces," among many others. Simply completing the survey gave me new insight into the vast array of difficulties the patients Eure treats might be experiencing, and made me feel fortunate that I really have no complaints.

The written instructions I had been given told me to shampoo my hair twice that morning, dry it completely, and use no styling products. They also informed me that I must forgo stimulants of any kind prior to the EEG recording. So I arrived at the Center with unruly, frizzy hair sprouting untamed from my head and a groggy somnolence from being denied my morning cup of coffee. I told Eure that she would need to turn the amplifier gain on her EEG machine up to the maximum, because I wasn't sure I *had* any brainwaves before my morning cup of Peet's. She said she had that covered. Her mentor and colleague, Robin Bernhard, would be arriving soon with coffee, and there would be a fresh cup for me to guzzle right after I finished my EEG recording. "I don't want you to suffer a gnarly caffeine withdrawal headache," she said with a sympathetic smile.

While we waited, seated comfortably in her office, she answered my questions and retraced the steps that led her to her work with Bernhard. Summing up her early experience in the mental hospital, she said, "I was watching people stay sick." In addition to the severely psychotic and homicidal patients in the ward, a lot of the patients were elderly people, living in nursing homes, who were depressed. "They are depressed and not taking care of themselves and getting catatonic," she said. Eventually, they'd be sent to the psych hospital for treatment. "They give them ECT and it gets them perked up for a couple of months," Eure explained.

"But then you are kind of back to where you were, but maybe with some memory loss. It was just really upsetting to me."

The despair she felt working there was a pivotal experience in Eure's life. Looking back on it now, she said, "I think being fresh out of college, I was shocked to go into this scenario, and I felt like, this seems pretty dire. I felt that ECT was so rudimentary in a way, just to create an electrical storm in the brain and hope that basically the brain forgets the state it was in before, so I started researching alternatives to ECT." Eure felt similarly disenchanted with some of the drug treatments that were being used: "I began to question what the purpose was—how were we actually healing people?"

"I was really interested in the intersection of mind and body. I decided to buy the equipment for neurofeedback and slap it on my own head and experiment. I bought my first EEG machine when I was twenty-two, and I'm thirty-eight now."

"What was the experience like when you first tried it on yourself?" I asked.

"I overtrained myself. I calmed myself down too much and woke up in the middle of the night in a panic attack [a rebound effect]. That gave me a healthy respect for its effectiveness. This is something that you can cause adverse reactions with, even though it is considered safe. I think that experience was good. I think everyone should practice what they do on their clients on themselves. I've done hundreds of sessions on myself at this point."

Contrasting her priorities and path to neurofeedback with those of her business partner Robin Bernhard, she said, "Robin is a psychotherapist first and a neurofeedback person second." In 2003, Eure became Bernhard's technician, responsible for hooking patients up to the neurofeedback devices. "I really liked it. It felt right—the intersection between clinical work and the science. It felt cutting edge . . . a nonpharmaceutical intervention that didn't cause harm and created real change."

So Eure went back to school to get her master's degree in counseling, eventually becoming licensed as a mental health counselor and nationally certified in neurofeedback. She and Bernhard have run the Center together since 2007. Coming to neurofeedback therapy from opposite directions has made them a complementary team.

At this point, Robin Bernhard arrived (bearing the coffee I wasn't yet allowed to drink) and joined the conversation. Having heard how Eure found neurofeedback through her interest in brainwaves and her experience in the mental hospital, I asked Bernhard how she had gone from standard psychological practice to using EEG analysis and neurofeedback on her patients.

"I fell into working with people suffering various types of trauma," she told me. "I started with hospice work and working with mothers who were dying and had young children. I started getting referrals for children who had witnessed violence, and I got into divorce groups for kids, and then I expanded into general PTSD."

She was using established techniques to help these patients, but found that it wouldn't take very much for them to regress. This was especially true for those who had suffered trauma very early in their development. Most traditional forms of psychotherapy focus on helping the patient consciously develop and apply insight, but this is difficult for such young patients who have not developed a cohesive sense of self. "I was interested in neurofeedback as another way of supporting these clients," explained Bernhard, "something where they didn't have to work so hard. Neurofeedback hooks up directly to your brain. It bypasses your mind."

"It is a physiological intervention at the unconscious level," Eure explained. "We are interacting with the reward centers in the brain and reflecting the brain's activity back to itself through auditory feedback."

According to Eure, about half their clients find them on their own, and half are referrals from other clinicians.

"For a lot of our clients, this is their last stop. They've been through the medical mill. They've tried every medical therapy; they have sometimes done electroconvulsive therapy, have been on multiple medications, and the drugs can have long-standing side effects, even when they are off of that medication. They feel hopeless. And we can say, 'Hey, there's something else you haven't tried and this can help you feel better,' and that's a *really* wonderful thing."

"It's a gift," Bernhard agreed. "When you can see someone come in a wreck, and go out feeling . . . sometimes remarkably better, and other times just a little bit better—but better!"

But early on Eure and Bernhard noticed that some people *didn't* get better after neurofeedback, and they couldn't determine why. At the time, they were only applying neurofeedback based on the patient's symptoms and weren't using medical-grade EEG to guide their neurofeedback. These devices worked by monitoring brain activity via EEG in real time and prompting the user when their EEG activity was changing in a way to increase or decrease activity of certain frequency components of the EEG, which are associated with general behavioral states, such as relaxation. They found that by giving patients a comprehensive nineteen-channel EEG analysis prior to neurofeedback therapy, they could identify subtle deviations in their patient's EEG that were likely to be associated with the cognitive and emotional issues they were experiencing. Surprisingly, they found that this analysis also frequently revealed what appeared to be clinically abnormal EEG activity. In such cases they referred the person to neurologists who could perform medical EEG analysis and other tests for diagnosis. Many of these disorders can be treated medically (and with neurofeedback) once they are properly diagnosed. This "quantitative" EEG analysis data (qEEG) allowed them to obtain objective measurements of EEG signals and apply more targeted neurofeedback to strengthen the specific weakness they found in the individual's brainwave patterns. The questionnaire I filled out is used to help localize which cortical regions are the source of the patient's difficulty. "Almost always the symptom clusters identified on the questionnaire are associated with the brain regions we end up training based on the EEG data," Eure explained. "So we'll see a lot of right temporal, limbic regions in the symptoms, and when we do the EEG analysis it will also come up and say that the right temporal lobe is unstable, and that's where we need to train." Sophisticated EEG analysis involves a lot of time, expertise, and expensive equipment, but it greatly increased the success rate of their neurofeedback therapy.

Now the time has come to record *my* brain in action. Eagerly anticipating that cup of coffee, I sit patiently while Eure wires me up, applying conductive gel and shirt-button-like electrodes to various locations on my scalp. She dabs a spot of sticky gel onto the bump on my skull that juts out behind my ear, and then presses an electrode atop it. "Your mastoids are very vertical," she says, struggling to keep the electrode from sliding off.

"No one has ever told me that before," I reply, not sure if her comment is intended as a compliment or a criticism. The knobby mastoid bone rises behind the ear from the petrous part of the petromastoid, a piece of the temporal skull bone that has the distinction of being the densest bone in the body.[9] DNA locked inside this ultradense bone is preserved as if in a vault, making it possible to recover genetic code from the dry, dead bones of humans who lived and died thousands of years ago. The petromastoid bone enables geneticists tracing the evolution of our species to probe the origins of humankind, but at the moment, mine is thwarting efforts to probe my brain.

Despite this, I soon have nineteen electrodes plastered to my head with clear goo. My hair is now bizarrely spiked like that of an insane punk rocker, with wire tendrils cascading down from the electrodes on my head and attaching to a remarkably small and simple-looking white

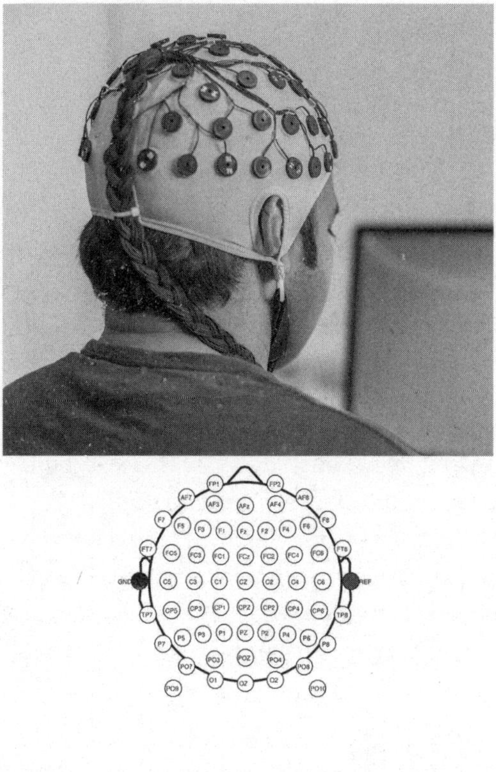


**FIGURE 16:** *Electrodes on the scalp pick up brainwaves for analysis (electroencephalogram, or EEG). The standard locations of the electrodes are noted in the figure at the bottom. Prefrontal (Fp), frontal (F), temporal (T), parietal (P), occipital (O), and central (C). Zero (Z) refers to the midline position, and the numbers refer to the lateral location of electrodes, with even numbers on the right and odd numbers on the left of the midline.*

plastic device about the size and shape of an elongated cigar box—the EEG amplifier that connects to the computer. Eure tells me that I can turn and look at the computer screen; I do and am stunned: Nineteen vigorously wiggling traces are sweeping across the screen. It is my electric brain!

The electrical traces respond instantly to whatever I do. When I blink or stiffen my jaw or neck muscles, the jiggling lines erupt in wild fluctuations. "Don't do that," Eure commands. "Stay still and relaxed." These disturbances are the electrical signals from my muscles; the contraction of any muscle fiber produces a large burst of electricity. To record the weaker electrical signals permeating my skull from my cerebral cortex, I must remain calm and stationary.

Looking at the voltage fluctuations being recorded by the nineteen electrodes, I can see that some are synchronized, tracing out near duplicate responses from two electrodes, while others appear uncorrelated. Synchronized signals result from two parts of my cerebral cortex that are functionally connected and working together, but this connection is dynamic, like two sections of an orchestra that may play together at one point and then break off to play individual parts on their own. The appearance of synchrony between two of the tracings can also arise when two adjacent electrodes are in fact both sampling the same region of my cerebral cortex. By carefully analyzing the synchrony of electrodes from precisely placed points on the skull, EEG researchers can triangulate the signals and locate their origin. This is called "source detection," and it allows us to map groups of neurons that are firing together.

Eure tells me to close my eyes for a moment. The jittery spikes of my EEG are suddenly quelled and replaced by slow, rolling oscillations—the low-frequency alpha waves growing in strength as the higher frequency waves subside. With vision—the main channel of information into the human brain—shut off, large parts of my cerebral cortex have little to do, and my brain is left to its own internal business. I see this sea change in my brainwave activity charted on the screen when I open my eyes.

What we now call alpha waves were originally called "Berger waves," and I have the profound sense that I am reliving that very first observation of human brainwaves in Jena, Germany, when Hans Berger asked his son, Klaus, to close his eyes, and saw the change in his son's mental

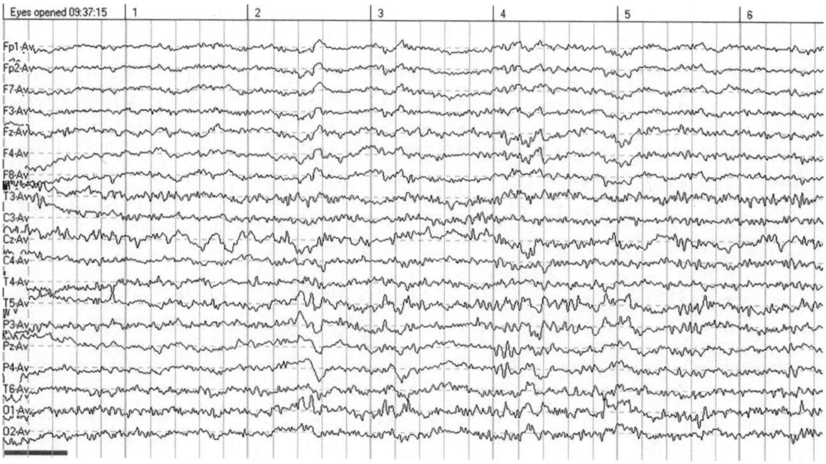


**FIGURE 17:** *A brief sampling of the author's EEG with his eyes opened. Each horizontal line represents the voltage changes recorded at each of nineteen electrodes placed in different locations on the scalp.*

state reflected in the physical manifestation of electrical energy. But I am also struck with a strange existential insight. There is no light in the brain. What we perceive as light is a fabrication of our brain concocted with brainwaves.

Eure records two ten-minute sessions of my EEG as I sit doing nothing; first with my eyes closed and then with my eyes open. Seeing my brainwaves gives me the peculiar feeling of catching one's reflection unexpectedly in a shop window. It's my brain! It's me! Me? I've never seen it before, but there it is, alive and changing; responding to my movements and thoughts, the physical underlying "me" that produces the outward representation of who I am. It is the man behind the curtain in *The Wizard of Oz*. It is a fascinating revelation—and a bit creepy.

They tell me it will take Jay Gunkelman and his colleagues at Brain Science International, in San Ramon, California, a week to analyze these recordings of my brainwaves. Once the results are in, I will return to Charlottesville to hear them, and then attempt to use neurofeedback to strengthen whatever weaknesses they reveal. As I sit, letting Eure extract my brain's secret transmissions, I'm a bit anxious anticipating the outcome. What will Jay find? With a sense of dread, I imagine I can hear

Gunkelman now—"Perfectly normal for your age . . ." The memory of Dr. Fayed in Zaragoza, saying those same words after reading my post-mountaineering MRI, pops into my head. Somewhere in the rippling electrical signals spilling out of my brain and into Eure's computer, I know that my recall of this memory is being reflected. The information is there; the challenge is developing the understanding and methodology needed to extract it.

We will return to Charlottesville and my brain's not-so-secret transmissions in chapter five, but first we need to dive deeper into what brainwaves tell us and how.

## DECODING BRAINWAVES

The electrical signals detected by EEG recordings are complex, fluctuating in a bewildering dynamic pattern like a chart of stock market prices. The stock market fluctuates over many different time scales, from the rapid minute-to-minute price fluctuations that some investors exploit using computer algorithms to make automated high-frequency trades at rates of thousands of transactions per second, to the hourly, daily, quarterly, and long-term changes that other investors monitor. All of these different scales of price fluctuation occur simultaneously, and so it is with our brainwaves, which fluctuate simultaneously at many different rates, from less than one to several hundred cycles per second (or Hz). Brainwaves are categorized into different types—which we'll detail shortly—based on the frequency at which they oscillate.

Just as stock prices fluctuate according to trading activity, brainwave fluctuations reflect the voltage activity around groups of neurons. But here is where our comparison with stock prices comes to an end. Brainwaves don't only tell us which neurons are firing; we now know that brainwaves themselves can directly cause neurons to fire. Brainwaves are a means of communication. Much like transmissions from different radio stations, brainwaves in different frequency bands radiate through the brain simultaneously. Each band carries distinct types of neuronal information that is transmitted and received by different groups of neurons located in different brain regions. You've probably heard the phrase "neurons that fire together, wire together"? Actually, brainwaves can couple populations of neurons that are not directly wired together into coordinated functional assemblies that fire together rhythmically, like soldiers marching in lockstep. Even neurons spread across distant regions of the brain can be coupled together by brainwaves, just as soldiers everywhere in a vast parade are synchronized by a drumbeat. "Neurons that wave together, behave together!"

To understand why this coupling of neurons by brainwaves happens, it's important to understand how neurons generate their electrical charge and how they are triggered to fire it. At rest, neurons are about one-tenth of a volt more negatively charged than the fluid surrounding them, creating a net negative voltage across the neuronal membrane. This negative voltage is created, just as in a battery, by separating positively and negatively charged ions—the neuron does this by accumulating fewer positively charged sodium ions inside the cell than outside, with the help of voltage sensors in the cell membrane that constantly monitor the difference between the voltage inside and outside the neuron. A neuron "fires" when it rapidly discharges the negative voltage created by the deficit of positively charged sodium ions inside the cell; this is the neuronal impulse or "action potential."

A neuron can be triggered to fire when another neuron releases neurotransmitters that carry a signal across the synapse, causing channels through the membrane of the recipient neuron to open briefly, allowing charged ions to pass through. The rapid flow of charged ions through these open channels disturbs the resting voltage, slightly reducing the negative charge in the recipient neuron. When the sensors in the cell membrane detect that the voltage has dropped to a certain point—say from minus 70 millivolts to minus 50 millivolts—they open further channels that allow positively charged sodium ions from the fluid surrounding the neuron to rush into the cell. The neuron's remaining negative voltage is completely discharged in a flash, generating an electrical impulse. A discharge of a tenth of a volt does not sound like much, but on the microscopic scale of the cell, this is equivalent to a lightning bolt.[10] If you divide one-tenth of a volt across the microscopic thickness of a cell membrane, you get the same result as dividing the thousands of volts discharged in a bolt of lightning by the distance the lightning bolt traverses from cloud to ground.

A neuron's discharge depletes the "battery" generated by the negative charge of the cell compared to its surroundings that was present before it fired the impulse. To fire another impulse, the negative voltage across the membrane must be re-established, just as a camera flash must recharge to fire again. To restore the negative charge after firing an impulse, different channels in the membrane detect that the negative voltage across the cell membrane has been depleted and open to eject positively charged potassium ions. These ions are accumulated inside cells by potassium pumps, which concentrate positively charged potassium ions and hold them in reserve to help establish and recharge the transmembrane negative voltage. When the potassium channels open after a neuron fires an impulse, these positive charges rush out of the cell. Like ballast dumped from a ship carrying too much weight, the ejected potassium ions restore the normal net negative charge inside the cell at rest.

To summarize, a neuron fires when positive charges (sodium ions) flowing into the cell drop the net negative charge across the cell membrane, triggering a rapid discharge of the neuron's negative voltage. Positively charged potassium ions then rush out of the cell, recharging its battery and readying it to fire another impulse. A neuron is like a banana floating in seawater—high in potassium on the inside, but surrounded by a sea of sodium chloride. The key point is that the firing of an electrical impulse is triggered when the voltage sensor in the neuronal membrane detects that the voltage across the membrane has reached a specific threshold point. Really, the concept of firing an action potential when the voltage drops to a specific point is no more complex than understanding how a toilet flushes. When the handle is dropped to a specific trigger point, the stored water in the toilet tank, like the stored negative voltage in a neuron, dumps out rapidly and then must be refilled.

You may have previously encountered all this information about how neurons fire in school, but you likely never learned that this textbook description is only half of the story—brainwaves are the other half. Because it is a reduction in the net difference in voltage across a neuron's membrane that triggers an impulse to fire, two different things can cause the voltage to reach the trigger point: Either the charge inside the neuron can become more positive, or the charge *outside* the neuron can become more negative. Either change will bring the voltage difference across the neuron's membrane closer to the threshold for firing an action potential. To think of it another way, you could flush a toilet—in a thought experiment at least—by holding the handle steady and lifting the commode a bit instead.

It should now be obvious how brainwaves, which are fluctuations in voltage outside neurons, can trigger a neuron to fire. If the strength of the electrical field surrounding a neuron or group of neurons becomes more negative, this can reduce the transmembrane voltage difference to the threshold point to fire an impulse. Conversely, if the voltage in the fluid surrounding a neuron becomes more positive, this will strengthen the cell's "battery," making the transmembrane voltage difference even larger, which will inhibit the neuron from firing. In this state, even if a neuron receives stimulation from another neuron, the recipient neuron may not fire an action potential because its voltage has moved so far below the threshold for activation. In other words, neurons are stimulated to fire when brainwaves make the transmembrane voltage peak, and inhibited from firing at the troughs.

The textbook description of how neurons communicate—by passing signals across synapses from one neuron to the next—is easy to understand and study, but brainwave activity is far more complex. Brainwave activity is so complicated that the scientists who study it are physicists and mathematicians using sophisticated methods of analysis. Fundamentally, this is why the science of brainwaves is outside the experience of most neuroscientists, and why very few nonscientists (the general public, reporters, and others) understand brainwaves themselves. Those who analyze brainwave activity are attempting to pinpoint the source in the brain of a weak electrical signal picked up by an electrode on the scalp. This is not as easy as it sounds: Electrical fields radiate in three dimensions; they change dynamically by constructively and destructively interacting with electrical fields generated by other sources; they propagate in three-dimensional vectors according to the combined strength of the fields interacting inside the brain and according to differences in electrical resistance. Such a complicated three-dimensional puzzle is exceedingly difficult to solve in a homogenous medium (like locating the source of a radio broadcast in the atmosphere), but the brain is not at all a homogenous substance. Brain tissue—and its electrical resistance—varies with the anatomy and microscopic structural differences of the brain.

Brainwaves are made up of a torrent of patterns and fluctuations in voltage, all churning together and changing constantly on millisecond to tens-of-seconds time scales, and, as a result, sophisticated mathematical methods are required to extract the relevant signals from the mix, much like the NSA (National Security Agency) extracting a single phone call of interest from the barrage of telephone traffic endlessly screaming around the globe.

Fourier analysis is a mathematical method used to decompose complex wave behavior into its multiple component frequencies. When the A key is struck on a piano, for example, most of the power generated comes from the string vibrating at a frequency of 440 Hz. But the string also vibrates simultaneously in many different frequencies to generate the distinct sound that we recognize as that particular note played on a piano and not, for example, on a guitar. Fourier analysis can determine the power of each of the different frequencies of sound simultaneously vibrating and plot out the results on a simple graph. The picture displayed will distinguish an A played on a piano from the same note played on a guitar.

The same process is used to analyze brainwaves. Like separating pocket change into stacks of pennies, nickels, dimes, and quarters, Fourier analysis breaks brainwave activity into its individual components of different frequency bands—alpha through theta. Combining Fourier analysis and other methods, brainwave analysis can identify the different frequency bands of brainwaves being broadcast simultaneously, how the amplitude, phase, and synchrony of those brainwaves are linked among different parts of the brain, and how these features change over time and in response to stimulation or cognitive tasks.

## NAMING BRAINWAVES

Brainwaves are categorized, like light waves, in spectral bands by their frequencies of oscillation. These different frequency bands are designated by the Greek letters alpha, beta, gamma, delta, and theta, but unfortunately, the alphabetic sequence of Greek letters does not correspond to the sequential increase in frequency, because some frequencies of brainwaves were discovered after certain Greek letters had already been used to name frequencies discovered earlier. Also, like colors, the cutoff frequencies for brainwave frequency bands vary somewhat in usage, just as the frequency of light waves that we call "yellow" is not sharply defined as it transitions into orange at the lower frequencies and green at the higher frequencies.

The scalp electrodes used in EEG recordings cannot pick up all the rhythms of brainwave oscillations taking place on the inside of the skull. Neurons operating deep inside the brain are too far away to be detected by electrodes on the scalp. The strength of an electric field rapidly decreases with distance from the source, so the electrodes used in EEG primarily detect sources of oscillating electric fields in the surface layer of the brain, the cerebral cortex. Neurons in the cerebral cortex are organized into an intricate pattern of six interconnected layers, but the cerebral cortex is only the 3-millimeter-thick "skin" of the brain. Still, being able to detect activity in the cerebral cortex is extremely important, because this is where sensory and motor functions are carried out

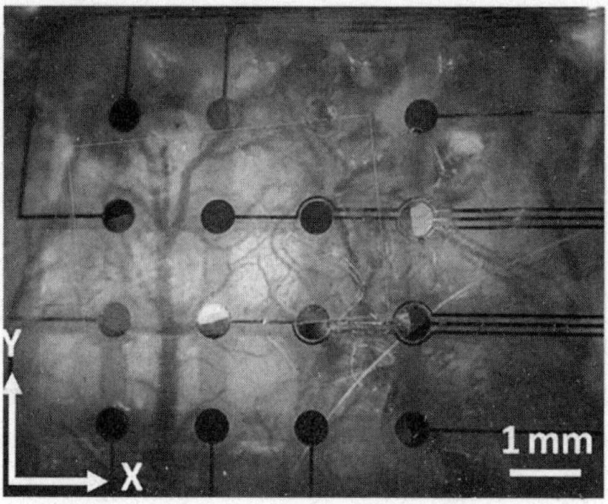


**FIGURE 18:** *Better recording of brainwaves can be obtained by placing electrodes on the surface of the brain, a technique called electrocorticography (ECoG).*

and consciousness arises, and thus almost anything that affects the conscious mind will, in some way, be reflected in activity taking place in cortical neurons. Those 3 millimeters are what makes us human.

Distance aside, the ability of scalp electrodes to detect various brainwave frequencies is impeded by the poor electrical conductivity of the skull and skin. Higher frequencies of brainwaves can be detected by opening the skull and placing electrodes directly on the surface of the brain, in a technique called electrocorticography (ECoG). Obviously, this is an invasive and delicate procedure, which is the main limitation of the method; although I was quite eager to have Jessica Eure and Robin Bernhard record my brainwaves, I wasn't about to open my skull to do it. ECoG requires a surgical team and a skilled neurosurgeon, but there are situations in which it is extremely useful, as during surgery for epilepsy, when it is used to assist in locating the source of a patient's seizures. Electrocorticography is also used in brain-computer interfaces. As will be discussed in part III of this book, these interfaces enable devices, such as artificial limbs, to be directly controlled by neuronal firing—that is, by thought.

The very best method of recording the electrical fields surrounding neurons is the use of the kind of sharp microelectrodes I described earlier. To record brainwaves instead of the firing of an individual neuron, the microelectrode does not need to impale a neuron or tightly push up against it, as I described doing in the rat brain at the beginning of this chapter. The sound of static waxing and waning that I heard as I advanced my electrode through the rat's brain tissue was the cacophony of dynamically fluctuating electric fields flowing around neurons and non-neuronal cells (glia) in a buzz, while the millions of neurons in the region of my electrode fired in their individual neural circuits. These extracellular electrical fields are called local field potentials. Neuronal oscillations of very high frequencies and very low frequencies, both of which scalp electrodes are deaf to, can be recorded with microelectrodes implanted in the brain.

Finally, multiple microelectrodes can be etched onto one sharp electrode thanks to the same printed circuit technology used to fabricate the complex microcircuits inside computer chips. With these multielectrode arrays, scores of recordings can be obtained simultaneously from one electrode. Such electrodes are commonly used by neuroscientists in experimental studies on animals, and they are also being used in humans for brain-computer interfaces. What implanted electrodes lack, however—and what EEG provides—is a global picture of brain activity. With EEG it is possible to study the brain operating as a system in which information is being processed locally and passed to other regions. This global view of the brain functioning as a complex network is transforming our understanding of how our minds work.

For the sake of completeness, I should mention that a complementary, and much more technically sophisticated, technique to EEG is magnetoencephalography (MEG). As the name says, electro*magnetic* fields have both an electrical and a magnetic component. The combination of electric and magnetic forces in electrical currents is the basis for electromagnets and electric motors. Extremely sensitive superconducting sensors, which are placed outside a person's head and cooled with liquid helium to approximately –269 degrees Celsius, can be used to monitor neural activity in the brain by detecting the magnetic component of brainwaves.[11] Finally, functional brain imaging (abbreviated fMRI) is a variation of MRI that is used to study the internal anatomy of the brain, as I experienced when I got my brain imaged after high-altitude mountaineering (see page 33). But fMRI can discern differences in local blood oxygen content inside brain tissue that are related to how active neurons are. While fMRI cannot monitor brainwaves, it can pinpoint areas of the brain that are more active than others during cognitive tests or even when the person sits quietly in the scanner doing nothing but letting their mind wander. More about this technology and the remarkable insights into brain function that can be achieved by fMRI, especially done in combination with EEG recording, will be in chapter nine.

The following is a list of the primary types of brainwaves as recorded by EEG, in sequence from the slowest to the fastest rates of oscillation, where the oscillation frequency is expressed in cycles per second or hertz (Hz).

**DELTA WAVES** oscillate very slowly, in the range of 0.5 to 3 Hz. These waves are characteristic of deep sleep, and they also develop in the unconscious states of coma or when loss of consciousness is induced by anesthesia.

**THETA WAVES** oscillate at about 4 to 7 Hz. They are generated during quiet periods of wakefulness, as in daydreaming and when drifting off to sleep. This frequency of neural oscillation has been found to be especially effective in strengthening synapses in the hippocampus, a part of the temporal lobe important for mapping the spatial layout of our environment and forming memories.

**ALPHA WAVES** oscillate at a frequency of 8 to 13 Hz. These brainwaves predominate when entering a very relaxed state, especially with the eyes closed, and in meditation. These were the brainwaves that Hans Berger first recorded in the human brain. They are found to be of larger amplitude (higher–voltage surges) at the back of the head.

**BETA WAVES** oscillate at 14 to 30 Hz. They predominate when the mind is busily engaged in mental tasks.

**GAMMA WAVES** oscillate very rapidly, at frequencies of 30 to 120 Hz. As a point of reference, the frequency of AC current in the United States is 60 Hz. This is so fast that the human eye perceives a tungsten lightbulb as generating a constant beam of light when in fact it is blinking at a rate of 60 Hz, which is too fast for our visual system to follow. Nevertheless, many circuits in the brain operate at these high frequencies (and higher). Gamma waves are associated with alert mental states and active information processing in the cerebral cortex.

It is important to understand that, contrary to the simplistic descriptions that are sometimes given in the popular press, each frequency band of brainwaves does not "do" a specific cognitive function. Although some mental states are associated with each of these bands, these are generalizations. And just as there are a million or more colors,

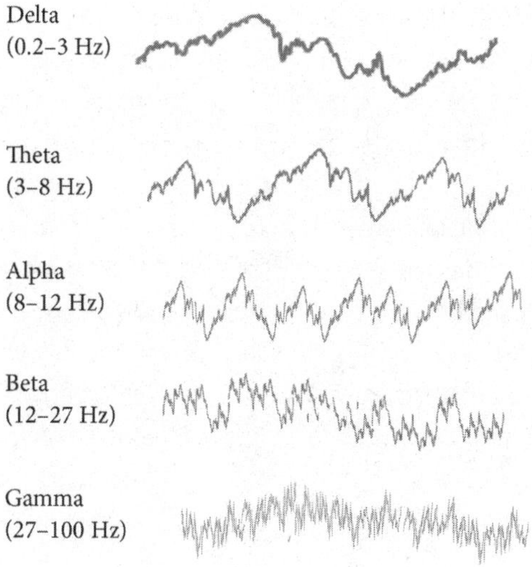


**FIGURE 19:** *A depiction of the primary types of brainwaves, categorized according to frequency of oscillation.*

not just the nine basic colors we name, so too do brainwaves come in nearly infinite variation and combination. Within brainwave bands, sub-bands are recognized, and different cognitive states and cognitive functions are associated with, for example, low-frequency versus high-frequency alpha waves.

In addition to these five broad characteristic patterns of brainwave activity, there are other patterns of brainwaves, which will be discussed later, among them several types of abnormal brainwave patterns. Many of these have useful diagnostic value for neurological disorders (and according to the latest research, perhaps neuropsychiatric disorders as well).

Every frequency band of brainwave oscillation is the result of multiple complex processes taking place simultaneously in millions of neurons locally and globally in the brain: Each brainwave frequency band is not generated by a single oscillator resonating at a specific frequency. Brainwaves are interdependent. It is fascinating to look out at the ocean as breakers roll in and crash on the shore, and we can sit and enjoy watching the sea like this for hours. Watching the surf break is spellbinding not only for its natural beauty, but because of its intriguing complexity and endless variation. There is a rhythm to the waves, but the rhythm is dynamic—it changes. A series of enormous breakers will roll in and then there will be a hush, and in the lull smaller waves splash and sizzle in the wet sand. Superimposed on every ocean wave are smaller waves and ripples, all of them the product of countless forces of nature: the wind puffing in gusts, blowing in forceful flurries, and raging in gales of prevailing wind patterns globally and locally; swirling currents; swelling and ebbing tides; the forces of fluids in motion; the frictional drag of the shallow sandy bottom on the moving masses of water; the reflection of a violent impact with rock—all of these are recorded and displayed in each wave as it approaches the shore. So it is with brainwaves. Wave riding on wave, every one is the collective product of countless interactions, like a rhythmic seascape vibrant with global forces leaving traces in salt and surge.

# CHAPTER 4

# Deciphering the Brain's Code

Today we are well on the road to understanding brainwaves, and that knowledge is fundamentally changing the way many scientists believe the brain operates. For me as a neuroscientist, the most fascinating aspect of this new research on brainwaves is the revolutionary impact this science is having on our basic understanding of how the brain works. This is science in action. Whether brainwaves are a breakthrough in understanding how the brain works, or instead are simply "electrical noise" generated by neurons communicating with electrical impulses, has deeply split the community of neuroscientists. As with all pivotal advances in science, many aspects of brainwaves continue to be mysterious and controversial. Scientists, and anyone who studies the history of science, know how easily we can go astray, attracted by an enchanting new idea or exciting discovery.

Any time science pushes the frontier of knowledge, naysayers and prophets will emerge to confront the strange new aspect of nature. There will be charlatans building fantasies on valid science that they do not fully understand or simply wish to exploit for profit. Already we have seen how research on the electric brain has progressed through countless twists and turns—from Berger's dabbling with psychic energy and mental telepathy, to Weinhold replacing the brain of a kitten with an electrical energy source, to Crosse recreating life with electrical sparks. Some feel this way about neurofeedback or reading thoughts by coupling brain and computer together; indeed, some experts feel this way about brainwaves in general. The truth is far more complex than the simplistic portrayals of brainwaves available in the popular media. If you believe the popular accounts that scientists understand the brain's code in the way we understand computer code, you are mistaken. The truth is far more remarkable and fascinating. But this is always the situation when science progresses from the known into the unknown. In the fullness of time the scientific method will prevail.

That method is to carefully examine the strange new thing that we have encountered. To measure it carefully, quantitatively, and objectively. To poke it (metaphorically speaking) and see what it does. To seek out others around the world for their insights. In this part of the book I will take you with me as I explore this intriguing area of brain research, going about it the way a scientist seeks answers. While I will distill this new information as best I can for the general reader and offer my synthesis and views, I will strive to be objective and present alternative opinions with fair consideration, because right now, the truth is before us, but we can't be certain what we are seeing.

## WATCHING THE BRAIN LEARN

We return to the scene that opened this book: Blue panel lights blink ominously on rows of black computers sealed inside a glass room. I walk past the supercomputer nerve center and proceed down an empty, dark corridor. Arriving at a door, I open it and enter a cavernous room painted entirely black. Twenty-four digital cameras stationed around the room pinpoint my every move. Two men stand up from behind a bank of computer screens . . .

I have come to the futuristic supercomputer center at the University of California, San Diego, to visit Howard Poizner, a neuroscientist. By using sophisticated instruments to eavesdrop on electrical transmissions flashing through my cerebral cortex, he claims that he can read my thoughts before I have them. He can see my brain learn. He can analyze the patterns of electrical signals flowing through my brain as my mind wanders with my eyes closed, and know what kind of brain I have. The most intimate details of my mind are his to see.

Professor Poizner's goal is to better understand the neurobiological basis of learning by exploring one of the most fascinating and complex of mental functions: how our memories are formed. As a scientist investigating the cellular basis of memory in my own lab, I learned of Poizner's research before it was released to the public, but it has now been published in prominent peer-reviewed journals.[1] The United States Navy, which funds Poizner's work, hopes to use these advances in neuroscience to help them train personnel and improve individual performance.

Learning theory rests on the century-old experiments of Ivan Pavlov and the dogs he trained to salivate at the sound of a bell. This simplistic model of learning through punishment and reward, called a "conditioned reflex," is fine for training pets, but is this really how human learning works? Whether recognizing a friend in a crowd, negotiating a new environment, picking up slang, or gaining insight into human nature by reading a novel, our brains are perpetually engaged in processing, organizing, storing, and recalling information—in short, learning—without the equivalent of a tasty cheese reward or punishing electric shock to guide us. Yet, if you ask most neuroscientists how this works, you'll get a shrug.

"Most of our learning is unsupervised," Poizner says. "Pavlov's learning experiment was a simple reflex. There was a specific stimulus (a bell) and a particular response (salivating) that was reinforced by reward (food). [Most of the time] you are just walking out in the environment and you are learning. You are learning the spatial layout, how to move in this environment, how to interact with it."

"But if this is the kind of learning we do all the time—and probably the most important kind, for humans—why has it not been studied more?" I ask.

"Well, it's difficult," he explains. The challenge of understanding how we learn to negotiate novel environments, for example, requires developing the technology to record how someone's brain responds as they are moving through new surroundings. Poizner's idea was to place people in a computer-generated virtual reality and watch their brainwaves as they explore.

This virtual reality is where I am headed. Poizner and his colleague Joseph Snider fit me with what looks like a swimmer's cap studded with seventy electrodes. The wires from the electrodes are bundled together into a long ponytail, connected to monitoring instruments inside a small black backpack I am carrying. Normally, to measure brainwaves, people must sit very still to eliminate the robust electrical impulses from bodily movements that can obscure the more feeble signals of the brain's activity, but Poizner's group has developed software to eliminate these unwanted muscle-generated signals and devised hardware that will wirelessly transmit my brainwave data as I move about freely in a virtual 3-D environment from the instruments in my backpack to the lab's computer for real-time analysis.

Poizner's colleague covers my eyes with large black electronic goggles. These generate a detailed three-dimensional scene across twelve miniature video screens on the lenses to provide me with a full-vision, 120-degree view of the virtual world. The two dozen video cameras arrayed around the room detect infrared diodes on a jumpsuit I'm wearing to track my movements and feed them into the computer. In response, the computer will adjust the three-dimensional virtual environment so that it moves and changes seamlessly as I explore. High fidelity speakers around the room create a three-dimensional sonic space to complete the illusion.

"The way we think of it is, we're putting you *in* the video game. It's not looking *at* the video game," Poizner explains. He's not kidding: The instant the goggles are switched on I'm stunned to find myself transported into a computer-generated reality, like a character in the movie *The Matrix*. I am no longer in a laboratory, but inside a storeroom on the US Navy aircraft carrier *Midway*. The illusion is totally convincing.

"That's absolutely critical," Poizner says. "It is not only critical for immersion [the virtual reality illusion]; it is critical for uncovering brain maps that are related to space." The complex integration of all the rich sensory information available to us—vision, sound, feedback from the position and movement of our body—enables us to build spatial maps of our environment inside neural circuits in our brains. The multidimensional maps our brains build as we move about the real world depend heavily on our motion, and, as Poizner notes, "If you are just moving a joystick or hitting a button, you are not activating the brain circuits that function in constructing spatial maps."

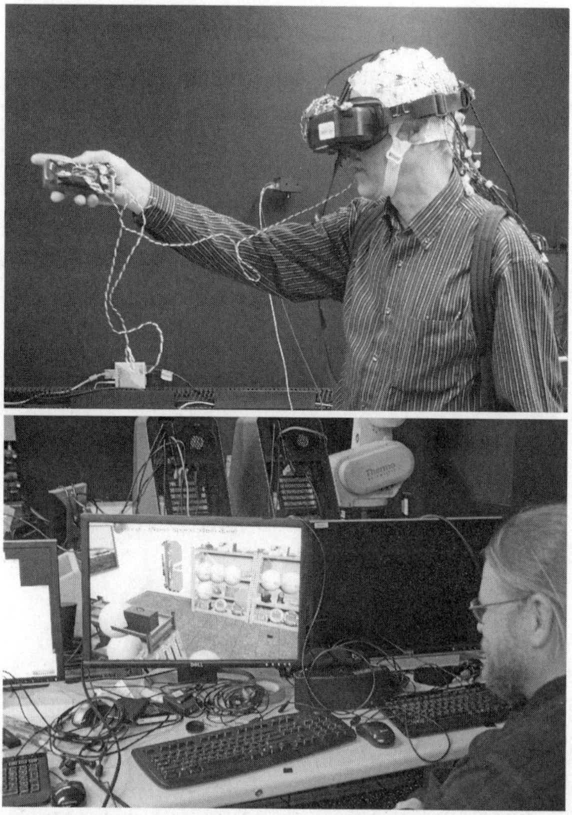

**FIGURE 20:** *The author experiencing virtual reality in experiments by neuroscientist Howard Poizner and colleagues at UC San Diego in which they monitor brainwaves to understand how we learn without being instructed or trained (unsupervised learning). At the controls, neuroscientist Joseph Snider monitors the virtual reality scene the author is seeing (inside the aircraft carrier* Midway*) as he explores the storeroom using virtual reality goggles. The task is to locate a green bubble on the storage rack (appearing gray in this black-and-white photo). The device in the author's hand is used to pop the bubble in the virtual reality environment, revealing an object hidden inside.*

Inside the virtual ship's storeroom, everywhere I turn I notice new, vivid details. Looking up, I see the steel triangular trusses reinforcing the ceiling that supports the flight deck above. Looking down, I see hideous government-issue blue linoleum. I wander over to an oval hatch and peer out on to the hangar deck where fighter jets are stationed in rows. I raise my leg to step over the high threshold. "Don't go out there," Poizner says. "You must stay inside the storage room." I retract my raised leg from the doorway—an observer would see me do this while standing in an empty black room, as if I were performing a pantomime, or sleepwalking.

Turning back to the room, I see beach-ball-size gray bubbles resting on storage racks. Poizner's computer screen lets him see what I am seeing and monitor my reactions to it. "You are looking for a green bubble," he tells me.

I search the racks and, finding the green bubble on the shelf among the gray spheres, I reach out and touch it. It pops! An object that was hidden inside is revealed—a fire extinguisher. I turn and locate another mysterious green bubble in the opposite corner of the room. When I pop this one, I see that it contains a wrench.

While his test subjects move about the room and explore, Poizner can see an increase in a specific frequency of wave activity in the parietal lobe of the brain—theta waves, oscillating at about 4 to 7 Hz—as the subjects build spatial maps. The parietal lobe is at the top-back of the brain, roughly beneath the part of the skull covered by a skullcap. Theta waves are known to be important in strengthening synapses in forming a memory. Strengthening a synapse means that the signal it delivers to the neuron it is communicating with is boosted. In fact, in my own research on the cellular mechanisms of memory, I stimulate neurons at the theta frequency to strengthen synapses in slices of rat brain that I keep alive in a dish. (Recall that a synapse signals by generating a brief voltage in the postsynaptic neuron. A synapse is strengthened by generating a larger voltage than before, and thus having a greater influence on whether or not the postsynaptic neuron fires a neural impulse.)

When analysis of brainwaves is tightly linked to a particular stimulus, it's possible to uncover powerful insights into cognitive function. A brainwave response that is initiated by a stimulus is called an "event-related potential" or an "evoked potential." These brainwaves evoked by a stimulus are much like the ripples caused by casting a stone into a pond. By studying the ripples produced by a particular stone—the brainwaves produced by a certain stimulus—we can develop an understanding of the relationship between the properties of the stone and the ripples produced. For example, doctors use evoked potentials (neural response to a stimulus) produced by test tones to determine whether or not an infant has a hearing defect. The newborn infant cannot verbally communicate, but the doctor will know, by looking for a brainwave evoked by a tone, if the child can perceive the sound. I used the same technique in experiments to discover whether or not strange deep-sea fish called chimaeras, which are distantly related to sharks, could sense extremely weak bioelectric fields that are generated by all animals in seawater.[2] The evoked potentials I recorded in the fish's brain as I delivered different frequencies and intensities of electricity near the strange-looking nose of the anesthetized fish proved that its brain received and processed this sixth sense, which humans will never know.[3] In my case, Poizner intends to use evoked brainwave potentials to tap into my brain and find out if I had learned anything during the virtual reality experience.

Poizner analyzed the evoked brainwave response his test subjects generated the instant they touched the green bubble and discovered that it held an object hidden inside. His analysis revealed a characteristic ripple in my brainwave activity that erupted 160 thousandths of a second after I popped the green bubble. "This is amazingly fast," Poizner tells me. "I mean, it takes 200 milliseconds just to make an eye movement. This is *pre*conscious perception that the brain is detecting something amiss. '*I can't tell you what it is, but something is amiss.*'" In other words, our brain responds to the surprising fact of a hidden object appearing in a bubble before we've consciously "seen" the object.

Analyzing these stimulus-evoked brainwave changes enabled Poizner's team to make some surprising discoveries—specifically, about the kind of automatic learning and memory formation we humans are engaged in all the time. When his experimental subjects revisited the virtual ship storeroom the next day, Poizner and colleagues had cleverly switched some of the objects that were concealed in the green bubbles. When a person popped a green bubble in a particular location and found that it contained a wrench where the day before the same green bubble had concealed a fire extinguisher, the evoked brainwave response was much larger than when they found an object hidden in the same location on the previous day. From these responses, it was clear that without any instruction, forewarning, or effort, the subjects had memorized the setting in detail while they were simply exploring. Faster than the blink of an eye, our brain knows when something has changed in our environment, and our *brain* knows this before our *mind* can comprehend it.

The navy is interested in tapping into these rapid preconscious signals for obvious reasons. Not only would these advances in technology and brainwave analysis allow them to identify quick thinking and help select the individuals best suited for the many demanding situations faced by military personnel, but they might also give the United States an edge in combat by coupling brainwave monitoring with computerized systems. Reading a pilot's brainwaves could enable a computer to take evasive action before the pilot is consciously aware of a threat. The quickest draw in such a gunfight wouldn't even know he had pulled the trigger.

For the military and civilians alike, the hope is that, someday, the information provided by brainwaves will be as valuable in assessing brain function as medical tests to measure the body's physical condition and performance are today. A United States patent has been granted jointly to Duke University and the Defense Advanced Research Projects Agency (DARPA) for "apparatus for acquiring and transmitting neural signals" with purposes including, but not limited to, "weapons or weapons systems, robots or robot systems." For the first time in human history, we are developing the capability to directly probe a person's thoughts and alter them. The ability to manipulate brainwaves has military application in interrogation and could be used to incapacitate enemy personnel by bombarding them with magnetic fluxes.

In grappling with this new reality, the Nuffield Council on Bioethics in England assembled a group of leading neuroscientists to consider the ethical implications of interfaces between the human brain and computers. In their report, published in 2013, they conclude that the Geneva Convention is woefully outdated and ill-equipped to deal with the potential misuse of neurotechnology in warfare, raising concerns from interrogation to brain computer interfaces enhancing fighter effectiveness, and noting that "the existing international conventions outlawing the use of biological and chemical agents in war do not cover the use of neurodevices."[4] In part III we will return to the ethical questions we now face by being able to monitor and change brainwave function, and the potential use of this capability for neuroweapons, but let's first return to the scientific question of how brainwaves could be involved in memory.

Poizner's research, and the research of others we will soon meet, suggests that brainwaves are fundamental in forming memories, because a memory requires knitting together myriad cognitive elements—sight, smell, emotion, temporal sequence, and preexisting memories—that are processed in different parts of the brain. How else could our brain create the layered experience that connects the spicy cinnamon and nutmeg aroma of hot apple pie with the love we feel for our mother who baked it, with the image of her favorite rolling pin as she rolled the dough thin and laid it over the top of the pie, and that Dad always takes his piece with a scoop of vanilla ice cream? The many and various aspects of a memory draw on millions of neurons from all over the brain, and to create a coherent scene, they must be edited together into the proper time sequence and context. By coupling activity in large groups of neurons across distant brain regions, brainwaves control the processing, encoding, storage, and retrieval of information. Like all electromagnetic waves that build or annihilate one another when they interact, these waves of electrical activity propagating through our brain tissue at various frequencies constructively and destructively interact in complex ways to filter and sort signals, carrying the enormous streams of data coming from our senses and drawing on stored information from many regions of the brain simultaneously to enable us to recall a memory in a flash or to learn without effort as we encounter the world around us. It is very difficult to understand how our brain accomplishes this integration and sorting of information from the textbook view of how the brain works.

## THE BRAIN AT WORK

The human brain has always been an object of wonder, and throughout history has been viewed through analogy to the most sophisticated technology of the day. Today the reference is the digital computer, but prior to computers, the functioning of the human brain was explained variously by analogy to hydraulics, light, and electric circuits. In the near future, the digital computer will be surpassed by quantum computers, and that more advanced subatomic technology will no doubt be tapped as a better analogy for how the brain works. But analogies are only models—mental crutches to help us grasp what may be too complicated for us to fully comprehend. The electrical circuit analogy of the last century that reduced nervous system function to the simplistic mechanism of an electric doorbell circuit is, of course, woefully inadequate. The phenomenal capabilities of the human brain cannot be achieved by simply wiring together more and more electrical "doorbell" circuits. But neither does the human brain function anything like an electronic computer. The most advanced information processing system known, our brain, works in far more sophisticated ways. It has to. Brainwaves, many but not all neuroscientists argue, are not an analogy; they are how the brain actually functions, the means by which it creates what we like to think of as the most complex network in the universe.

From this glimpse into how we learn achieved from the virtual reality and EEG experiments in San Diego, let's move on to visit other laboratories around the world to see how researchers are finding brainwaves to be a window into thought. Some of these researchers are finding brainwaves to be the basis of consciousness. Others are using brainwaves to link brains and computers together and to transmit information from one human brain to another. There is perhaps no better example of how brainwaves are coupled to mental activity than sleep; indeed, the stages of sleep are defined by the systematic changes in our brainwave activity throughout the night. Brainwaves reveal mental and neurological health, and EEG analysis can empower us to change our own brain wiring to overcome our individual cognitive weaknesses. Yet, the nub of the controversy persists: Are brainwaves the power or the by-product of mental activity?

French neuroscientist Karim Benchenane and his colleagues at ESPCI (École Supérieure de Physique et de Chimie Industrielles de la Ville de Paris) are doing fascinating research to try to break through this logical dilemma. In the course of this pursuit, Benchenane and his colleagues have, among other things, implanted dreams into a rat's brain by using a brain-computer interface and erased a fearful memory from a rat by introducing oscillations at just the right frequency into the frightened rat's brain. But still he is queasy about accepting what he believes in his gut to be true—that oscillations are an important new dimension of information processing in the brain. "I would not like that someone in five or ten years should demonstrate that what I did was an artifact," Benchenane told me when we met in his office in Paris. (When scientists speak of an artifact they are referring to an erroneous result that is unintentionally produced by the experimental procedure.)

## BEAUTIFUL ILLUSION OR SECRET CODE?

Founded in 1257, the Sorbonne has endured nearly eight centuries of the world's turmoil and transformation; predating Christopher Columbus, the printing press, and the Renaissance by two centuries, the institution persists, and indeed flourishes, as a testament to mankind's greatest strength—learning. In the same era, in the year 1163, construction of another enduring testament to the other pillar of humankind's strength and character, spirituality and religion, began as the first stones were placed in the foundation of the Notre-Dame Cathedral, situated on a spindle-shaped island in the middle of the Seine River that slices through the center of Paris, flowing in the direction of the sun, east to west. Only a short walk from the towering cathedral, passing over the river across either of two bridges, brings one to the Sorbonne, now a cluster of colleges and universities on the south (left bank) of the river in the Latin Quarter of the city, so named because of the ancient, universal tongue of scholarship that once permeated the cobblestone streets surrounding the renowned academic institution.

A small plaque is affixed to the side of a weathered, six-paneled brown wooden door. Its shiny rod-shaped brass handle stylishly, but ineffectively, sits at the center of the door, which explains the badly scuffed bottom rail, chipped and split from countless toe kicks to open the portal through the massive tan colored stone walls faced with stately quoins. The inscription commemorates the momentous events that occurred here when a young woman changed history with her scientific discovery, and this former governess became the first female to be honored with the Nobel Prize.

"You can't go into that lab," a scientist who works at the Sorbonne told me. "It's contaminated still. All her notebooks are still highly radioactive." The half-life of radium is 1,600 years, which means that the intensity of the radioactive contamination inside this laboratory has hardly diminished from the days when Marie Curie worked here and carried around vials of the highly radioactive element in the pockets of her lab coat. Her discovery has saved the lives of millions by the application of X-rays in medicine, but her accomplishments came at the cost of her own life. Unknown to her or anyone else at the time, the invisible high-energy particles radiating from the curious elements she studied ransacked DNA in the nucleus of her bone marrow cells, sterilizing them and terminating their ability to divide normally and make new blood cells. It is possible that the physical remnants of her work, in the form of radioactive contamination that is too difficult to expunge, could outlive both of the stone monuments to knowledge and religion built by men in the center of Paris.

I wandered through a busy laboratory with students in lab coats scurrying about with glassware and notebooks in hand. They paid me no mind as I roamed the hallways looking for Benchenane's office. I finally came to a hallway where the doors to every laboratory were not numbered but instead had names like "amygdala," "prefrontal cortex," and other parts of the brain. I knew I was getting close. I finally met Benchenane, and we sat down for a conversation. Benchenane sported a seemingly perpetual five-o'clock shadow and a butch haircut. The small gold ring in his left ear gave him the look of a jolly marooned pirate.

I got straight to the point, beginning with the accepted facts and outlining the controversy: Oscillations of electrical activity in the brain (brainwaves) are a fact, unquestioned by anyone, and obvious to anyone with an electrode to measure them. These oscillations in voltage can be stationary, like a standing wave, rising and falling like a child jumping rope, or they can sweep through the brain, just as all waves can propagate or remain stationary. But the problem is that almost all of the data we have about the connection between brainwaves and information processing, cognitive function, and behavior are correlations. Alpha waves increase when eyes are closed, for example, but how is this correlation produced? What significance does the correlation have for neural circuits and behavior; and is the correlation the cause or the consequence of an underlying process?

"For me I do believe oscillations are important," Benchenane told me in French-accented perfect English. He noted that oscillations are everywhere in nature, and they have many important properties that make them well suited for information processing in the brain. In particular, he said, oscillations build and transmit power very effectively. He gave the example of people clapping hands. Many people clapping in synchrony can rock a stadium with thunderous sound; however, using the same amount of energy when everyone is clapping, but not simultaneously, will not have the same effect. We see this effect dramatically when a skyscraper resonates with the frequency of vibrations in an earthquake and energy transmits efficiently and compounds, collapsing the entire structure.

"The problem is demonstrating the importance of oscillations in the brain. There are not many results that clearly demonstrate that oscillations are important," he said. To prove causation, one must break the correlation. The way to do that in the case of brainwaves is to change the oscillation in some way and see whether or not the correlated neural event or behavior also changes in the predicted manner. Benchenane and his colleagues set out to determine if they could change a rat's behavior by changing the rate of its brain oscillations by using optogenetics to drive neural activity in a critical neural circuit known to mediate a behavior.

Optogenetics is a new method of driving electrical activity in neurons by using laser light to excite ion channels that are inserted into specific neurons via genetic engineering. The advantage of optogenetics is that only the specific neurons that are genetically modified are stimulated by the laser light, whereas electricity delivered by a microelectrode spreads through neural tissue, acting on many neurons, including neurons in regions far from the electrode that have axons passing in the vicinity.

The behavior Benchenane and his colleagues chose to study was fear. If they could disrupt the fear memory in a rat by driving oscillations at the right frequency in appropriate neural circuits, this would demonstrate the importance of oscillations in neural function and open the possibility of new treatments for anxiety disorders, such as post-traumatic stress disorder (PTSD).

Learning to fear a specific stimulus, called fear conditioning, requires communication among several brain regions. The hippocampus encodes spatial information, the amygdala and limbic system evaluate potential threats and regulate emotional responses, and the prefrontal cortex integrates the complex information available to the brain to give the event the proper context to correctly interpret whether the event is a sudden danger or not. For example, our prefrontal cortex provides us the context to fear a knife in the hand of a stranger in a back alley, but not in the hand of a stranger in a restaurant kitchen.

To study this fear-learning process, rats are first trained to fear the sound of an alarm by delivering a mild electrical shock to their feet through the wire floor of their cage immediately following the tone. The rats very quickly learn to freeze in fear when they hear the tone, instantly freezing when they hear it even without the foot shock. If electrical activity from the amygdala and the prefrontal cortex of a rat is recorded during fear conditioning, oscillations at a frequency of 4 Hz in the prefrontal cortex and amygdala suddenly become phase locked (synchronized) when the animal freezes in response to the tone. This resonant locking of theta-wave oscillations in the prefrontal cortex and amygdala could be how the functional connection and transfer of information between these brain regions is made to evoke the fear response. The sound, transmitted to the brain's threat-detection center, the amygdala, becomes coupled together with location, coded in the hippocampus with its connections to the prefrontal cortex, where memories of past experiences in this test cage are brought together and interpreted by the prefrontal cortex. The rat will not freeze at the sound if it is presented in another place or context. The resonant coupling of theta waves between amygdala and prefrontal cortex package together all of this information distributed from across the brain and the synchronized cortical-amygdala feedback of theta waves instantaneously unleashes fear and freezing. Meanwhile, other brain regions oscillating at different frequencies independently carry out their distinct neural functions (vision, locomotion, and so on), like different radio stations broadcasting at different frequencies are picked up only by radio receivers tuned to the appropriate frequency.

To test this correlation, the researchers generated 4 Hz oscillations in the prefrontal cortex of rats by using optogenetic stimulation, and the rats froze in fear, strong evidence that the 4 Hz oscillations cause the behavior. If they increased the duration of 4 Hz stimulation, the rat freezing behavior was increased, and if they shortened the duration of 4 Hz stimulation, the animals froze less frequently.[5] By simply generating and manipulating the 4 Hz oscillations synchronizing these two brain regions, the researchers could regulate the rat's fear and behavior in this situation.

Benchenane maintains the disciplined skeptical objectivity of a careful scientist. No matter how beautiful the theory and elegant the results from an experimental test of a hypothesis may be, experience shows that conclusions can be wrong. So scientists try to knock down their beloved deductions by trying to imagine every possible alternative explanation. This is the difference between belief and proof. To understand how this seemingly iron-clad test of causation between oscillations and behavior in this experiment might be wrong, it is important to consider the fundamental question of how information is coded in the nervous system in the first place.

There are two fundamental ways that information is coded by action potential firing in the brain: 1) rate coding and 2) temporal coding. "What is important?" Benchenane asked rhetorically. "Is it the precise time of the spike, or is it the number of spikes?" Clearly, both are important. The intensity of light, for example, is coded in the frequency of neural impulse firing—"rate coding."

My earliest experience seeing rate coding firsthand was as a graduate student, recording visually evoked responses from horseshoe crabs, squid, and frogs, as well as studying electric fields that are sensed by special sensory receptors in sharks, rays, and chimaeras, which was the subject of both my master's[6] and PhD theses.[7]

We were patiently schooled in this technique by my adviser, David Lange, professor of neuroscience at the University of California, San Diego. Working in a dark room, my partner and I used fine glass needles to fray apart the frog's optic nerve under a microscope. We suspended fine wisps of nerve fibers on a pair of silver chloride wires that we bent into miniature meat hooks using watchmaker's forceps. While the anesthetized frog was positioned in front of a screen, the wires were connected to an electronic amplifier to pick up action potentials racing from the frog's retina to its brain.

We could see the action potentials flashing on our oscilloscope and hear them pop by feeding the signals into a speaker. The brighter the light on the projection screen, the faster the rate at which neurons in the retina fired action potentials to the brain, ranging from a rattlesnake-like rattle in dim light to a screaming buzz, like a deep-sea fishing reel screeching when a monster fish is hooked, in bright light. We could plot the frequency of action potential firing against the intensity of light we shined on the retina, and with that plotting we had an accurate biological light meter. We could determine precisely how bright the light was, simply by referencing the frequency of action potential firing.

I'm leaving out a lot of fascinating neurobiology here to stay on point. But for accuracy, it is important to add that neurons in the frog retina (and those of most animals) don't only report the intensity of light; their discharge rates also depend on the shape of the light spot, the contrast and orientation of its edges, its movement, and whether the spot of light has flashed on or flashed off. However, all of these features of visual information are conveyed to the brain as spike rate codes.

Also, the neurons do not maintain their firing constantly in a single steady tone; the firing rate gradually slows down, just like the sound a fishing reel makes when it drops in pitch from the initial force of a fish strike. This gradual decrease in action potential firing rate while holding a stimulus constant is called adaptation, a common feature of sensory systems. Walking out into bright sunshine momentarily blinds us until our visual neurons adapt. Our retinal neurons fire at maximal rates when the bright light first hits the eye, but the firing rate subsequently declines as we adjust to the light. Similarly, we feel the cool cotton cloth of a t-shirt when we pop our head through it in the morning, but if it stays in place, we don't feel the t-shirt throughout the day. Those neurons in our skin responding to touch have adapted to the constant stimulus. The nervous system is most highly tuned to *changes* in our environment, like a sudden movement in the bushes. In fact, frogs are blind to objects in their visual field that do not move, because those visual neurons stop firing if they do not detect motion. That may sound strange, but blindness to stationary objects suits the frog biology well so they can quickly zero in on buzzing flies and sense the creeping shadows of predators. An image of a fly projected on the screen that fades away to nothing in the frog's brain if the fly does not move is no more a deficiency in the frog sensory system than being unable to feel our t-shirt throughout the day is for us.

In other instances, it is not the rate of action potential firing as we saw in frogs, but instead the precise time that an impulse arrives that conveys the critical information, which is called "temporal coding." For example, we localize the source of a sound in our environment because the brain can process the minute difference in spike time arrival at a relay point in the brain that receives input from both ears. If the source of sound is to our right, the sound waves will arrive at our right ear slightly before they arrive at our left ear and our brain is able to discern the difference in spike time arrival in a neural circuit that clocks precisely when an impulse arrives and, by the difference in spike time arrival, identify the location of the sound.

Oscillations allow for a third possibility of coding information in neural circuits: "phase coding," or the point in time when a spike arises with respect to the phase of oscillation in local field potential or rhythmic firing of a neuron. A spike occurring at the peak of an oscillation may have a different effect than a spike occurring at the trough, just as the force of a punch to a tether ball very much depends on the precise timing of the punch with respect to the ball's oscillation around the pole. In the case of Benchenane's rat experiments, phase coding is what would encode fear by the coherent 4 Hz oscillation between prefrontal cortex and amygdala.

It's possible to interpret the results of the fear conditioning experiments as a clear demonstration that oscillations are functionally important, but, cautioned Benchenane, "The problem is that you've got more spikes after the oscillations are imposed, so you can still interpret the result as just basic rate coding." Driving 4 Hz oscillations increased spiking, and the longer the oscillations were induced, the more spikes were generated.

Another sticky problem is that brain oscillations also track with oscillations in bodily functions, such as heart rate and breathing rate; even arm movements are oscillatory. We've all seen our hands shaking. Do the body's systemic, physiological oscillations impose oscillations on the brain? The pulsing of blood flow and rise and fall of oxygen could have some influence on neural circuit function, for example. Such bodily oscillations picked up by electrodes in the brain (or by EEG) could also simply present a rhythmic background that becomes an artifact in electrophysiological studies. It may be that the physiological changes that are induced by a behavior also change respiration or heart rate. From a neural circuit perspective, that would make the brain oscillations generated by the change in cardiac function an irrelevant consequence of the behavior. The feedback loop between the behavior and oscillations in breathing, for example, could simply cause a change in oscillation picked up by electrodes in the brain or on the scalp in response to the startling tone. Your respiration is very much affected by a sudden threat, for example, so an alternative hypothesis could be that the sudden change in brainwaves could be the consequence of a change in respiration and not directly regulating the neural circuitry controlling the behavior.

Input from the senses and rhythmic motor function could have a similar effect on brainwaves. Many meditation techniques control breathing to alter cognitive states, which are also accompanied by changes in brainwave frequencies (discussed in chapter eleven). Strobe lights and rhythmic sounds ripple waves of activity through the brain that alter and synchronize brainwave activity, in some cases to the extreme of inducing massive increases in cortical oscillations that result in an epileptic seizure. Physical movement can align with wave frequencies; for example, when a rat runs, its head moves back and forth at the theta frequency. Rats also sniff and whisk their whiskers back and forth at the theta frequency when they are exploring their environment. What are the consequences of all of that rhythmic sensory input to the brain from motor activity accompanying any behavior? The olfactory system has strong connections to the cortex and hippocampus, thus breathing introduces theta rhythms in those brain regions. However, it could be that rhythmically synchronizing information processing in neural circuits in time with the natural rhythms of walking and breathing could be beneficial, because it couples neural processing in the brain in sync with the rhythms of sensation and motion of our body. "And that's where we are stuck right now," Benchenane said, because they can't prove causation.

Probably the best example of the importance of brain oscillations and phase coding is in the hippocampus. Cells in the hippocampus, called place cells, fire when an animal moves to a specific location in its environment. The spot in its environment where this neuron fires is called the "place field" for that cell. Cell A will fire when a rat is in the right arm of a Y-shaped maze; cell B will fire when the rat is at the junction; and cell C fires when the rat is in the left arm. This is a typical rate code: no spikes in cell A when the rat is in the left arm, and high-frequency spiking in cell A when the rat enters the right arm. However, there is also a theta rhythm in the hippocampus, and if the spike fires at an optimal time with respect to the phase of the oscillation, spatial coding is improved to more precisely code space. When the rat moves toward the center of the place field, the phase of the theta oscillation advances. This is called phase precession. This phase shift in theta waves provides two types of information about where the animal is in its environment. The phase shift carries additional information from the rate coded, because the shifting phase of theta oscillation as the animal moves signals information about where the animal was, where it is, and where it will be.

Others argue that this phase coupling and phase precession are simply a coincidence. If you have prolonged neuronal firing on top of a theta oscillation in local field potential, you will find apparent phase coupling and phase precession as a consequence of individual spikes occurring at a slightly different interval than the period of the theta oscillation, especially when the theta oscillation changes as the animal moves. More perplexing is that some animals, such as bats, have very good spatial localization, with place fields in their hippocampus, but they do not have theta oscillations at all. "If you are totally honest, you have to be puzzled," Benchenane said.

There is a bigger technical problem in studying coherence of oscillations, because electrodes measuring rhythmic changes in local field potential sample electrical currents flowing through tissue around large populations of cells that are firing. An electrode recording in one brain region, the prefrontal cortex, for example, might pick up electrical fields generated by neurons oscillating in a different brain region like the amygdala as a measurement artifact. It's possible this could be responsible for the coherence seen in the experiments studying 4 Hz oscillations that apparently join these two brain regions in fear conditioning. The apparent oscillations and coincidence of oscillations could simply be due to one of the electrodes picking up signals from the other brain region as well. This inability to weed out the source of signals is an especially difficult problem in EEG recording. "The reason that we can do EEG is that electrical fields propagate. That's good because you can record brain electrical activity with a remote scalp electrode, but you have the exact same problem when you put an electrode in the brain," Benchenane said. "I think for oscillations, it is so beautiful that sometimes we go a little bit too far."

One of the fascinating outcomes of studying place cells is in understanding how sleep promotes long-term memory, a subject that will be considered in greater depth in chapter seven. Sleep learning is relevant here, because Benchenane used it in a clever experiment to determine if these cells truly code an animal's location. The hippocampus is necessary for encoding information, but that information has to be transferred from the hippocampus to the prefrontal cortex for long-term storage. György Buzsáki, Biggs Professor of Neuroscience at New York University School of Medicine, found that as animals solved a Y-maze task, there was strong coherence between the local field potential oscillations in the hippocampus and in the prefrontal cortex, indicating a possible transmission of information via phase coupling between two oscillators. When recordings are taken from place cells in a rat that is moving down a long hallway, different place cells will become activated in sequence at different locations along the way as the animal progresses through the environment. Recordings performed when the rat was asleep found that the same place cells became reactivated in sequence, just as they had during the day, indicating that at night the rat was replaying the experience of moving through that pathway in its mind. In 1989, György Buzsáki proposed that information is encoded during theta oscillation and transmitted to the prefrontal cortex, where it is replayed mostly during sleep and consolidated into long-term memory.[8] Recently, Buzsáki's group has shown similar reactivation of emotional memories during slow-wave sleep in rats, by finding reactivation of coordinated activity between the same neurons in the hippocampus and amygdala as the animal sleeps that were activated when the animal was trained.[9]

As compelling as these findings are, they are still correlations. In an effort to test these lingering alternative explanations for how oscillations encode and transmit information in the brain during sleep, Benchenane and his colleagues devised an elegant and imaginative experiment.[10] If, by stimulating neurons while it slept, they could implant a dream into a rat's brain about being in a certain place and then see if upon awakening the rat acted on that dream, this would be a clear demonstration that place cells encode memory for space and that lasting memory is created during sleep. This ambitious experiment would eliminate the many confounds, because the researchers would not record or stimulate any part of the brain after the animal was awakened. There would be no possible artifact or contamination of recording and stimulating electrodes—they would simply watch the rat to see if it acted upon the dream they had implanted while it was asleep.

To achieve this, they combined brain-computer interface with studies of hippocampal place fields. They recorded from place cells that became reactivated during sleep, and when a particular place cell fired, this triggered the computer to deliver an electrical stimulus to make neurons fire in the animal's reward circuitry (the medial forebrain bundle, known to give reward in both animal and human studies). Then they woke up the animal and put it in the test arena.

"In a sense we asked the rats, did you dream about this location?" Benchenane explained. The rat answered with its feet. It ran directly to the spot in the test environment that the scientists had rewarded the animal for reaching in a dream. "That was the first time we could demonstrate that a spike in a neuron means a location in the physical world."

"How did you feel the first time the rat went to that location?" I asked.

"The first time, I didn't believe it . . . the effect was very strong," he said. He immediately began searching for some bias in the experiment that could have caused the rat to run to that spot. "The second time you say, *Oh!*" After that, the researchers were silently and anxiously rooting for the rat to do it every time, particularly because these were extremely difficult and long-term experiments. But there was no question: Stimulating the reward pathway in the rat's brain when a particular place cell fired spontaneously while the rat slept had induced goal-directed behavior in the rat to go to that very spot the next day.

Oscillations are everywhere in nature, and they are beautiful. We see them in rippling cloud formations, in waving fields of grain, in the blaze of color on the horizon opening and closing every day. While Benchenane and others believe neural oscillations are important in the coding and transmission of information in the brain, the challenge is to provide unequivocal proof. The oscillations are there. Even if they arise in the brain as a consequence of how oscillations everywhere in nature arise when fluctuating forces interact, and not because the brain creates them for information processing, the brain had to evolve a way to live with these oscillations. Mechanisms devised to avoid the interference oscillations would inflict on information processing in the brain are necessary to comprehend, so brainwaves, through design or accident, are intimately involved in brain function.

Now, equipped with the necessary background understanding of brainwaves, we return to Charlottesville to learn the results of my own EEG analysis. The results will reveal some surprises that would have perplexed Hans Berger—my brainwaves deviate from normal.

## CHAPTER 5

# Brainwaves, a Window into the Mind

I had waited a month to get the results of my brainwave analysis, a time punctuated by bouts of curiosity and concern; after all, it was *my* brain being scrutinized. (The delay was in large part due to my schedule. The turnaround time is much faster for their clients.) Today, I am about to be exposed, and there is no turning back. Whatever the EEG recordings reveal, I will need to live with it. I imagine the worst: *They will probably find epileptic activity or evidence of a brain tumor.*

It was a muggy, sweltering summer day when I visited the Virginia Neurofeedback Center a month ago, but on this visit storm clouds are soaking the streets of Charlottesville, tires of cars hissing as they pass, and rain obscures my windshield in defiance of my wipers frantically beating back and forth. I am driving through the fringes of Hurricane Harvey that submerged Houston in a devastating flood a few days earlier. Despite the weather, the Neurofeedback Center is bustling with activity. Several wet umbrellas sit open and drying on the front porch like an array of satellite dishes. Inside clients sit in the waiting room, and as I wait, the sounds of people talking in several rooms upstairs and downstairs penetrate the walls. Jessica (we are now on a first-name basis) greets me and escorts me upstairs to her office to meet with Robin. A computer monitor on Jessica's desk is cued up for a live video conference with Jay Gunkelman at Brain Science International in California, who will reveal my brainwave results and explain what they mean.

Brainwave analysis is a complicated process, requiring a team of several people with different specialized skills and expertise to distill the wobbly lines of voltage seeping through my scalp into a comprehensible synthesis. My raw brainwave recordings had been sent to EEG analysts who combed through the recordings to edit out artifacts (muscle and eye movement) and use several techniques to analyze the nineteen channels of EEG recording from my brain. Their processed data were then sent to neurologist Meyer L. Proler, MD, in Houston, who studied them clinically for any abnormalities. Proler's report and the EEG data were then sent to Jay Gunkelman in California at Brain Science International for further analysis. This included a statistical analysis comparing my EEG with a database of 1,000 healthy controls from the Human Brain Institute (the result of a collaboration between researchers in Switzerland and St. Petersburg, Russia). Based on deviations from age-matched controls, Gunkelman outlined recommendations for a program of specific neurofeedback techniques, which Jessica Eure and Robin Bernhard would provide.

I am handed a neatly compiled report with my name on the cover page. Inside I find Dr. Proler's analysis written in succinct paragraphs, snippets of my EEG traces, along with the scores and plots of my EEG power spectrum relative to a large database of normal EEGs. The data are also displayed in graphs and in a pictorial manner with the relative power of different frequency bands of my brainwaves (delta, theta, alpha, beta, and gamma). Color maps of the quantitative EEG analysis are superimposed on a cartoonlike drawing of my head, viewed from a perspective of looking down at the top of my skull. These color-coded maps show various aspects of my brainwaves in different parts of my brain. The cartoon character's head is blotched in pseudocolor much like the weather map of Hurricane Harvey on a map of the United States. Other figures in the report show blue and red lines connecting different pairs of the nineteen electrodes on my head, each line representing the degree of coherent activity or coupling between regions of my brain, and finally the report ends with Gunkelman's summary and specific recommendations for neurofeedback training.

The instant I glance at the power spectra I see that my brainwaves are not typical. They are missing the characteristic peak of alpha power that rises when the eyes are closed. So as not to bias the report I was about to receive, I do not mention what I saw.

"This is a normal, clinical EEG," Gunkelman says, summarizing the findings of Dr. Proler, "ruling out epilepsy and encephalopathies." That's a relief! I recall Jessica telling me that one third of their clients have clinically abnormal EEGs. This high rate of abnormality was completely unexpected, because these people suffered from no obvious seizures, but EEG analysis of their clients who had been diagnosed with a wide range of psychological disorders showed that many were in fact having micro-seizure activity in certain spots inside their brains—a clear indication that their psychological disorders had a physiological component. This was critical information to have to prescribe appropriate drugs to treat mental conditions as some medications could exacerbate the micro-seizures.

Feeling as if I have dodged a bullet, I wait for the next shoe to drop. "We've got a low-voltage EEG," Gunkelman says. Circling his mouse pointer around one of the EEG traces, he says, "We do see a little bit of activity at 8, 9 [Hz], but there's no persistence to the alpha."

"The background alpha, most people think of it as the brain twiddling its thumbs doing nothing; in fact, alpha is enhanced when the brain is resting, but ready," Gunkelman explains. "When you don't see a big alpha, it actually decreases the efficiency of sensory packeting. When alpha goes electronegative the thalamo-cortical gate is open and information flows to the cortex." (By "sensory packeting," Gunkelman is referring to how information is often sent in a series of short bursts, rather than streamed continuously.) In referring to the thalamo-cortical gate, I understand that he is talking about the thalamus, a major relay point inside the brain that shunts sensory information to the cerebral cortex where consciousness arises.

"When it [alpha waves] goes into its electropositive phase, it briefly turns off," Gunkelman continues, referring to the peaks as opposed to the troughs in the alpha-wave oscillations. "Information presented during the positive dip is actually not perceived. So you packet information with the alpha wave. Your alpha frequency is your basic sampling rate of the outside world. Slightly faster alpha, higher resolution; slower alpha, lower resolution."

Although the terminology might not be familiar, you have experienced the effects of thalamo-cortical gating from everyday experience. When you hear that guy yakking into his cell phone in a restaurant oblivious to everything and everyone around him, you are seeing the thalamo-cortical gate in action. To enable his mind to focus on the cell phone conversation, sensory input from the outside world, and his awareness of the outside world, is shut off from the conscious mind. In this case, the thalamus throttles the flow of sensory information to our cerebral cortex, making us oblivious to what is going on around us. Normally, much information in the brain is pulsed through neural circuits in waves and oscillations, to coordinate and packet streams of sensory information up to our cerebral cortex where it is analyzed and pieced back together into a meaningful experience. Like a movie, which appears to us to be a continuous fluid motion but is really discrete single-image frames flashed rapidly at twenty-four times a second, much the same process of quickly shuttering and shuttling bits of information in cortical circuits normally goes on inside our brain. This packeting of information splits up the extremely complex processes of decoding sensory information into meaning and coordinates it with other sensory information and cognitive functions that would be too difficult to perform in a continuous stream. This is much like the buffering we see when large amounts of complex video information being downloaded into our computer is interrupted by abrupt breaks to process input.

For an example of how packeting information works, studies have shown that if an audio recording is made of a person talking, and a researcher cuts the audio track into very short bits that are only a fraction of a second long, then reverses each bit, stitches them back together, and then plays the entire audio string again, people do not perceive gibberish of backward speech; they perceive normal speech.[1] (I can verify this phenomenon personally, having heard this demonstrated in a lecture by David Poeppel, of New York University, when he presented his experimental results at a scientific meeting.) If this fragmented auditory information is presented in packets of optimal frequency, the neural circuits doing the auditory processing and interpretation are simply doing what they normally do in receiving bursts of neural activity flowing through the cerebral cortex at appropriate brainwave frequencies—they make sense of the input and splice it back together properly into the information stream. The length of the recorded fragments is critical, however (they must be less than 50 milliseconds), because the frequency at which our auditory system decodes packets of information is very precise.[2] (Interestingly, much the same thing works in reading, as your brain automatically unscrambles scrambled words to make sense of sentences.[3])

"Also the sweep of alpha across the cortex helps orchestrate processing," Gunkelman says. Recall that waves of any kind can be stationary—that is, rising and falling in one place like a jump rope spinning between two people—or waves can propagate by sweeping over a distance, like the ripple of a fly-fishing line whipping in a curl from rod to tip. "So when alpha is chaotic [like mine, apparently] there is an inefficient packeting and processing stream."

Despite what my EEG analysis shows, I don't feel as if I have any difficulty processing information. I don't think I could have succeeded in my academic and scientific career if I had such difficulty. This is what makes EEG analysis so interesting, because a nonstandard EEG does not necessarily mean that there is any cognitive impairment. This is because individual brains are different. A brain with a weak or nonexistent peak in the alpha frequency, for example, likely has offsetting differences in other power bands that enable the brain to operate efficiently (or even superiorly in certain respects) to the typical brainwave pattern. The increased power of beta, theta, and delta waves in my brain, compared to the average population, are likely compensatory mechanisms, Gunkelman concludes.

In general, there are certain traits that can be associated with EEG patterns that differ from the norm. These generalities are not necessarily applicable to a specific individual, in the same way that cigarette smoking is associated with lung cancer but many smokers never get cancer and live to old age.

"Low-voltage EEGs with no apparent alpha [like my EEG] are a normal variant," Gunkelman says, indicating that it is seen in about 10 percent of the population. "Meaning that there is nothing medically wrong with that. However, a person with a low-voltage fast EEG is much more likely to have overarousal complaints—anxiousness, nervousness, or being tense. It is easy to get harried or scattered with a low-voltage fast background."

"Now your alpha is essentially blocked, which means it is essentially gone with the eyes open. Additionally, we see a peak that is in the slower frequency range. A little bit more in the spatial than the verbal hemisphere, but it is ubiquitous through the sensory integration area [of my cerebral cortex]. It is not in the primary visual field; it is in the sensory integration area."

He goes on to point out that there is increased higher-frequency (beta wave) activity in many regions of my brain. "Usually when we have slow content and beta (presenting together), the beta is thought of as compensatory for the slow content—the brain making more fast beta to compensate for slow (alpha) waves. For example, elite athletes and high-level business executives have more central beta than is normal [as defined by the large database of EEG recordings he uses for reference]. To a certain extent it is a drive mechanism for them. Commonly the elite athletes and high-level executives can have sleep issues." Gunkelman explains how this happens due to the changes in brainwave activity that occur during different stages of sleep (a subject we will examine in chapter seven).

There is, however, another factor that affects the power of EEG recorded from electrodes placed on the scalp. "Skull density affects amplitude. If you have a low skull density with less calcium, it does not attenuate [reduce the brainwave voltage] as much. When we see something like this across the board we have to ask, well . . . is this just a thick skull?" Comparing the relative power of different frequencies of my brainwaves to each other, rather than the absolute power, can help to correct for the muffling of EEG signal by thicker skulls. By analogy, two photos might be very different in brightness of the colors, but if the colors are compared with each other within the same photo, an accurate adjustment of the image content can be made, even if one photo is faded. Gunkelman's analysis suggests that there is evidence of such signal damping by my skull, but, still, the basic findings are clear.

During the hour-long conversation, Gunkelman goes through all the data carefully, explaining in great detail all the information, and answers all of my questions. In comparing my EEG to the average EEG of large numbers of people in their database, my EEG is a bit unusual. "Most people don't have a theta peak, as is witnessed by the theta/beta ratio in the back of the head," he says, pointing to a color-coded map comparing my brain's EEG to a large database of comparable individuals. "The age-matched healthy cohort does not have that theta peak."

Turning to another analysis, he explains that the red lines drawn between many pairs of recording electrodes indicate higher than normal coherent brainwave activity between these two points in my brain. In general, my brain is hyper connected, especially in the frontal lobe regions and in the slower brainwaves. (Frontal lobes, behind the forehead, carry out higher-level executive functions.)

Summarizing, Gunkelman says, "Most of the time we view low-voltage fast EEG as a marker for overarousal," and he recommends neurofeedback training to increase my alpha-wave activity and decrease my delta/theta- and beta-wave activity.

But given how responsive the EEG is to brain state, I ask him how an EEG recording of a few minutes could give information about the brain's basic wiring or pattern of activity. I was highly aroused and intensely interested when Jessica and Robin were recording my EEG. No wonder I had low alpha power—I was certainly not in a meditative alpha state. But Gunkelman replies that he is confident from the thousands of EEGs that he has analyzed that the resting-state EEG of a person does reflect intrinsic features of their brain's operation and wiring. In fact, in the early days of his career, needles were stuck into the scalp to record EEGs. "They were a lot more aroused than you ever would be, because we used to have to stick twenty-one needles into their head."

Gunkelman says he performs his EEG analysis without first reading the patient's medical history or the initial complaint that motivated them to have their brainwaves recorded and analyzed. In my case I lost hearing in one ear several years ago,[4] and when I tell him this, Gunkelman says he can see how my brain's response to that sensory deprivation has altered connectivity and brainwave activity in my auditory cortex.

According to Gunkelman, another risk factor for people with low-voltage fast EEG is developing dependence on alcohol. "It is very common for someone with a pattern like this to have run into something that winds them down, and they like it. Now with a low-voltage fast EEG, if you get home in the evening and you have a couple glasses of wine, you now have alpha," he says. This occurs because alcohol acts on inhibitory neurons in the cerebral cortex, and these neurons are very important in determining the frequency of brainwave oscillations. Alcohol, activating GABA neurotransmitter receptors, slows brain oscillations, thus synchronizing and amplifying the lower-frequency alpha waves. Fortunately, though, I do not have alcohol dependence issues, and Gunkelman adds that this, too, is reasonable, because there are many other cognitive and social regulators on drinking alcohol to excess.

He then outlines a detailed program of neurofeedback training for me. The first technique will strengthen the sensory motor rhythm, to increase my brain's alpha power and decrease beta and delta power. Then they will switch to alpha/theta training to change the relative power of alpha and theta waves. These techniques use neurofeedback to strengthen specific frequencies of brainwaves, relative to others, at specific locations of the brain—that is, at different electrode recording locations.

It has been a fascinating hour, hearing Gunkelman talk about my brain and looking at the data on the electrical activity going on inside it. At the end of our consultation I summarize Gunkelman's review: "The bottom line here is that my wife's right. I have a thick skull, and the doctor recommends that I drink more."

Waves of high-amplitude laughter erupt with robust intercoastal coherence.

## NEUROFEEDBACK SESSION: CHANGING MY MIND

Following the video conference with Gunkelman, Jessica begins attaching the electrodes to my head for my neurofeedback session. She attaches one wire to my earlobe and two other electrodes to two points on my skull designated Pz and C4. These designations represent an electrode over the back of my brain at the midline (Pz: *p* for posterior and *z* designating the midline) and an electrode over the right parietal/central region of my brain (C4). As she applies the sticky goo that affixes the electrodes to my scalp and conducts the electrical signals from my head, she senses my concern that changing my brain through neurofeedback might not be the best idea. Just because something differs from the norm, it does not necessarily mean it is bad. It might even be that the absence of alpha power in my brain is what makes me—me. It could be that my brain, the way it is, is not only just fine; it might in some way be the key to my success as a scientist.

"A lot of our clients are exceptional in some ways," she says, trying to put me at ease, "but will have complaints in other ways." She goes on to tell me that I shouldn't worry about neurofeedback taking away exceptional qualities, if I have them. "I will say it's like a gearshift. If you have a sports car and it's always in fifth gear, when you go to park in a parking deck, it's going to be really challenging. We all need to be able to park. We all *need* to be able to go into reverse. We need that flexibility of all the different gears." Neurofeedback training would enable me to do just that.

Robin explains the neurofeedback plan. "What Gunkelman's plan is that you are going to train up SMR at C4 [the EEG electrode location on the right of the midline of my skull, at a point between my ear and the top of my head referenced to Pz, which is an electrode at the back of my head on the midline]." SMR refers to sensory motor rhythm, a characteristic wave at 12 to 15 Hz in the sensory motor cortex. "Training the sensory motor rhythm will help you feel more relaxed and focused."

I'm ready to go. Jessica tells me to relax and avoid muscle tension. "Also, take stock of how you are feeling now, because especially initially, changes are going to be really subtle. So just feeling your body; noticing if you have tightness or tension anywhere. Just kind of take stock of how you *feel* in terms of your arousal level."

Looking at the computer monitor, I see three green vertical graph bars bobbing up and down and turning bright red when they rise above a horizontal line in the middle of the screen. Above these bars there is a single oscilloscope trace of my live EEG recording dancing across the screen. The wiggly line sweeps left to right leaving a wobbly track, and when it reaches the edge, that part of the screen goes blank and the quivering line starts up again as if drawn by someone with very shaky hands while riding in a car with no shock absorbers on a rough road. This is the raw electrical signal in my brain recorded between the two electrodes. The bars reflect the analysis of how much power is being generated by the different brainwave frequencies in my brain happening in real time.

"You can watch this [computer display] if you *want* to," Robin says, "but it may take your attention away from your body, and get you into your brain—your thinking brain. We're actually trying to slow you down. You could have your eyes closed and let the auditory (tone indicator) work for you," she says. (She is referring to tone signals that will sound when my brainwave patterns are shifting during neurofeedback in the direction they want to see.) "But if you are curious and want to peek, that's fine too." Jessica adds, "Some people find this display overstimulating, but usually when I'm first starting with somebody, I have it on."

The tones begin. They sound mystical; deep resonant chimes, like a new-age jazz musician tuning up or freewheeling on a marimba, searching for inspiration before launching into a tune.

"What am I supposed to be trying to do?" I ask.

"You don't need to . . ." Robin halts her reply to my question and turning to Jessica, she says under her breath, "Did you see what he said to me?"

Rephrasing my question, I ask, "How do I interpret that screen?"

"You see, that's what I *knew* would happen," Robin says in a tone conveying good-natured, restrained frustration. Jessica giggles knowingly. "You thinker," she scolds. "So the cool thing is that your amygdala knows exactly what is going on already. The conscious mind is really way too stupid to do anything to affect it." She and Robin chuckle. They've already cracked into my brain with their machine. They knew I would take a left-brained, analytical approach to neurofeedback, and that tack would not be of much help.

"*However*," Robin says with emphasis, conceding to the fact that I'm not going to resist peeking, "in case you are interested . . ."

Bing, bing, chong goes the bell.

"This is the reward frequency," she says pointing to the bouncing green bar in the middle of the screen. This [the bar on the left] should be your delta slow wave that you are trying to suppress, and this [the bouncing bar on the right] is the high beta that you are suppressing."

She also explains that eye blink and muscle tension affect the different signals so they must be distinguished from true electrical activity inside my cerebral cortex.

The mysterious tones continue, and I try to make sense of them. *Interesting,* I think to myself. *Maybe I'm supposed to make the pitch change in a certain way? Possibly the tones represent the different EEG frequencies. I should probably inhibit the base note chimes and the treble ones and try to emphasize the midrange chiming pitch. But how?*

I cross-examine myself: *Maybe that's not it. The tones could be meaningless, just random variations. Yeah, that's probably right. Single tones or simple notes would get boring. No reward in that. Right, but they* could *be signaling something?* I counterargue with myself: *My amygdala will figure it out. How can my amygdala figure it out? No one told it what to do. Yeah, but* you *know and so your amygdala does too. You can't hide that from your unconscious brain. Maybe I should try to organize the sounds—play them like a musical instrument. Can I develop a rhythm here? How about a melody? Can I play a scale? Quarter notes? Ring the bells in rapid succession?*

I ask aloud, "These tones . . . what are they?"

Jessica explains, "You'll notice these two bars are changing from red to green." I see the left and right green bars bouncing up and down in height turn red the instant they cross above a horizontal line. "This is the threshold setting," she says, pointing to the line that makes the bars turn from green to red when they cross it. "If you'll notice these two [bars]; we are trying to inhibit them, and we are rewarding the middle." That is, the bar in the middle represents an increase in the SMR power and the other bars represent a drop in delta/theta and beta/gamma activity. "When the bar goes over the threshold it turns red. When it is red you are not meeting criteria, when you are in green you are. You'll hear a tone of bells all ringing fast when you see all green on the screen."

"So I'm trying to make it green? All channels?" I say to confirm my understanding.

"Yes, yes," they both say.

"And I'm trying to do what to the bells?"

Jessica responds, "Well, you are not really *trying* to do anything." "Exactly," Robin emphasizes.

This is not how I expected neurofeedback would work. I assumed it would be something like a video game where you are rewarded by racking up points for doing well. "You wouldn't tell your clients what they are trying to do?"

"No," Robin replies.

My puzzled expression no doubt registers my surprise. "A lot of other neurofeedback practitioners are coaches for their clients, to engage them in the process a little more actively," Jessica says. "Because of the clientele we tend to see—they tend to be trauma people, and people with ADD and other issues—they often feel that they have been bootstrapping it, and pushing, and pushing, and working and working. We take the attitude of, 'this is a physiological intervention, and it is going to work whether they do *that* or not.'"

I get it, I think. She's talking about a type of implicit learning that Poizner was researching in the virtual reality experiments I experienced in the aircraft storeroom. No one told me I needed to learn which bubble contained the fire extinguisher—my brain figured that out automatically and unconsciously. Still, this comes as a surprise to me. I had assumed neurofeedback worked by trying consciously to do the right thing to my brainwaves, which was prompted by the tone, like a video game. I'm a bit confused by this, so I press them.

"So I'm not conscious of anything happening?" I reiterate, puzzled and a bit slow to accept what I hear them telling me.

"You don't *have* to be," Robin answers. "And really—especially for what you need—you might want to try the next three minutes trying to loosen all your muscles, close your eyes, and let go. See if you can track sensation in your body that will give you information that you are relaxing."

*OK, mindfulness*, I think to myself. *I'm not much into that. People like it—I'm sure it's great, but I don't have the time. I'll try.*

"What I like to do with my clients is help them experience *sensation* in their body while they are being helped to get there with the equipment. And that might make it easier for you to replicate later."

The three-minute recording and neurofeedback session ends, and Jessica clicks her computer keyboard to analyze and display the results on the computer screen. The bar graphs are replaced by a graph with several lines of different colors. They are zigzag, jagged lines, as if drawn by a child on an Etch-a-Sketch. I realize that this is a plot of the voltage of my brain's EEG over the three-minute neurofeedback session we just completed. I'm not sure what the different colored lines represent, but I presume they are traces of the amplitude of different frequencies of my brainwaves.

"Dark blue and light blue lines are the delta and theta plotted over time," Jessica explains. "Dark brown and red are higher frequencies. That beta activity we want to inhibit," she says, pointing to the red line soaring above the others.

Robin points out a prominent spike, like a mountaintop in the traces midway through the experiment. "You can see when Jessica started talking to you right there."

"Yeah, you've got a lot more beta and gamma," Jessica says, pointing to the peak. "You've got this [mental] state change that happened in this area."

"He might have been interested in what you were saying," Robin suggests. The peak occurred at the moment during the neurofeedback session when they were explaining to me how to interpret the computer display and the strange chimes.

"Yeah," Jessica agrees.

"Your eyes might have been blinking a little more while you were listening," Robin adds, suggesting that this sudden burst of signal also could be muscle artifact.

This first taste of neurofeedback was just a trial run to introduce me to the method and to confirm that the EEG recording equipment, analysis, and visual/auditory rewards were functioning properly. "In the next session just gaze down, not looking at the screen and with no talking," Robin instructs.

"Off again," Jessica says as she taps the keys on her computer keyboard to start another session.

The bells, bells, bells start ringing again.

*OK, don't think about it,* I think to myself. *Ring the bells. Just ring the bells . . . Bells, bells, bells. That's Poe. Stop it! Don't think . . . feel.*

Bing, bing . . . chong . . .

*There, you did it!* I praise myself internally. *What did I do?* I ask in my internal dialogue. *I don't know, but do it again? OK . . .*

Nothing.

The voices in my head begin again. *That's a really long period of silence. You are not doing it. You can't think it. You have to feel it. I can't stop thinking, though. No one can.* My thoughts ricochet back and forth inside the walls of my cranium. *Well, think about being calm . . . OK, I'm calm. Yeah, but you can't be calm when you are doing this as an interview. Yeah, but if you don't do it right, the technique will not work and the session will be a failure. Right. OK, don't think . . .*

Bing, bing . . .

*OK, that's it! . . . but it stopped. You have to do what you did just then again to ring the bells . . . Poe was crazy. It was his mental illness that gave him his genius. Tried to commit suicide. Tragic. Neurofeedback might have helped him.*

*Stop it! Clear your head!*

*OK, OK . . .*

Bing, bong, chunggg . . .

Now the image in my mind of the jazz musician striking the wooden keys of a marimba with marshmallow-tipped drumsticks suddenly morphs into a calypso steel drum player. *Yeah, it sounds kind of like a calypso drum. Nice rich chime tones . . . Those drums are amazing. How do they get such resonant sounds out of an oil can? It's amazing. I heard some calypso drums last week on my trip to the Caribbean. The skin diving was fantastic. Remember that six-foot shark you came nose to nose with when you swam into the mouth of that cave? Startling, but I knew it was a nurse shark. And those remoras were amazing.* The image appears of the long black fish swirling around the big sleepy shark right in front of my mask. They were so close I could see the ribbed sucker disk on their head that they use to attach to the shark. *The sucker is really a highly modified dorsal fin. I'm really glad I can hold my breath so long. I don't really get how those rich resonant sounds come out of the lid of a dented oil drum. I wonder how they ever discovered that? Human ingenuity is pretty amazing. I wonder if I could make one? The calypso drum player is dark-skinned ebony black with dreadlocks down to his waist. He's wearing a red headband, and he has a faraway, spacey look in his eyes. Bob Marley is standing behind him bobbing his head, waiting to join in.*

Chime, bing, gong . . . bong . . . a long pause.

*But no tune, no rhythm. He's just diddling the drum with a stoned look in his eyes. Too much ganja, probably. No chiming . . . He freezes as if struck by narcolepsy. Or maybe he's dumbstruck by a drug-induced stupor. No bells are ringing . . . He's catatonic. Probably not marijuana, probably heroin would do that.*

*Stop it! You are not supposed to be thinking!*

*Right.*

Dead silence is followed by a couple of chimes.

*. . . I don't think Marley would be there. He's too good to just stand there while this guy tries to get it on. Right. Marley's got rhythm. He's dead now. I wonder when he died? Can't remember. That's reggae anyway, not calypso. Came from calypso. Cool sound. This guy is an amateur. He's stoned . . . Probably can't even play the drums.*

*No one is there! You are not supposed to be thinking! You are supposed to be making your mind blank. Don't think about the sounds at all. Yeah, but you have to make the reward sounds or nothing happens. You have to ring the bells. But you can't think about it. Right . . .*

A bell rings. Then a series of three more chimes in rapid succession.

*That's it! I wonder how I did that? You have to do that again. You can't think it. You have to feel it. But what did I feel just then? Didn't feel anything . . .*

Bing, bing, kachunggg . . .

*Maybe it is more like a marimba.*

The Jamaican calypso player vanishes and the jazz musician is back at the marimba. *One of those glossy varnished wood types. It is amazing how resonant some hardwoods can be. Like the rhythm sticks I got in Australia. I'd love to make a guitar out of that hardwood. Australia is nice. Steel drums are similar to this bell sound, but they have a metallic edge to them. Yeah. Not quite a steel drum but close. Or what about a wind chime with really resonant tubes of hardwood. Right, more like that . . . and that would explain why there's no rhythm. It is a very pleasant tone. I'm glad it's not just a sine wave tone or something. I do like the sound. People like wind chimes. They find them peaceful. Yeah. Imagine it is a wind chime and relax. Don't think about it; just listen . . .*

Bing, ba-bing . . .

*I made a wind chime once out of copper water pipe cut to different lengths. Yeah, that was really interesting. It was interesting cutting the pipes to the right length to make them ring in the proper pitches of the pentatonic scale. Did I make it with a pentatonic scale? No, I might have made it with a chromatic scale. Not sure I used that many pipes. That wouldn't have sounded like music, if I used a chromatic scale. Probably a pentatonic scale. You wouldn't have had to use all twelve notes and they still could have been in the chromatic scale. I don't think you had twelve pipes, more like six. Maybe chord progressions, 1-3-5, 1-4-5, or something. I don't remember. Major scale, minor scale, Dorian scale. A minor scale would be perfectly restful and peaceful. Maybe you did. Yeah. A minor seventh or diminished, that's really peaceful. I think my favorite chord is a Cmaj7. I really like all of those jazz chords. Did I build in any harmonics? Did I span octaves? I can't remember. It is amazing how Pythagoras figured out how to make pipes and strings the proper lengths to make all the notes. Yeah, and he did it all with geometry. Right, but there is an equation, too, but I can't remember it. I used it, though, to calculate the lengths of the pipes on my wind chime. Yeah, but the geometric method is more fun anyway. It is amazing. All you need is a protractor. Melanie is waiting outside. Where will we go for lunch? I'm not that hungry now. You take the length of the string and make a triangle from the end to the nut and then just swing your protractor down and that's the first fret. Then you just keep doing that all the way down. Amazing. I wrote it up for the child's science magazine* Odyssey. *Yeah, but they didn't publish it.* Odyssey *went out of business. Great magazine, though, and Beth was a wonderful editor. Yeah, but the wind chime didn't sound very good. I know. It sounded like junk crashing in the wind. That's because that's what it was. But you made them the proper lengths, according to Pythagoras. Yeah, but it didn't work right. Funny how certain frequencies of sounds are pleasant to the human ear. It's all ratios. It probably has to do with the physical world, how the pitch changes when the lengths of things change in meaningful proportions. Sounds accompany everything. Like breaking a stick or striking a hollow tree limb, and when the object breaks into exact proportions, halves, thirds, or whatever, that accompanying sound is rewarding and the human brain favors it and finds it rewarding. You aren't going to get rewarded because you are thinking! The bell's not ringing! You shouldn't be thinking. Stop it!*

Bing, chong . . .

*. . . Or it could be the physics of the middle ear that makes certain ratios of sound pleasing. It's just the way our nervous system is built. The way it stimulates the cochlea. All of that sensory information that had some reward aspect to it for our nervous system led to music. Maybe. It could be a bit of both. My wind chime was fun, but it really didn't sound very good. Copper was probably not the best material. Too soft. Needed to be brass or something. I wonder what happened to it. All the strings got tangled up. Copper is really expensive now. People steal it. I made that whiskey still out of copper. Tasted like Lagavulin. It can't really be rooted entirely in physics, because it's cultural. Asian music has only seven notes and Western music has twelve. Right. Asian music sounds bad. I mean it's fine to appreciate because it is exotic, but it's weird. The twelve-note scale is better. Yeah, but you are being ethnocentric. Too bad. That's not nice. At least I get to hear rock and orchestra music instead of Kabuki string plucking. That's bad to say. I got to like Asian music when I was in Japan. Remember you bought all those CDs of Asian music when you were there? Yeah, but I haven't played them in years. I'm not sure where they are. I don't play CDs anymore anyway. Pandora now. Pandora is too repetitive. I'm getting bored with it. I like Japan, everything is different there. The food, art, the people. Symmetry. They hate symmetry and we love it. It is weird, though. Melanie is waiting outside. Where will we go for lunch? I'm not that hungry now. Western music is nice. People listen to it all around the world. Rock, folk, blues, jazz, country before it went soap opera, bluegrass, opera. Opera makes you cry. Why does it do that? Rap stinks. The Beatles, amazing. No matter what country I'm in I hear Western music. It is always playing on the taxi radio. Yeah, but it is cultural. People all around the world hear it now because of global communication and the dominance of Western civilization, politics, and business. The Jihadis hate us. A culture clash caused all the problems and the 9/11 attack. Their music is strange, all that uvula undulation, belly dancers, semitones, off key, and what not. Remember the Fado music we heard in Portugal? It was really nice. Beautiful and sad. Such a memorable time we had at that café in Lisbon with Morgan. She talked the waiter into selling her a bottle of Portuguese brandy as a surprise gift for me because I enjoyed it so much after dinner. She wasn't even twenty-one. He sold it to her in secret. It was*

*so good. It's all gone now. I'll never get any more. Melanie is waiting outside. Where will we go for lunch? There are the blue notes in the blues that are off scale and they sound great. That one note built an entire genre of music and somehow that particular frequency and sequence of notes conveys an emotional response that transcends language. It's visceral. That's what you are supposed to be doing now! Feeling! You are not supposed to be thinking now . . .*

Bong . . . bong . . . Time passes.

*I've got to focus. Make the bells ring. Turn the bars green. I don't really feel anything, though . . . Melanie is waiting outside. Where will we go for lunch? Cultural experience wires the brain, no doubt. Like many of Jessica and Robin's clients who have horrible traumatic experiences early in life and suffer long-term scars because of how that tragic experience altered their brain wiring. I feel sorry for them. Yeah, but they can help these people by using neurofeedback to rewire their brain. I'm glad I don't have any mental issues. Maybe you do, who knows? I don't or I would feel bad and want to change things. I'm happy with my mind. I'm happy with my life. You are lucky. Yeah. I feel sorry for people with depression and it is such a shame we can't cure schizophrenia. That's such a tragic disorder. Remember that psychiatrist in Winnipeg, Canada, who treats schizophrenics and his patient committed suicide the day you were there? They wouldn't call it suicide. They said he 'completed,' because there is no such thing as a successful suicide. He was heartbroken and he told me all about the guy. It was so sad. You're* not *supposed to say schizophrenics. You are supposed to say "people with schizophrenia." In and out of jail, no friends, horrible hallucinations. I really admire psychiatrists like him caring for people with such severe psychiatric disorders when we don't have the basic understanding of what's gone wrong in their brain. "Yes, but I learned a long time ago that being a physician is often about* care *not* cure. *That's what I try to do," he said. I'm glad I don't have schizophrenia. He was a nice guy. I hope he is doing well now. He told me he had early signs of Parkinson's; I hope he's OK. You could get it. Never know. One out of 100 people do, but no one talks about it. Anyone could get schizophrenia. You could get depression or suffer some horrible tragedy in life. A car wreck or murder or something horrible. So far, you've been blessed. Yes. Everyone in my family is blessed. That's not the case for everyone. No. But you should not be thinking about that, you should be zoning out. Relaxing and listening to the reward tones. Focus. Imagine they are a wind chime peacefully playing in the wind. OK, I'll try that . . .*

Bonggg, bonggg . . .

*Jessica and Robin are really nice. They've been so generous with me and the work they do is so important. They are a great team. Robin's speech is calming, slow, deliberate, almost hypnotic, and precise in choosing words carefully so as not to offend and to be clear. Jessica's exuberance, enthusiasm, quick mind, and openness are endearing and both approaches put you at ease because they are so generous, caring, and friendly. A great team. I really want this neurofeedback to work. I might not be able to do it right now, but later and with practice I'll bet I could. I need to try harder not to think . . .*

*I wonder if the bars are green more of the time now. I can't look. Don't peek. Don't peek. That's cheating. You'll spoil Christmas. Just listen to the bells . . . Wind chimes never sound that peaceful for long. They do at first but then it gets kind of annoying and repetitious and the chimes always get wrecked in a windstorm. You have to throw them away. You shouldn't buy them anymore. I don't buy them. I have gotten some nice ones as gifts. Yeah, from Morgan and Dylan, and I gave Morgan and Lydia wind chimes made out of white translucent soapstone cut in the shape of butterflies and birds that I picked up in Mexico last year when I went to Cuba through Mexico. Hope I can go back to Cuba again, but it will be hard now. Trump changed everything. A bully stomping on sandcastles. Morgan loves butterflies. She always has. Yeah, but you've probably overloaded her with butterfly stuff. Really wonderful decorations at her wedding, and they have such special significance for us. Remember the letter I wrote to her about butterflies when she left home to go to college. But I'll bet the butterfly wind chime I gave her sounds crappy too. They don't ever really get the scales right. And the strings get tangled. Bach changed the scale that Pythagoras made. It is true. A guitar is never in true pitch when you move up the neck and play in different keys. Right. It is a pity. No way around it, though. Banjos are never in tune. That's why the original stringed instruments had movable frets. That's why real orchestra instruments like cellos and violins don't have frets. Kelly is a wonderful cellist. I love the sound of a cello. I can't play that thing at all. Why don't modern guitars have movable frets? We wouldn't have to make them out of catgut now, they could be really high-tech. It would be slick. I hate putting frets into the guitar necks. I don't know how other builders do such a nice job. It is really hard to get them flat and smooth and to have the neck perfect. Aaahhh, a perfect neck. It's all about the neck in the end, isn't it, with guitars? The shape, feel, and the angle. Yeah, remember the feel of that guitar neck you picked up at the county fair on San Juan Island back in 1980 when you were at Friday Harbor Labs with Melanie? It had this deep vee shape. Yeah. The instant I touched it, I felt it. Melanie is waiting outside. But it is very personal. Different people like different neck shapes. They have different shaped hands, that's why, and they play different kinds of music. The Martin guitar neck is the best. I need to make my necks like that. Most of my necks are a bit too wide. Yeah, but you used to like it that way. Yeah, but not so much now. I need to finish the guitar I'm working on for Kelly. Yeah, that's taking forever. She wanted all that inlaid pearl on it, that's why it's taking so long. I hate inlaying pearl. Yeah. It's really tedious and I don't think I'm that good at it. Yeah, you kind of botched up the inlay around the sound hole on her guitar. I think I can cover it up with some careful black stain. She might not notice. She'll notice. Other people will notice. It will look amateurish. Yeah, but there's not much I can do about it now that it's there. I'll have to fix it. I could start all over. Make an entirely new top. I don't think you want to do that. It is taking too long as it is. Stop thinking and relax. Make the bell ring. Right.*

The silence of the room was erratically punctuated by the chime of bells.

*How can I make these bells ring? Imagine I'm dozing off to sleep or something? That's when alpha gets stronger. I'll try that . . . Berger recorded the first alpha waves—on his son and mental patients. He was a Nazi. I'd like to go back to Jena. Maybe I could play this neurofeedback machine like a musical instrument and make it play a tune—"Jingle Bells" or something. No, not "Jingle Bells," that would sound too childish. Maybe Clapton's "Tears in Heaven" or something nice like that. They would love it. Wouldn't Robin and Jessica be amazed to hear me playing "Tears in Heaven" with my brainwaves! You probably can't do it because you haven't figured out what the tones mean and what scale they play, let alone how to make them play the right sound when you want them to. Yeah. Anyway, the sounds are arbitrary probably. But I'll bet I can do it. Wouldn't they be amazed! It would be like that Spielberg movie* Close Encounters. *You can't do it! I'll bet I can. Here goes . . . I wonder if any of their clients can? Maybe that's what I'm supposed to do. Figure out how to play these sounds or at least play them in a rhythmic pattern. No! It's just a reward and you can't do it by thinking. I'm not going to think. I'm going to let my unconscious brain do it. Get a little calypso beat going here. "Shave and a Haircut," maybe . . . "Shave and a Haircut . . . Shave and a Haircut." Not working. "Shave and a Haircut . . ." You have to stop thinking! I don't think anyone can ever stop thinking. But then, you've never been inside anyone else's head. Jessica and Robin have. Still I think everyone has these internal dialogues going on constantly. Dialogues with who? I hate that who/whom business. My prefrontal cortex, I suppose. Yeah, but talking to what? Your association cortex? That doesn't make sense. Default-mode network. Frontal, parietal, temporal, cerebellum. PCC, mPFC. And who am I talking to? I'm not talking to myself. I'm talking to* you. *Who* are *you? It's like there are two people inside my head. And* you *are always bossing me around, telling me how to behave, what to do. It's not* me *talking to* me, *it's* you *talking to* me; *not talking really, but* telling *me what to do. You should stop it. Stop thinking. Relax. OK, I'll try. It's hard. You have weak alpha and hyper connectivity especially in your prefrontal cortex, that's probably it. Yeah, but I seem to be doing just fine without alpha. You need the alpha to package and coordinate the flow of information. Yeah, but too much synchrony is bad too. Higher entropy equals higher information. You have higher bandwidth if your brain can oscillate independently to hold and process different bits of information and perform computation independently and simultaneously in different regions and then send on the information to the next stage in the network. The faster brainwaves probably do it faster than when constrained by slower alpha waves. Maybe, maybe not. Yeah, and you have high-power theta. Theta is great for enhancing synaptic plasticity. You stimulate synapses in rat hippocampal slices with theta in LTP studies. I'm so glad we got Olea's LTP paper published. Low power–high frequency EEG is probably an indicator of dementia. I'm getting Alzheimer's probably. Jay won't tell me. Just like Dr. Fayed in Zaragoza. "Perfectly normal for your age." Hate that. Nothing lasts forever. Told them I'd have low EEG power without my morning coffee. Your brain is going. Don't think so. You might not be aware of it. Cognitive reserve may be covering it. Happens with age. Nothing you can do about it. You are getting old. I don't feel old. "No one gets out of this world alive," Mom says. Melanie is waiting outside. They say you need to reduce both theta and beta, and maybe you'll find that mental state even better. It could make you hyper focused. I don't have any trouble focusing. I can sustain concentration for hours on end. It's that high dopamine in my prefrontal cortex caused by that gene variant I found I have when I did 23andMe. I'm 3.8 percent Neanderthal. 23andMe was great but it is lousy now. The FDA ruined them and now they are exploiting the data. I want to see what it is like to have higher alpha power. Could be just an artifact of the EEG recording—low recording sensitivity. I might really have normal alpha and EEG power. Or just the transient state, not innate wiring. These EEG data are real. I want to boost alpha and see what it is like. But you have to stop thinking about it and let the bells ring . . .*

Ding, dong, ding . . .

*That barred C chord in "Tears in Heaven" is hard. I'm always a bit slow to get into it. You need to practice more. You don't practice enough.*

*You are thinking! Stop and feel your body.*

*I don't get who you are. It is not* me *talking to* me, you *are telling* me *what to do. Maybe it is my two cerebral hemispheres debating with each other? Is that what happens? Left fighting the Right? Rational versus analytical? I don't know. Maybe others don't have this sort of internal dialogue. Sure they do. Keeps them up at night. Maybe with themselves but* not with you. *How would you know that? Are you my conscience? No, that is visceral, the unconscious mind. It is your mother. Right. She is never wrong, but it is easy to miss it sometimes. Your conscience does not talk very loud. It is a feeling. Have to listen for it. Yeah, but you are not my conscious; you are demanding. You are* telling *me what to do. Stop thinking, dammit, and ring the bell! You are not doing it right.*

Bing bong, bing bing.

*Hey! I got a string of notes going there. Cool! The steel drummer is back. Chugged down a swig of Red Stripe and he's started up again. That's good. Just let him play whatever he wants. Still stoned, though. Can't keep it up. Bells are not ringing. How do they clean those dreadlocks? Clay had dreadlocks, but had to cut them off to get his climbing helmet on. Jack*

*Kerouac wrote his entire book on a scroll of paper using stream of consciousness. That's really how the mind works, I guess. His mind anyway. I thought it was sort of interesting but not really such a big deal. Anyone could have done that. Somehow that book became a million-copy seller and a classic. Wish I could figure out how that happens. Books are a lot of work and I wish mine would be successful. Probably won't be. But you never know. No one knows. That's marketing. No one really understands marketing or everyone would do it right. It is pretty random. A crapshoot. The bells are not ringing. Stop thinking and relax.*

The three minutes are up.

"So what was that experience like for you?" Jessica asks.

There is a long pause as I try to encapsulate my experience of the last three minutes and convey it accurately. I can't do it. "It was interesting" is all I can choke out with a laugh. It is a laugh of embarrassment, and I instantly hope that they don't interpret my laugh as dismissive, so I try again to communicate something of the experience I just had. "I guess I was wondering what the different tone pitches meant . . . I guess they mean nothing."

They explain to me that the changing pitches of the tones had no meaning. "The older machines just had one tone and it just drives you bananas," Jessica explains.

"Were you able to attune to your body at all?" Robin asks hopefully.

I have to be honest. "No," I exhale in defeat as my shoulders slumped. If only they knew what mental frenzy was going on inside my mind during those three minutes of silence.

"You were thinking," Robin pronounces, as if she'd caught my hand in the cookie jar.

"I was trying to play 'Jingle Bells,'" I confess.

They both burst into uncontrollable laughter.

Looking at the results, Jessica points to a point in time on the graph on the computer screen and says, "So here's what happened that time. We still have a lot of delta . . ."

Clicking the keyboard, she pulls up other brainwave frequencies. "This is very interesting! You *did* get a little relaxed."

"Yeah," Robin agrees.

"Right around in here. Because the green, your alpha and lower alpha, light blue and green, peaked up over the top of the amplitudes of your other frequencies." I could see those colored lines rising on the chart. "Then it went back down."

Robin adds, "The other thing that's good is that we were trying to suppress that theta. And the theta went down. That's the dark blue line on top."

I think that this change was at a point in the session when I was most attending to suppressing thinking and trying to relax. I just wasn't able to sustain that and at the end of the session I ended up bickering and going *mano a mano* with my internal mind about who was in control of my thoughts and behavior. That's when the alpha lines crashed back to baseline.

"What is his SMR doing?" Robin asks, referring to the sensory motor rhythm. "It's hard to change SMR," she continues, seeing that my results were not positive. Jessica replies, "His SMR is—" Robin interjects, pointing at the descending line. "It went down."

So the session was not entirely a failure in their eyes, it seems, but the changes were very subtle. After some helpful coaching from Robin—she suggests thinking of a heart-opening thought, "Something you really love to do, like rock climbing. Or someone you love, or a pet. That tends to increase SMR"—we start another three-minute session, this time with my eyes closed.

But as I sit there quietly with my eyes closed trying to attend to the neurofeedback reward bells and to dwell on a pleasant experience, no one could possibly tell that my internal mental experience was just like the previous sessions—my usual mode of rapid-fire thoughts and memories swirling around inside my mind. My mind is always abuzz. It's a blizzard of thoughts from which I normally pick one to develop into a coherent train of thought or to express verbally. Thinking is like catching snowflakes. In addition to the verbal thoughts, there are images, and probably more important than either of these is nonverbal and nonvisual thinking—thinking by knowing. For example, automatically approaching a car from the correct side to drive it. Unconscious thinking constantly directs your behaviors. You notice it if you are in England and find yourself opening the driver's door when you intended to open the passenger door, which is now on the "wrong" side of the car.

At the end of the third three-minute neurofeedback session, the EEG results are much the same as the previous two. My EEG is not changing as much as desired. I ask if I can try it one more time. Robin suggests that I take a break and maybe come back another day. We have done enough for one day and neurofeedback takes time. I insist on trying it again. "I would like, this time, to see if I feel any different when the bell's ringing," I say. "This is all new and stimulating, and I can't just veg out [to boost alpha waves], but I'll see this time if I feel differently when the bells ring."

"All right, let's do it," Jessica says. Robin whispers to me as if sharing a secret, "It doesn't work that way." Everyone laughs. "It's fine, though," Jessica says. "You can experiment with it."

So we begin the fourth session, again with my eyes closed.

I try to think of the wonderful sensation of weightlessness I felt while skin diving in the Caribbean last week. Floating on the surface, hyperventilating, and then jackknifing, bending at the waist and kicking my outstretched legs into the air, driving me deep into the water. The image of the six-foot nurse shark inside the cave comes into my mind. *Don't think about the shark. That's too exciting. Think about the sensation of floating underwater. Drifting with the ebb and flow of the currents; swaying in sync with the colorful schools of fish; drifting past waving fronds of coral; diving down deep at the edge of the coral reef where it drops off into deep blue open water a thousand feet deep, and flowing through the buoyant fluid effortlessly like a seal rolling and tumbling. I dive deep beneath the surface down, down, down, below where the stabbing lances of sunbeams penetrate, descending into the dark silent solitary indigo universe of the deep ocean 60 feet below, the vertical coral reef wall on one side and infinite open blue water below and behind me. I'm a visitor in another world. I'm in outer space.*

The bells stop ringing for a long time before I notice and realize the session must have ended. "Is that it?" I ask.

"Um-hum," Jessica says.

I don't remember much else that happened during that session.

The results pop up on the computer screen and they are starkly different from the previous sessions. The graph plummets from a peak at the start of the recording, decaying exponentially to a low flat line. "The delta had a learning curve," I say.

"*Yeah,* it did!" Jessica replies. "You had a nice decrease in delta, and that's great, that's great."

Scanning through the other frequency bands of my EEG, she says, "The different frequency bands are much cleaner and they don't get tangled up. They are more stable . . . I like the look of this; it is an indication to me that your brain was starting to get the idea, to stabilize and calm."

There before my eyes were the data. My brain had changed itself. My brain had altered the waves of electrical activity flowing through it exactly as we sought to do. I don't know how it did it—the conscious brain can't do it alone—but my entire brain figured it out, got together, and changed itself—mind over matter.

## SECOND OPINION

Feeble alpha waves! Who knew? Should I continue neurofeedback, which began tuning up my electric brain to bring it into normal specs after only three sessions, or should I leave my brain's brainwaves to their own unusual mode of vibrations? The following week I met with a leading authority on EEG, who runs a large research program at a world-leading institution and who wishes to remain anonymous, to show him my EEG results and ask about the implications of having an EEG with low alpha power and higher-power beta, gamma, and delta wave activity. The expert was dismissive of my EEG report. He rejected the EEG data and discounted the approach of using EEG in general to identify psychiatric conditions or personality traits, as well as using neurofeedback to alter brainwaves for therapy. He cited a researcher from forty years ago who was widely recognized for using EEG for these purposes, but he stated that this body of work was unsound, and "it gave EEG a bad name." (The researcher he was referring to is still active in EEG research, and well known, so I will not identify the individual by name.) In his expert opinion, the EEG recordings in these early experiments in the scientific literature were not sufficiently reliable and the experimental results were not well replicated.

My expert did not accept that I had low-power alpha waves and he rejected entirely the coherence data. "You have nothing to worry about," he said, trying to reassure me about my apparently abnormal electric brain function. The EEG data and the analysis were simply not adequate, he said. Moreover, he did not accept that these recordings of my resting-state EEG were indicative of fundamental properties of functional connections in my brain, but rather they most likely reflected the state of my brain activity during the recording session—not its innate state.

I pressed him on this, considering the large body of studies in the scientific literature on resting-state EEG correlating with innate brain properties, functional connectivity, and associations with psychological disorders and other traits. (We will examine these studies in greater depth in chapter nine.) He agreed that there was indeed good evidence for this, but that more research needed to be done before resting-state EEG could be used clinically to diagnose psychological disorders, or neurofeedback used for therapy.

This expert is a world-renowned working scientist, not a therapist, and scientists are skeptics by absolute necessity. As a scientist myself, I can attest to the fact that we've all been burned by accepting our own hypotheses too soon, only to find with additional experiments that Nature has outwitted our feeble attempts to figure her out. (Hopefully this revelation happens before publishing the results in a scientific journal.) Working with that uncomfortable uncertainty leaves a queasy feeling in one's stomach, as if traversing a narrow ledge of a cliff face, but that is where the action is—the exploration of the unknown. So until a hypothesis is fully supported by rigorous experimental data using the proper scientific method, and independently replicated by others, any conclusion remains tentative. In the view of this expert, there must be strong evidence-based data that the method works and is reliable before it can be clinically adopted. Clearly, this rigorous scientific viewpoint is why EEG analysis and neurofeedback treatments are not currently approved and covered by many health insurance companies. This cutting-edge research is being actively pursued around the world, including in this expert's own laboratory, but this is neuroscience research in progress. No matter how promising or exciting these new studies are, this expert feels that it is premature to apply EEG diagnostically or therapeutically until the conclusive scientific evidence proves that EEG can be reliably used to diagnose and treat mental illness.

He agreed that the current scientific evidence strongly supports that resting-state EEG can provide the type of information on the innate state of functional connectivity in a person's brain and other features that are characteristic of certain mental illnesses and individual traits and behaviors (to be discussed in chapter ten), but that to obtain such information from EEG, very stringent and sophisticated procedures must be used. The state of the individual during the recording session must be carefully controlled. The quality of the recordings must be very high. The methods used to analyze the data must be appropriate. Coherence measurements in particular, he said, are very hard to do accurately and they are especially error prone. Conducting coherence measurements properly requires sophisticated mathematical methods to reach valid conclusions, he warned. In his research, he uses 128 EEG recording electrodes, in comparison with the nineteen electrodes used to monitor my brainwaves, and he uses powerful computers running sophisticated algorithms, supported by a team of expert scientists and mathematicians, to distill the enormous volume of complex brainwave data streaming in from those 128 channels and make sense of the information. He said that two ten-minute-long recording sessions of my EEG were not adequate, and there were too many artifacts in the records from muscle contractions and electrode movements in my relatively short recordings. "You can do it right," he said, referring to using resting-state EEG to diagnose cognitive function, but you have to use far more sophisticated equipment and analysis. He cited studies that accurately assess cognitive function in people with Alzheimer's or senile dementia, Parkinson's disease, and other brain disorders, and assess the brain's innate pattern of connectivity using resting-state EEG or fMRI together with advanced graph theory methods of analyzing brain networks.

He agreed that neurofeedback can enhance cognitive function, memory, sleep, attention, and relaxation, as well as have therapeutic benefit in some studies of psychological disorders, but that more basic research is needed before it can be applied for diagnosis and treatment.

It is important to appreciate the different perspectives of a therapist using EEG and neurofeedback to alleviate a person's debilitating psychological problems, and of a scientist at one of the premier institutions in the world doing advanced research on the brain using EEG. The objectives of a therapist and a scientist are completely different. The equipment and resources this researcher has in his laboratory for experimental research are the best available, vastly more sophisticated than the instruments used for my brain's EEG analysis. While Jessica and Robin at the Neurofeedback Center have the best equipment, team of experts, and quantitative EEG analysis that they can obtain, they are not a research institution, and they do not claim to be scientists. As long as it helps their clients, that is all they need.

Commenting on the problem of the very large voltages caused by muscle contractions, which always contaminate EEG recordings and which must be carefully removed for analysis, Karim Benchenane, when we met in his lab in Paris, told me that this was a concern for neurofeedback, because relaxation of the body in an alpha state, for example, will also change signals from muscles, but that some researchers have used changes in muscle tension to drive neurofeedback, and it works very well. (Note that biofeedback, in contrast to neurofeedback, works by monitoring any number of bodily responses—heart rate, for example—to enable a person to self-regulate mental state or other bodily functions.)

"How can you be sure that what you're doing is really controlling brainwaves and not a by-product [of relaxation or tension in muscles]?" He cited a specific example: "SMR [the sensory motor rhythm] is used in neurofeedback to relax, but when you control the experiments rigorously, you get exactly the same neurofeedback response with EMG (electromyographic recordings of muscle contraction) that you do with EEG." (To record EMG, researchers simply turn down the gain on the amplifiers on their EEG device so they pick up only the much stronger electrical signals from muscles and do not detect the weaker electrical activity in neurons.) While this is very important from a scientific perspective, it does not undermine the benefits of neurofeedback. "Do you want to cure or do you want to understand?" Benchenane asked.

We will return to the question of EEG analysis to diagnose an individual's cognitive strengths and weaknesses, as well as psychological and neurological dysfunctions, in part III of this book, plus look in detail at the latest research into how neurofeedback can be applied to modify brain function. But in preparation for that, and because we now find ourselves at the crossroads of what is known and what is unknown, let's examine at a deeper level what neuronal oscillations are and separate what oscillations are *known* to do from what many neuroscientists believe they *may* do.

## REALITY CHECK

The two-story cottage-style house of white clapboard trimmed with black shutters is where Albert Einstein lived on the grounds of Princeton University from 1935 until his death in 1955.[5] It is a private residence now, but according to my cab driver, ogling tourists sometimes still pester the homeowners. Passing the nineteenth-century home and traversing the quiet residential fringe into the heart of the university is like journeying through time as the massive green glass and silver monolithic building of the Princeton Neuroscience Institute, built in 2013, comes into view. Timothy Buschman, assistant professor of psychology in the Institute, is investigating how our brain makes decisions and guides our actions toward a goal—that is, the behavior of executive control.

"To me, *this* is thinking," he says. He goes on to say that we have any number of thoughts running around in our heads, and we have the ability to pick one, think about it some more, and act on it. Unlike the simple conditioned reflex of Pavlov's dogs responding in a stereotyped way to an external cue (the dog's dinner bell), executive control is the result of the brain's *own* internal cogitation or information processing that decides what to think about, and what to act upon.

This vital task of executive control is carried out by the prefrontal cortex, located in the forehead region (the lobe severed from the brain in prefrontal lobotomy), but no brain region can accomplish this complex task alone. Decision-making requires integrating diverse types of information over large regions of the brain that extend across a brain-wide network from the prefrontal cortex (higher-level integration and decision-making) to the parietal cortex (important for knitting together sensory information of all kinds) and the basal ganglia deep inside the brain (the hub of the brain's reward circuit that is responsible for that puff of elation we feel when we accomplish a goal). The parietal lobes are in the upper back part of the cortex. Other brain regions are called upon as well, such as the hippocampus (to provide memory and spatial mapping) and the amygdala (for threat detection). Neuronal oscillations and brainwaves are what Buschman and his colleagues suspect are critical in tying together activity of neurons across distant regions of the brain into a dynamic, flexible association to make quick executive decisions.

Buschman, despite his high-profile research demonstrating the importance of synchrony of neural oscillations in attention and decision-making,[6] considers this cutting-edge research a strong "working hypothesis" that requires further proof. "Even within the neuroscience community there is a debate of whether neural oscillations are playing a functional role," he says. Neuroscientists tend to agree that in certain places in the brain neural oscillations play a fundamental role—for example, in the hippocampus where interactions between theta and gamma oscillations are clearly important for the computations and information processing for memory encoding—but this cannot be expanded into a new information coding mechanism in the brain in general without much more research.

"The counterargument is that this [neural oscillations] is just reflecting some process resulting from the true underlying mechanism—the exhaust fumes of computation," he says. The reason for this uncertainty is that most of the current evidence is based on correlation—observations that certain brain rhythms are in sync with neuronal firing (coherence) and behavior. But correlation does not necessarily mean causation. (I always awake at sunrise, but I can't take credit for producing that spectacular celestial phenomenon.) Experiments using new methods, such as transcranial magnetic stimulation (TMS) and optogenetic rhythmic stimulation of neuronal firing, are testing causation. "TMS has effects," Buschman says. "If you stimulate people with TMS at the frequency at which their brain naturally resonates, you get nice enhancement effects, but this is not something that works 100 percent of the time." (In part III we will see how TMS is being used to boost learning and performance in other mental tasks.)

Better evidence of causation is possible in animal experiments, primarily with rodents and nonhuman primates. Buschman and others are using TMS and optogenetics to induce different brain rhythms into the brains of animals and see how altering rhythmic firing influences neural processing and behavior. For example, in experiments stimulating the somatosensory cortex of a mouse to enhance gamma rhythms, the data show that the animal exhibits better perception of sensory input that is presented at certain phases in that rhythm, he says.

It is a fact that a very common feature of neuronal firing in neural circuits is its rhythmicity, which begs the question of why should neuronal firing oscillate rhythmically if this serves no purpose? "Oscillations are incredibly ubiquitous," Buschman says. "You see them across species. You see them across brain regions, but you also see them being disrupted in a lot of neuropsychiatric and neurodevelopmental disorders." Oscillatory phenomena are everywhere in the natural world and can be very pronounced in many cases. From the perspective of physics, however, many natural phenomena exhibit oscillations—a pendulum swinging from the constant pull of gravity or the vibrations (and sound generated) by striking a ruler against a hard surface. Oscillations are the bane of many sensitive electronic devices, such as high-gain amplifiers, where the force of electricity meeting a resistance sets up a back-and-forth oscillation resulting in "amplifier hum" or the deafening audio feedback from a PA system in an auditorium when the speaker's microphone gets too close to the loudspeaker. In numerous instances, that oscillatory behavior is vital to the operational mechanism of a process or device, like a pendulum in a grandfather clock, for example, but not always. Are neuronal oscillations the clockwork of the brain or the screeching audio feedback of a PA system?

This is a fundamental question with which neuroscientists around the world are grappling. Dr. Oscar Herreras Espinosa, at the Cajal Institute in Madrid, Spain, for example, studies the shape and dynamics of electrical fields created around neurons when they fire, called local field potentials. As we have seen, these extracellular potentials are the substrate of brainwaves. During my visit to his laboratory in Madrid, Herreras Espinosa pulled up on his computer a record of electric fields that he recorded in the rat hippocampus. "The oscillation is there," he said, pointing to the wavy lines moving up and down across the screen. However, Herreras Espinosa said, "The important thing to determine is, what is making the neurons oscillate?"

Often oscillations are caused by the push and pull of excitatory and inhibitory connections collectively stimulating and dampening a neuron's firing. Excitatory neurons stimulate neural activity in other neurons, and inhibitory neurons reduce neural firing—the "gas and brake pedals" that make signals go at the proper rate in neural circuits. However, Herreras Espinosa cautioned, just because two populations of neurons are observed oscillating in synchrony, it does not necessarily mean that the two are united in the same functional network. "The input causing this oscillation here," he said, pointing to a recording from one location in the hippocampus, "may be totally different from the input to this other region," which is oscillating in synchrony with it.

"But of course, there are also many oscillations that we don't see because they don't create field potentials," he said, indicating another important confound. Recall that the electric fields surging around neurons like ocean waves are the result of many factors and interactions among electric fields created by other neurons. These interactions can cancel out the oscillations in a field potential detected at a certain point in the brain, although a neuron in that location might be firing rhythmically if one used an intracellular electrode to monitor its firing directly. He warned that missing data in recordings of local field potentials and EEG recordings, like missing pieces of a puzzle, can corrupt the overall picture formed from piecing together the data we do have.

Reflecting on Herreras Espinosa's cautionary words, I now listen as Buschman says that a primary cause of neuronal oscillations is the necessity of a neuron to maintain itself in a balanced state of excitation to operate optimally. If a neuron is too excitable, and firing at maximum rate, it cannot signal a higher level of activity, and, likewise, if its activity is too depressed, it cannot signal a decrease. Oscillations help the neuron find and maintain this optimal state of balance between excitation and inhibition. In this balanced state, a neuron can encode the maximal amount of information.

The oscillatory activity of a neuron waxing and waning to achieve and maintain the optimal balance point in excitability could then lead to oscillating electric field potentials that couple other neurons together in synchrony. This synchrony may be what binds relevant information across the brain into a transient functional network to record, integrate, process, and retrieve information, in complex cognitive functions such as making executive decisions. But too much synchrony is also a problem. "Imagine if all the neurons were firing at the same time—there is not much signal there, right?" Buschman asks. "You might as well have just one giant neuron—if they are all doing the same thing, then why have a million of them?"

"I think the brain is trying to maintain its excitation/inhibition balance, but it is a biological system, so you have a lot of noise in biology (unlike if it were an engineered system). I think what oscillations might be doing is essentially acting as a way to sample [or test] that excitation/inhibition balance. So you have this fluctuation of relative excitation and inhibition, and then in certain periods of the oscillation you are more balanced, which allows the network to be more dynamic," and thus process the maximum amount of information efficiently. "Then you increase the inhibition, otherwise the network would go crazy—sort of an epileptic state." Summing up, Buschman says, "Oscillations enable the network to sample and express different forms of dynamics." It's like balancing on a surfboard—oscillations are like the adjustments you make to keep the board from becoming overbalanced.

Sharing his current unpublished research, he says that his findings are supporting a causative role of oscillations in coupling together the prefrontal cortex with other brain regions in decision-making. "There is synchrony that is established between prefrontal cortex and basal ganglia at certain times during learning, but it is unknown exactly how—what specific cell types are driving these interactions. There are two pathways from basal ganglia [to cortex] that are antagonistic to one another, that might allow the push-pull that we were talking about before. Different rhythms allowing you to positively or negatively select ensembles [of neurons] in the prefrontal cortex." By this, Buschman means that sets of neurons in the prefrontal cortex that come into "resonance" with neurons from the basal ganglia become coupled functionally, while other sets of neurons that resonate at a different frequency are excluded. "The strong oscillations that you see in parietal cortex and prefrontal cortex—these beta oscillations—seem to be playing a strong role in top-down information processing. [Top-down processing is behavior that is willful—controlled by higher-level cognitive function in the cerebral cortex, as opposed to bottom-up processing, the stream of information into the brain from our senses, which also grabs our attention and can direct our behaviors reflexively.] Those beta oscillations are also very strongly represented in the basal ganglia. This is the rhythm that becomes pathological in Parkinson's disease." Parkinson's disease impairs control of voluntary movement because of the death of specific neurons that control the excitation/inhibition balance in the basal ganglia.

In his current research, Buschman is trying to crack the question of whether oscillations are correlations or causations in binding neurons into functional networks. "We are putting in different types of rhythms [using optogenetics to drive neuronal firing] and looking at how they flow through the brain." He is also doing these experiments in genetically modified mice that are animal models of autism to see if this disorder really causes disruptions in normal brain rhythms. The data thus far support the hypothesis that these oscillations are instrumental in coupling neurons into functional ensembles in executive decision-making.

In other pioneering research, which is still under way and not yet published, Buschman and his colleagues are performing experiments that attempt to decode what the brain is thinking, and then feed that information back into an animal's brain to transplant the specific thought and command a specific behavioral response, a fascinating subject that we will explore later in this book (see chapter nine).

But to sum up what we have learned in our quest thus far to understand the strange discovery that oscillating waves of electromagnetic energy radiate out of the human brain, easily detected with a wire touching the scalp, using even the most primitive equipment available when mankind first began to examine the mysterious force of electricity, let me take you back to San Diego.

## GAMMA WAVES IN THE BRAIN—FUMES OR FUNDAMENTAL?

I arrived early to get a good seat, but when I turned the corner and peered down the hall, I was shocked. An overflow crowd of disgruntled scientists was jammed at the entrance, unable to get into the packed auditorium. Those in the back of the mob bounced on tiptoes trying to peer over the heads into the room where muffled sounds of a public address system originated.

I was attending a session at the Society for Neurosciences meeting held in San Diego in November 2018 on the subject of gamma oscillations. The presentation, titled "Gamma Waves in the Brain—Fumes or Fundamental?" featured two experts taking opposite sides in a debate. The program was inspired by a recent pair of articles in the *Journal of Neuroscience* examining the controversy over whether gamma waves are fundamental to how the brain operates or if they are instead an irrelevant by-product, like the hum of an electronic amplifier.[7] Brainwaves, which had for so long been an esoteric specialization, had suddenly exploded into a major issue, drawing in neuroscientists of all kinds and dividing them into two camps. These neuroscientists from around the world were swarming to the auditorium because the most fundamental question of how information is coded and processed in the brain is being re-examined.

Standing among the crowd in the hall, I could hear but could not see the speakers making their opening statements at the podium. I slowly oozed my way through the throng and managed to squeeze through the doorway into the room. Every seat was taken. The aisles were packed with people standing in the overheated room—a fire marshal's nightmare scenario. Inching my way forward, I worked my way through the standing-room-only crowd, compressing myself between people packed along the walls and in the center aisles as if in a fully loaded elevator.

The moderator and two speakers laid out the issue: The textbook explanation that information is digitally coded by the firing rate of neural impulses (the spike code) is well established, but possibly insufficient. High-frequency waves of electrical activity are also detected in the brain, the speaker remarked, oscillating at about the same frequency as AC current in the United States (60 Hz). Unlike neural impulses, which are digital, with every impulse being discrete and identical to every other one, there is also an analog signal in the brain—brainwaves. Brainwaves, like radio waves, can encode information by the phase, frequency, and amplitude of the oscillation in the same way radio waves encode and transmit information in AM and FM (amplitude modulated and frequency modulated signals). Seen as oscillations in neuronal voltage much weaker than what is required to spark a neural impulse, and from rhythmically fluctuating electric currents flowing around them, and also in the rhythmic firing of neural impulses at this frequency, gamma waves ripple through brain tissue at a frequency of about 60 Hz—but why?

The argument was put forth that gamma oscillations could be boosting information processing in sophisticated ways to provide cognitive capabilities that cannot be achieved by a simple spike code. With spikes in neurons, information can only be coded by when a spike occurs and by how rapidly spikes fire in succession. (The spike rate and spike time coding of neural information is discussed in chapter four.) But because neurons are driven to fire at the peaks of gamma waves and are inhibited from firing at the troughs, gamma waves would strongly influence when a spike occurs (as brainwaves of other frequencies would also do). In theory, the rhythmic oscillations could sort, filter, and organize the flow of information through the brain with the proper timing and synchrony required for thinking, perceiving the world through our senses, focusing attention, and making and recalling memories.

Other frequencies of brainwaves have the potential to influence when a neuron fires on different time scales, but the frequency of gamma waves is a perfect match to the time frame over which synapses change their strength during learning and information processing. In 1949 psychologist Donald Hebb proposed the fundamental rule of learning at a cellular level. He hypothesized, and subsequent research verified, that when a synaptic input onto a neuron fires simultaneously and repeatedly along with the firing of the postsynaptic neuron, that synapse will become strengthened. Recent research has precisely defined the time window for what is considered "simultaneous" firing. If the synapse fires precisely when the postsynaptic neuron fires or up to 20 milliseconds before the postsynaptic neuron fires in response to other inputs it receives, biochemical mechanisms are kicked in to make that synapse stronger. The synapse will generate a larger voltage change in the postsynaptic neuron when it fires in the future, and thus have a greater influence on the operation of the neuron it excites. However, if a synapse fires late, that is, right after the postsynaptic neuron fires a neural impulse (within 20 milliseconds), it loses out like anyone late to the party, and it will be weakened. Similarly, a synapse that fires prematurely also gets weakened. This 20-millisecond time window for strengthening and weakening synapses matches the time between peaks of oscillations at the gamma frequency. Oscillations at the gamma frequency could have a commanding influence on strengthening or weakening synapses during learning. Gamma-wave activity in the brain increases during heightened attention and cognitive processing. In contrast, slower oscillations of alpha and delta waves are associated with a mind at rest.

Secondly, as Benchenane in Paris noted (see page 117), waves transmit forces very efficiently. This is in part why Tesla's AC current was destined to win over Edison's DC current as the most efficient means of distributing power through the nation's electrical grid. The combined force of people pushing on the bumper of a car may be inadequate to roll it out of a rut, but if the same people deliver their pushes in carefully timed oscillations, the force of each shove adds to the rocking car's momentum, rapidly building up and propelling the stuck car out of the rut.

Waves of all kinds in nature undergo complex interactions. Just consider the countless complex interactions and consequences of the simple fact that our planet spins on its axis in a twenty-four-hour period. The consequences of the diurnal cycle of the Earth span from weather to biorhythms.

In situations where there are multiple centers of oscillation, complex interactions result. Two or more waves will interact constructively or destructively to amplify or quash signals and change their phase relationships. Higher-order structures are also possible when waves of different frequencies interact. Waves of high frequency can become nested within waves of lower frequency, further coordinating and structuring signals in sophisticated ways, just like quarter notes are organized into measures of a regular number of beats—three to a measure or four, for example, to create either a waltz or a march. All of these features of waves that we see around us in nature would apply to the waves of electrical energy in the brain.

Resonant coupling between two populations of neurons oscillating at gamma frequencies could unite neurons together into functional ensembles in the same way the violin section of an orchestra is joined in time and rhythm with the drum section to create music. Resonant coupling at appropriate frequencies might combine all the rich elements of a memory—place, sequence, emotion, and so on—that are processed in different neural circuits into one coherent, recalled scene.

Wolf Singer, a neuroscientist at the Ernst Strüngmann Institute in Frankfurt, Germany, and a pioneer in the field of gamma oscillations, rose after the debate to offer his view: "It seems unavoidable that microcircuits oscillate, but once you start to couple oscillators then you get very rich repertoires of behaviors" that greatly expand the capacity and complexity of information processing in the brain.

The discussion swung back and forth. There is no dispute over the power of oscillations to contribute a new dimension to information processing in the brain, but definitive experimental proof is lacking. Critics of brainwaves as being fundamental to brain function pointed out that oscillations arise everywhere in nature. The fact that neural circuits can oscillate does not mean that oscillations are part of the mechanism of neural function.

"Right now, these are a lot of beautiful theories without a lot of experimental testing," Jessica Cardin, a neuroscientist at Yale School of Medicine, countered. She noted that present evidence is largely correlational, not a rigorous test of causation, and that synchrony will be observed by chance between any two oscillators. Stronger evidence is necessary to test whether gamma waves contribute to information processing.

Vikass Sohal, a psychologist at UC San Francisco who took an opposing view, argued that gamma waves are important for information processing, but he respects Cardin's critical perspective and admitted that "science works because we are all skeptics and that's essential."

Regardless of which position one favors (gamma waves are fundamental or gamma waves are fumes), there was wide agreement among scientists that gamma waves are often abnormal in neurological and psychological disorders. "I am a psychiatrist and one of the reasons I am interested in this is because gamma rhythms are clearly disturbed in a number of psychiatric conditions, most commonly schizophrenia and autism," Sohal said to me later when the speakers, Wolf Singer, and I met privately to continue the discussion after the session ended. Whether altered gamma waves are a marker of dysfunction or are responsible for the mental health disorders is unclear. However, Singer said that his research and others' have shown that people can learn to control the power of their gamma waves in specific regions of their cerebral cortex by using neurofeedback, which might be therapeutic. Neuroscientist Elizabeth Buffalo of the University of Washington, Seattle, who moderated the debate, added that "there are a couple of small studies on autistic kids with biofeedback that have promising results."

Other research is beginning to test whether rhythmic electrical brain stimulation (with electrodes implanted in the brain or by beaming electromagnetic pulses through the skull) to alter gamma waves (and other frequencies of neural oscillations) can provide an effective treatment for a wide range of disorders, including chronic depression, autism, schizophrenia, and others. Cardin cautioned that this therapeutic approach may not work by correcting abnormal brainwaves; such stimulation could simply disrupt neural circuits by overwhelming the brain with excessive stimulation or inhibition. This is what deep brain stimulation is believed to do in treating Parkinson's disease. In the extreme, electroshock or electroconvulsive therapy (ECT) is effective in treating depression, but this does not necessarily mean that the high-voltage brain stimulation is treating the disorders by correcting abnormal oscillations. No one really knows why ECT works.

Apart from the possible therapeutic potential of manipulating brainwaves by electrical stimulation, this same criticism over the uncertain effects of the stimulus undermines experimental efforts to test whether oscillations are fundamental to brain function or just fumes of the cognitive engine at work. Transcranial stimulation, deep brain stimulation, and optogenetics (genetically engineering neurons so that they become light sensitive and can be stimulated by laser) are being used to alter the oscillations, which does affect behavior, but these methods impact brain activity too severely to provide a convincing test. "If the whole brain becomes entrained in rhythm, the spike code is probably going to die," Cardin said.

The consensus reached at the end of the debate was that rather than using wholesale perturbations of brain activity produced by current methods, we need new technology to manipulate the phase of oscillations and the timing of neural impulses very precisely in specific neurons. In other words, the proof we seek to settle this dilemma may be far in the future when new technology develops. Sohal agreed, but he suggested another more likely possibility. There may be no single "silver bullet" experiment that proves beyond a shadow of doubt that gamma waves are fundamental to brain function. Instead, like the theory of evolution, countless experimental observations over time will accumulate to build an irrefutable case that the theory must be correct.

After this session ended, I raced to see other scientific presentations on other subjects, but the controversy and impact of the debate lingered. "I could not get in, even though this is my field," Phillip Gander, neuroscientist at the University of Iowa, told me, referring to the overflow crowd at the debate, "but they definitely are *not* fumes." He shared with me his own new experiments on gamma oscillations in memory. He performed his experiments on patients who had electrodes implanted in their brains for diagnostic purposes prior to undergoing surgery for epilepsy. His experiments show that when a person remembers a tone, gamma waves are kicked up in the idling brain during the interval between when a test tone is sounded until the subject is asked to recall it. The gamma waves surging through the auditory and frontal cortex—while there is absolutely no stimulation to generate a code of spikes—are holding that sound in working memory, Gander said.

That conclusion, he admitted, is a belief, not a proof. But belief—and disbelief—are what drive science into the unknown.

Whatever belief you may have formed at this point, there is more intriguing evidence to add to the deliberation. In addition to memory, which we have already explored, perhaps the two most intriguing mysteries of brain function are consciousness (our awareness of ourselves and the world around us) and sleep (the baffling state of altered consciousness in which we spend one third of our life). Brainwaves are the best measurable indicator we have of mental function in these two brain states. Let's closely examine brainwave research during sleep and consciousness, not only to illuminate these two great mysteries of brain function, but to collect further evidence to reach a conclusion in the fundamental debate raging in neuroscience over what brainwaves do, if anything.

# CHAPTER 6

# Consciousness, Riding on Brainwaves

Consciousness has long mystified philosophers and scientists. What is conscious awareness? Do animals have it? Consciousness touches on the most fundamental question in philosophy, psychology, and biology, of how the brain creates the mind. What we now know with certainty is that consciousness arises when there is appropriate brainwave activity in the cerebral cortex. Too much brainwave activity (seizure) or too little (anesthesia or sleep) results in the loss of consciousness. This insight illuminates what consciousness is and how it arises from neural networks. Regions beneath the cerebral cortex are also key to understanding how our mysterious unconscious and preconscious minds operate, but here we are concerned with the relation between consciousness and brainwaves, and EEG detects cortical, not subcortical, electrical activity. In a coma, brainwave activity is unusually tranquil and it oscillates abnormally. When brainwaves cease entirely, we are gone, regardless of whether or not our heart is beating.

## COMA AND VEGETATIVE STATES

On Friday, July 13, 1984, thirty-nine-year-old Terry Wallis had just celebrated the birth of his daughter when he lost control of his truck and plunged through a road barrier into a creek. When he regained consciousness, his daughter, Amber, was nineteen years old.

People can persist in a coma for years. When patients are in an unresponsive state, it can be difficult for doctors and family to know if the person has any awareness of themselves and their surroundings. In such cases, doctors often cannot predict when or if the person will ever regain consciousness, leaving loved ones and caretakers in agonizing limbo. Monitoring electrical activity in the brain with EEG provides doctors with critical insight, and, for our present purposes, the research on brainwaves and consciousness adds evidence for those who believe that brainwaves coordinate neural activity to achieve higher-level cognition.

In another case, Ryan Finley's wife, Jill, had a heart attack in 2007, caused by a congenital condition. At the hospital, doctors succeeded in restarting her heart, but Jill was left in a deep coma from lack of oxygen, which damaged her brain. After sitting by her bedside for eleven days, reading the Bible and doing what he could to comfort her, Ryan came to an agonizing decision. Jill would not want to be kept alive through heroic measures just to linger indefinitely in a vegetative state with no hope of recovering. So he solemnly signed the papers authorizing doctors to disconnect his wife from life support.

When doctors pulled the plug at 6 PM that night, Jill did not die immediately, but Ryan was determined to stay with her until the end. At 11:45 PM she started to stir. The last gasps before death—it is very common, the doctors told him. "Get me out of here!" Jill demanded suddenly. Then she added, "Take me to Ted's and take me to the Melting Pot." Ted's was a Mexican restaurant the couple enjoyed, so Mexican food and fondue, apparently, were what Jill felt she needed after two weeks of being fed through a tube.[1] She fully regained consciousness and was not upset with her husband for pulling the plug. After rehab, she is enjoying time with her husband.

After a person suffers severe head trauma, as can occur in a car accident or violent attack, the level of consciousness cannot always be determined with certainty by a neurological exam. How much can the comatose person perceive? Do they have any consciousness at all? Are they drifting peacefully in an alternative dreamlike state, or suffering agonizing pain and despair? A person may be completely nonresponsive because they are suffering from trauma-induced Locked-In Syndrome. In this state of absolute horror, the person can be fully cognizant of their situation and the people around them but powerless to let anyone know. The scientific literature suggests that 40 percent of patients judged to be unconscious are misclassified and are indeed conscious but unable to respond.[2] Moreover, it is difficult to know if a person will come out of a coma, and, if so, when? Every brain injury and every patient is different. If the person awakes from a coma, what will be left of their brain function, personality, and quality of life as a result of their injury? The most profound decision about terminating life support of a loved one hangs in the balance. Compounding the agonizing decision that must be made based on the patient's present dire state is the fact that people can sometimes remain in a deep coma, unresponsive for years, and suddenly recover.

Methods to monitor electrical activity in the brain can be extremely helpful in assessing what cognitive processing is occurring in patients with severe traumatic brain injury, but this research is also yielding surprises and providing insight into how consciousness arises.[3] In many cases, EEG recordings on unresponsive patients in intensive care show electrical responses in the cerebral cortex in response to the patient hearing speech or music. In 2017, researchers Brian L. Edlow and colleagues at Massachusetts General Hospital published their investigation in the journal *Brain*, using fMRI and EEG to assess consciousness in patients in ICU suffering from severe brain trauma.[4] Because the patients in this study could not move, talk, or otherwise respond to questions, the doctors asked their inert and apparently sensationless patients to imagine performing a specific task. They were asked to imagine squeezing and relaxing their right hand, for example, while EEG readings were being taken and also while they were in an fMRI scanner. Doctors also looked for brain responses to sounds of speech or music taking place in the higher-order cerebral cortex, where the brain processes and makes sense of sounds, not in the lower regions of the brain involved in automatic sound detection. Of eight patients who were classified as being unable to respond to language in bedside examinations, the brain activity recordings yielded clear evidence of consciousness in four of them, including three who were classified in the direst vegetative state.

But 25 percent of healthy control subjects failed to show electrical brain responses to these stimuli in these studies, as did one comatose patient who had a remarkable recovery six months later. The reasons for this may be due to technical failures, but the absence of evidence is not evidence of absence, even when assessing unconsciousness as measured by the failure to find electrical responses in the brain. Certainly, if electrical activity is detected in the cerebral cortex in response to language or other stimuli, those circuits are processing information and the person may have some cognizance of their situation and surroundings. But even with this information, it is difficult to make a prognosis. Overall, the data did not show a reliable connection between early electrical responses in the brain to these sensory stimuli while patients were in a comatose state and their long-term outcome. But the research does indicate that even if a person is in a deep coma and entirely unresponsive, their brain (the cerebral cortex where consciousness arises) can in some cases be working and aware.

Better information about cognitive function in comatose patients could be obtained by more sensitive means of measuring brainwaves than through electrodes placed on the scalp as in EEG, but sticking electrodes into brain tissue is riskier. In part III of this book, a new noninvasive approach to monitoring electrical activity in the brain will be explained. The new method is experimental, but clinical trials are about to begin. But there is a far more common comatose state that many of us have or will experience personally—general anesthesia. New research on electrical activity in the brain as people are rendered unconscious with anesthesia and awaken from it provide clear evidence of how consciousness is linked to brainwaves in the cerebral cortex.

## ANESTHESIA

Recently, I underwent a routine medical diagnostic procedure, done under general anesthesia. As I lay in my birthday suit on a gurney, cloaked in a ridiculous faded blue cotton gown, open in the back with two useless ties, I was eager to engage in a conversation with my anesthetist about the miraculous ins and outs of anesthesia, consciousness, unconsciousness, coma, GABA type A receptors, which I knew the anesthesia would act upon, and especially brainwaves, but the doc seemed to take my probing inquiries as questioning her authority. She politely and professionally indulged me as she mixed up a concoction in a 12-cc plastic syringe, white as milk, without really responding to my questions. *She's not going to squirt that white crap into my veins, is she?* I thought to myself. *Yes, she is,* I answered inside my mind. My thoughts flashed: *I don't like it, it's white.* She stuck the hypodermic needle into my IV line and pressed the plunger. Instantly a pleasant sensation of falling into a deep, dreamy sleep overwhelmed me, and I was out like a boxer tagged on the chin. I admit some envy at her ability to terminate an unwanted conversation with an annoying person by simply pressing her thumb on a plastic plunger. What a sense of power that must be!

I awoke immediately. Strangely, she had vaporized in the midst of our conversation, and now only my doctor and a nurse remained beside my bed, muttering as they watched my pulse oximeter report my pulse and the oxygen levels in my blood. I was now fully alert, but I had been out for over half an hour. I had absolutely no sensation of time having passed. That in itself is weird, because when we awaken from a deep sleep, we usually know that time has passed, even if we can't say exactly how long we've been sleeping. Anesthesia, however, is not sleep, and brainwaves make this clear.

Propofol, the anesthesia I was given, works by inducing delta waves. The anesthesia does this by upsetting our normal brain rhythms by interfering in the balance of excitation and inhibition in neural circuits in the cerebral cortex. The neurotransmitter GABA is used by inhibitory neurons in the brain, which are the neurons that put the brakes on other neurons firing. Propofol boosts inhibition of neural activity by activating GABA receptors on neurons. When that white "milk of amnesia" hits your veins, the effect is as if every inhibitory neuron in your brain suddenly slams on the brakes. The strong inhibitory drag slows the normal busy oscillation of electrical activity in cortical circuits from its delicately balanced inhibition and excitation, and brainwaves grind to slow, low-frequency delta waves, causing the person to black out.

Emery N. Brown, MD, PhD, anesthetist at Massachusetts General Hospital, and professor at MIT, together with his colleagues, is using anesthesia, not only to render patients unconscious during surgery, but to explore the very nature of how consciousness arises in the brain. Dr.

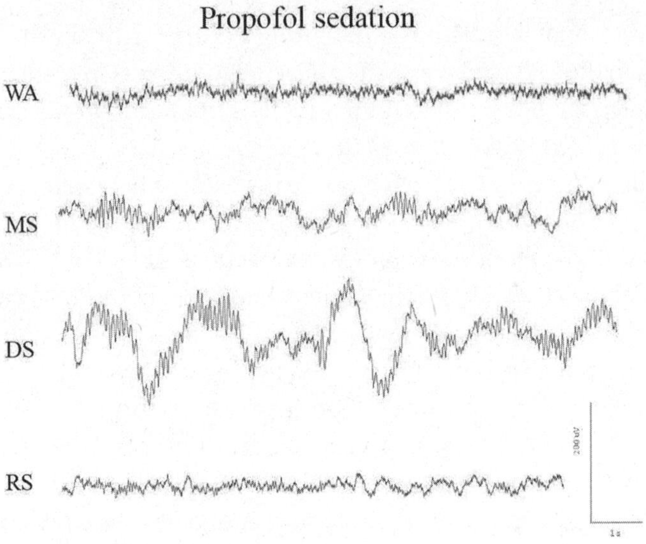


**FIGURE 21:** *Brainwaves and consciousness are intimately related. The effects of the anesthetic propofol on electroencephalogram activity are shown as the patient loses consciousness: wakefulness (WA), moderate sedation (MS), deep sedation (DS), and recovery (RS).*

Brown's approach is unique in that he uses EEG to monitor electrical activity inside the brain as anesthesia takes effect and consciousness slips away. It is surprising to me that such experiments were not done in earnest until about 2010.[5] Brainwave research has been conducted separately from neuroscience in general, and from medicine as well, except for special fields such as epilepsy research and treatment, where EEG recordings are used to localize seizures in the brain. This late-to-the-table application of EEG brain science and technology is especially concerning because general anesthesia carries the risk of serious brain injury and death,[6] and because anesthetists gauge the dose and mixture of drugs they apply to keep their patients unconscious by monitoring consciousness through indirect bodily functions, such as heart and respiration rate, rather than knowing what is happening inside the person's brain in terms of conscious awareness. Too little anesthesia and you can be subjected to the unimaginable horror of being awake on the operating table but locked in, paralyzed by drugs, and unable to move or scream while you are fully conscious of doctors cutting up your body, chatting, and inflicting excruciating pain. Too much anesthesia, and you may never wake up. Everyone's body is slightly different in how we react to and metabolize different drugs. Redheads, to cite one example, require higher doses of Novocain, often used by dentists, to have the same pain-relieving effect.[7] So an anesthetist's skill, as much as pharmacology, determines how well anesthesia works.

Equally fascinating is how neuroscientists, who have long pondered and pontificated about the nature of consciousness, never thought to study this connection between mind and brain by using anesthetics—drugs that move the brain between consciousness and unconsciousness at will—in combination with electrophysiology, monitoring the electric activity in the brain. Taking advantage of his skills as an anesthetist, Brown began doing such experiments on people, by simply applying EEG electrodes to their heads and watching what happens to their brainwaves while under anesthesia.

I am sitting in a cavernous lecture hall among thousands of other neuroscientists who have come to hear Dr. Brown give a fascinating presentation of his research at the Society for Neuroscience meeting in 2016. As he narrates his PowerPoint presentation, Dr. Brown describes what's happening to the brainwaves being recorded from a sixty-year-old woman undergoing anesthesia for thyroid surgery. Videos of EEG recordings and pseudocolor maps of brainwave activity superimposed on the brain flash across the projection screens positioned all throughout the auditorium. The pseudocolor brainwave maps look just like the ones I obtained of my brain at the Neurofeedback Center in Charlottesville, with different brainwave frequencies represented in different colors on the head, but he is showing video animations, so the audience can see the wash of rainbow colors sweep over the brain, reflecting how brainwaves are changing with anesthesia and consciousness. He describes that in studies in which he asked test subjects to respond to a series of prerecorded words or clicks by pressing a button while he administered propofol, he and his colleagues were able to determine exactly when the

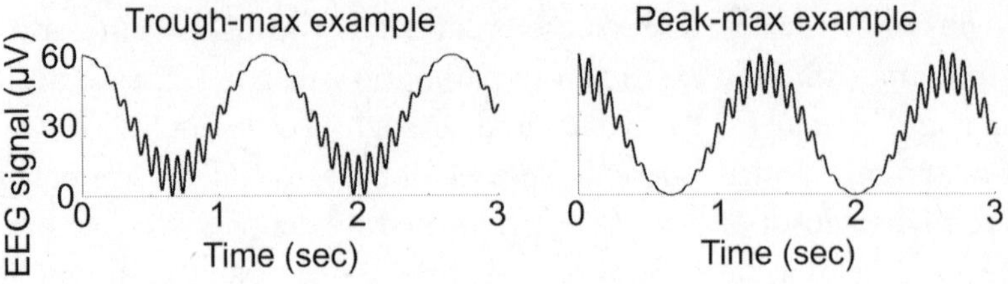


**FIGURE 22:** *When the anesthesia propofol is used as an anesthetic, loss of consciousness occurs when alpha waves (8–12 Hz) and slow-wave/delta oscillations (0.1–1.5 Hz) become coupled in phase such that the peak of the alpha waves occurs at the trough of the slow waves (trough-max), as illustrated in this drawing. At deeper levels of anesthesia, the coupling reverses such that the peak of the alpha waves occurs at the peak of the slow waves (peak-max). When anesthesia is stopped, consciousness returns once the trough-max brainwave pattern develops.*

person lost consciousness as well as when they regained it. Studying the EEG recordings, they were able to pinpoint a telltale change in brainwave activity the instant the person loses consciousness or regains it after anesthesia is stopped.

During the state of unconsciousness, very slow, less than one-cycle-per-second oscillations (slow delta waves) undulate everywhere across the cerebral cortex asynchronously. The recordings show that populations of neurons in different parts of the cerebral cortex are no longer firing together rhythmically: They are oscillating, but disconnected and uncoordinated from one another. Like watching the picture in a beautiful sand painting dissolve on a table vibrating in an earthquake, the picture our mind paints of the world disappears as brainwaves in the cerebral cortex interfere with each other and oscillate destructively.

He explains that when he and his colleagues compared how the alpha waves and slow-wave/delta oscillations interacted, they noted that during unconsciousness, the peak in amplitude of alpha waves was in synchrony with the peaks of low-frequency delta waves, something Brown calls "peak-max." It is as if someone in a boat at sea, riding slow-rolling waves, waves his hand at the peaks of the ocean swells. But during transition into and out of unconsciousness, maximum alpha wave amplitude becomes 180 degrees out of sync; that is, alpha wave amplitude peaks during the troughs of the low-frequency waves (which Brown calls "trough-max"). The peak-max brainwave pattern resembles the slow brainwave oscillations that have been studied in deep sleep, which we will discuss in chapter seven, but there are other differences that make brainwave activity during anesthesia very different from sleep, he says.

Consciousness, then, must arise from the coordinated, precisely timed activity among many different populations of neurons in our cerebral cortex, linking together different cognitive processes. If the complex network of intracerebral communication is disrupted, so that populations of neurons throughout the cortex fire rhythmically but in an isolated and uncoordinated fashion, consciousness is lost, as happens during propofol anesthesia. By analogy, if we were considering sound waves recorded from an orchestra onstage, rather than electrical waves from the brain, it would be like the difference between the cacophony of noise and chaos heard as the orchestra tunes up before the symphony and the sound created when they play music together. Likewise, if all populations of neurons in the cortex fire absolutely in synchrony, which is what happens in a seizure, consciousness is lost, because there is no interchange of information across the cortical expanse of the brain if all neurons are firing together. No music is possible if every instrument in the orchestra strikes a note simultaneously with the beat of the kettle drum.

Although we commonly refer to anesthesia as "putting someone to sleep," Dr. Brown says his research monitoring brainwaves during anesthesia shows that the brain in sleep is nothing like the brain under general anesthesia. Sleep is an altered state of consciousness, not strictly the absence of consciousness. Instead, as he promises to demonstrate in a moment, brainwaves under anesthesia are like those seen during loss of consciousness caused by traumatic head injury. "General anesthesia is a drug-induced reversible coma," Brown says. Patients, understandably, prefer the sleep euphemism to the truth that the stranger fiddling with their IV line is going to put them into a coma—even if only temporarily.

Brown explains that typically, a combination of drugs is used during general anesthesia. Benzodiazepines (such as Valium) or other hypnotic drugs are used to cause amnesia. Most people do not want to hear or feel any aspect of their surgery, he notes. Propofol or barbiturates are used to render the person unconscious, and ether-like gases or nitrous oxide are often given to maintain the unconscious state. Anticholinergic drugs, like curare, the paralytic toxin that South American natives paint on the tips of their poison blowgun darts, are injected to paralyze muscles and prevent the body from twitching and writhing in response to the surgeon's knife cuts. You really don't want the doctor attempting delicate surgery on a moving target (oops!). Other drugs are used to maintain blood pressure, kidney function, and oxygen/$CO_2$ levels within healthy limits, and finally, drugs to reverse the action of these agents are given to bring the person back out of the unconscious state. Even for a simple procedure you will receive about ten drugs during anesthesia, according to Dr. Brown. Any drug can have adverse side effects, but rendering a person unconscious, pain-free, and untraumatized by preventing the formation of horrible memories requires bathing the brain in a pharmacological stew, not just giving one medication. (What could possibly go wrong?)

"I'm going to show you how her EEG changes over time," he says, showing his EEG recordings of the sixty-year-old woman about to have surgery on her thyroid gland. At first, the woman's EEG is jiggling actively in the typical energetic mix of waves of electrical activity that one records from the brain in the awake state. Flashes of enormous seismic disturbances periodically erupt, which are caused by her eye blinks. Muscles situated just beneath the thin skin generate electrical discharges that are huge, compared to the weak signals from neurons seeping out of the brain and through the skull.

When propofol is injected into her veins, he says, "It's going to get real, real noisy," and it does. All the tracks in her EEG record start to scratch jagged lines frantically. "That's because, when you inject propofol into a small vein, it burns like crazy. She's tensing up." (Muscle contractions cause large-voltage fluctuations in EEG recordings.) I don't remember feeling any burning pain when my anesthetist squirted that milkweed-looking stuff into my veins. But of course, I wouldn't remember. My memories of anything that happened while I was under the influence of propofol were erased or, to be more specific, never formed.

"Now watch, you are going to see when the drugs take over . . . Right here!" The noise disappears and the EEG traces begin to roll in slow oscillations like the sea gently rocking in a harbor. "What you have now is a very regular oscillation, it's a beta oscillation at about twelve to sixteen cycles per second. This is the oscillation you would get if you took a sleeping medication—if you took Ambien, or you took one of the benzodiazepines, you would get a beta oscillation. You wouldn't get sleep; you would get this beta oscillation," he explains, narrating the EEG recording streaming out across the display screens. (When we consider brainwaves during sleep in the next chapter, it will become apparent what Dr. Brown means here. Sleep is a multistage, complex process. While beta oscillations are characteristic of some stages of sleep, and Ambien boosts these waves, many more factors and additional brainwave responses are required to achieve sleep.)

"Now watch what happens next, just a few seconds later—large slow oscillations." Extremely low frequency waves of less than one cycle per second—delta waves—begin to rise in the woman's brain. Brown explains why there is a sudden sea change in her EEG. "So the blood is traveling through the basilar artery now, and going to the brain stem, and it's hitting these GABAergic circuits that come out of the preoptic area and it's inactivating [this part of] the brain stem and you see these slow oscillations." (The brain stem, the stalk at the top of the spinal cord, upon which the rest of the brain is affixed, controls the general level of arousal throughout the brain. With the excitatory circuits in the brain stem squelched by the strong inhibition from the anesthetic drug, the brain is rendered inert in its lowest possible state of arousal. Nothing but very slow, rhythmic fluctuations in EEG activity is possible.) "It is like slow-wave sleep, when your brain is inactivating, and that's what's happening here." Now the woman is completely flaccid; all muscle tone is gone, her eyes are closed, fixed at the midline, as if she were dead. But this stage of anesthesia is only a transitory one resembling the state of deep sleep. Just a few seconds later her EEG pattern changes abruptly. "She's very deep," he says. "She's in burst depression . . . There's the depression right there, and there's the burst!" He points. "This is a very deep state like you would see in hypothermia, or someone who has had traumatic brain injury," he explains. Essentially the woman is in a coma. "Between the bursts she's actually isoelectric [flat line]."

The spasmodic bursts that erupt in the flat-line EEG trace are like cannonballs, creating enormous waves in the otherwise tranquil sea. The normal EEG of a person who is awake is about 5 microvolts in amplitude. A 1.5-volt flashlight battery is 3,000 times stronger. Brown explains that "as you move under [transition into] anesthesia, the EEG becomes highly organized and very, very regular, and you have these large oscillations that occur. So it goes from something that's about 5 microvolts to something that is about 10 to 20 microvolts, or even as large as 500 or 1,000 microvolts." Such brainwave activity is never seen in an awake, normally functioning brain.

Next Dr. Brown shares an EEG recording he made of a nineteen-year-old woman under anesthesia, but this time the raw EEG traces have been processed into spectrograms, which break down the brainwaves into their component frequencies and displays them on the brain in pseudocolor, showing how brainwave activity changes over time and sweeps over the surface of the entire cerebral cortex. "There is another very important dynamic that happens every time you anesthetize someone. There is a very important spatial dynamic," he says, meaning a change in how the brainwaves sweep through different regions of the cerebral cortex. Describing alpha waves, which Hans Berger first discovered, Dr. Brown explains that when you close your eyes, EEG oscillations at the back of the head in the alpha frequency (about 10 Hz) increase. "Pretty much everyone has them [alpha waves], but we tend to lose them as we get older. Maybe 5 to 10 percent of people don't have them." (Apparently, I am one of them.) It has been known for at least forty years from experiments on monkeys that under ether anesthesia, the alpha oscillations at the back of the head move to the frontal cortex. They return to the back of the head as the monkey comes out of anesthesia. In this experiment, Brown and colleagues are about to do the same thing to a human volunteer, who agreed to receive propofol injections in different doses to systematically uncover how brainwave activity changes with the loss of consciousness, and whether the huge sweep of alpha-wave activity at the back of the brain that moves to the frontal cortex also occurs in people. If it does, unlike a monkey, a person can describe what is happening to their mind as they begin to go under.

"This person is sitting there with a very strong eyes-closed alpha oscillation," he says, pointing to the slowly oscillating EEG trace. "Now, drug is injected . . . It is starting to break up at the back." The alpha waves become irregular and start to diminish in strength as they begin to build in the frontal cortex. "It's starting to appear at the front and now you can see these really strong alpha oscillations at the front."

The spectrograms allow one to see what's happening to the other brainwaves, which are shown in different colors. Delta-wave activity is strong and surging at its slow oscillation of less than 1 Hz throughout the brain. Also, by comparing the timing of the waves recorded at each electrode, an important effect is seen: The alpha waves at all the electrodes in the frontal region are oscillating in synchrony. "In other words, it [alpha wave activity] is moving in phase across the front of the head, whereas the slow oscillation across the brain is totally incoherent." When Dr. Brown stops administering the drug and the patient begins to regain consciousness, the alpha waves in her frontal lobes subside and build up again in the cerebral cortex at the back of her head.

"So what's happening here?" he asks, as he begins to explain the neural circuitry that is responsible for the changes in alpha waves. Research on experimental animals, mathematical modeling, and data from epileptic patients who have electrodes implanted in their brains to help doctors locate the origin of seizure activity indicate that a feedback loop of activity between two brain regions generates the large-amplitude, coherent alpha-wave oscillations in the frontal region. Neurons deep in the brain (found in the thalamus) that send and receive connections from the frontal cerebral cortex are oscillating at about 10 Hz, setting up back-and-forth waves of electrical transmission in lockstep between the frontal cortex and thalamus. If all the neurons in the brain's frontal lobes, which are responsible for complex coordination, processing, and interpretation of all activity received from other brain regions, are stuck in a phase-locked pendulum swing of synchronous activity, it is as if the entire frontal lobes had only one neuron. One neuron is not going to give you consciousness.

Remember that the brainwave oscillations are changes in the electrical field around neurons. Because the firing of neurons, and thus transmission of information between neurons, is facilitated when the neurons are depolarized (that is, at the troughs of the waves), but their firing is inhibited at the hyperpolarized peaks of brainwaves, then the transmission of information is greatly impaired when the millions of neurons in the frontal cortex are locked in phase at 10 Hz. Like a sputtering engine firing on only one cylinder, the brain cannot function. Only very low-frequency oscillations, which can wax and wane over a much slower course of time, will be possible. Consciousness, the recreation of our internal and external state of mind by receiving and analyzing sensory input, integrated with our memories and other cognitive functions, is impossible. Brainwaves are key to consciousness.

But the results thus far are correlations and don't prove cause and effect. To determine whether changes in brainwaves during anesthesia are responsible for consciousness or simply associated with it, Brown begins to describe experiments he and his colleagues performed on mice in which they could drive brainwave activity by optogenetic stimulation. Optogenetics uses genetic engineering to insert ion channels into neurons that will open in response to specific colors of bright light. Then the researchers can drive activity in these neurons (stimulate or inhibit it) by beaming laser light into the brain through a fiber-optic cable. The audience members and I watch the video of the mouse on the projection screen. As soon as they stimulate neurons in the circuit between thalamus and cortex in the alpha-wave frequency, the synchronous alpha waves knock the mouse unconscious, without anesthetic.

There are many different types of anesthetics, and they do not all produce the same effects on brainwaves as propofol. In every case, however, brainwave activity is disrupted in a way that impairs communication among populations of neurons in the cerebral cortex, leading to loss of consciousness.

Brown says he finds the current standard of practice of anesthesia irrational. "We are obliged to monitor heat rate, respiration, temperature, oxygen, carbon dioxide, but we are not obliged to monitor the brain. The irony is that the drugs are having the most profound effects on the brain, but we monitor all the physiological responses besides the brain!" He advocates that EEG should be standard practice in anesthesia to monitor a patient's level of consciousness. By following his patient's EEG while under anesthesia, he has been able to give one third the standard dose of propofol, and still know that the person is unconscious by following their brain oscillations. Additionally, these patients recover from the anesthesia much more rapidly and with fewer side effects. Typically, a person receiving propofol anesthesia is not permitted to drive or engage in other cognitively demanding activities for twenty-four hours, while the drug is being broken down by the liver and the brain gets back into its normal swing. I remember that in my case, my wife drove me home and I took the rest of the day off, using the opportunity to write. But almost nothing I wrote survived the editing process when I read it the next day. The sentences were properly constructed and they were clear and coherent, but the choices I made about what to write were poor, and I stripped them out of this book the next day.

Brown and his colleagues have taken this line of research a step further to develop a most remarkable experimental technology, which

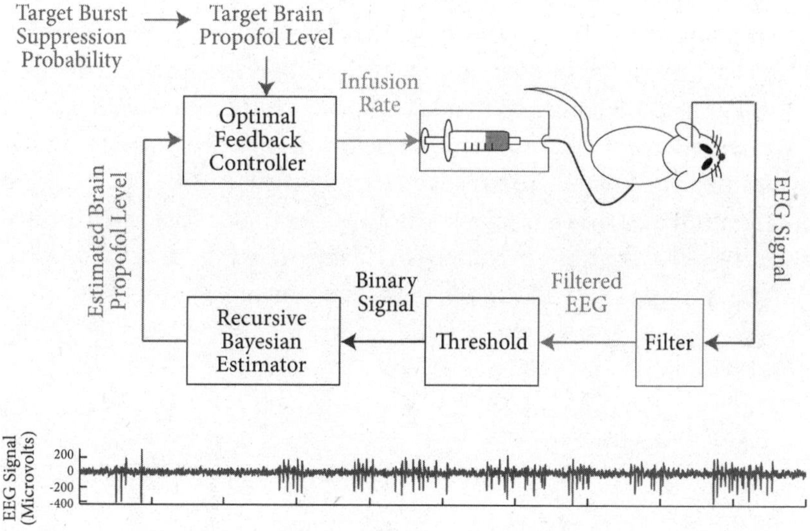


**FIGURE 23:** *Brain-computer interface to automatically maintain anesthesia or a medically induced coma by computer monitoring EEG activity and delivering anesthetic to maintain burst-suppression EEG activity. Burst-suppression EEG activity are brief bursts of high-voltage brainwaves that appear against a quiet, flat-line EEG, seen only during deep anesthesia or coma. A one-minute recording is shown.*

could eliminate the need for an anesthesiologist. In place of an anesthesiologist, a brain-computer interface (BCI) monitors brainwaves of the patient on the operating table, and it regulates the delivery of anesthetic drugs to keep the brain constantly and safely in a state of unconsciousness as determined by the brainwave patterns throughout surgery. By automatically dispensing drugs into his or her own body to keep their own brainwaves in the state of a deep coma, the patient's brain, hooked up to a computer, becomes the anesthetist, and renders itself unconscious by direct action. From Dr. Brown's experience observing changes in EEG activity in his patients going in and out of consciousness during anesthesia, there is no doubt in his mind that brainwaves are what drive consciousness or obliterate it.

General anesthesia is something that not everyone experiences, but all of us lose consciousness every night. The most mysterious state of mind is the complex phenomenon that takes place in our brain while we sleep. Sleep is perhaps the area of human behavior where brainwave analysis has had the greatest success. In fact, the various stages of sleep are now identified by the characteristic brainwaves that take place in the five stages of sleep, from dozing off, to deep slow-wave sleep, to REM sleep and dreaming, as we will explore in the next chapter.

# CHAPTER 7

# A Sea Change in Brainwaves While We Slumber

What is sleep and why do we do it? It is fascinating to realize that although sleep is an essential requirement for humans and most animals, and that we spend so much of our time on Earth in this bizarre altered state of consciousness, we do not fully understand *why* we sleep. Through brainwave research, however, we have learned a great deal about *how.*

Before delving into the brain's electrical activity in sleep, let's begin by examining some fascinating aspects of the peculiar state of consciousness and unconsciousness we call sleep. Then we will explore how brainwave analysis reveals what is happening inside our brains as we slumber and dream and when sleep is abnormal. The most fascinating part of sleep is dreaming. What are dreams and why do we have them?

## DREAMS: OUR NIGHTLY EXCURSION INTO MADNESS

We spend a third of our life in a completely altered state of consciousness, which can indeed be considered madness. Dreaming is a descent into what would otherwise be a severe form of psychosis. These nighttime hallucinations can be exhilarating, insightful, meaningless, nonsensical, or terrifying. Nearly everyone has experienced them. What's in your dreams—especially dreams that revisit night after night? Recurring dreams are especially curious and often disturbing. Are you flying high above the heads of other people in gleeful euphoria? Are you sharing a tender moment with a loved one in beautiful surroundings? Or are you terror-struck, running for your life from a monster or murderer? Maybe you are helplessly tumbling through the air or are badly injured. A recent study revealed the most common content of recurring dreams and found very different hallucinations in the dreaming minds of adults versus children.

The study of several hundred boys and girls between the ages of eleven and fifteen conducted by Aline Gauchat, psychologist at the Université de Montréal, and colleagues, found that most recurring dreams are not pleasant.[1] They are most often terrifying confrontations with deadly threats of some sort. But the recurring dreams of kids, adolescents, and adults differ in interesting ways.

Here are the most common themes in recurring dreams according to Gauchat's research:

**Being chased:** The dreamer is chased but not physically attacked.
**Physical aggression:** Suffering a threat or direct attack on one's person or character, including sexual aggression, murder, or being kidnapped or sequestered.
**Falling:** The feeling of falling in midair, off cliffs, or from another elevated object.
**Car accidents:** The dreamer or another character is involved in a car accident.
**Contact with strangers**
**Death of the dreamer**
**Death in the family**
**Confrontation:** The dreamer is confronted by monsters, animals, zombies, or similar creatures.
**Dreamer is injured or ill**
**Stranger entering the dreamer's house:** A stranger is breaking into the dreamer's house or trying to enter it.
**Being stuck or trapped**
**Flying or being able to control/withstand natural forces**

Consistent with what we all know about children and nightmares, the study found that only 9 percent of the recurrent dreams of children between the ages of eleven and fifteen were positive experiences. The most common recurrent dreams in children involve serious threats, such as confrontations with monsters, animals, or zombies. The next most common theme in recurrent dreams of children involved threats of physical aggression, falling, and being chased. In 87.9 percent of the children's dreams, the dreamer is the target of the threat. Themes involving car accidents played out in recurring dreams of 6.9 percent of boys and girls, but gender differences are found. The threats to girls in recurring dreams were twice as likely to be related to being chased (11.3 percent versus 6.5 percent). All the other common dream themes listed above were reported in only 6 percent of the recurrent dreams of children.

It is interesting to compare these results on children with a 1996 study on adults.[2] As with children, most of the recurrent dreams of adults were disturbing (77.3 percent). In contrast to children, where 45.5 percent of the threats in dreams involved aggression and violence, this harrowing theme dropped to third place in the top five themes of adult recurrent dreams. Escapes and pursuits were the most commonly experienced recurrent dreams of adults (25.9 percent). Interestingly, dreams of physical anomalies were common in the recurrent dreams of adults, but this terror was absent from the sampling of children's dreams.

The most common recurrent dreams of children in this study were:

1. Aggression and violence (45.5 percent)
2. Accidents and misfortunes (28.8 percent)
3. Escapes and pursuits (22.7 percent)
4. Disasters (4 percent)

Dreams about failures and physical anomalies were not reported by children.

The most common recurrent dreams of adults were:

1. Escapes and pursuits (25.9 percent)
2. Accidents and misfortunes (19.7 percent)
3. Aggression and violence (19 percent)
4. Physical anomalies (17 percent)
5. Failures (6.9 percent)

Dreams that were absent from the childhood recurrent dreams were the common adult dreams of losing one's teeth, being unable to find a private toilet, and discovering or exploring new rooms in a house. One striking result is that friendly interactions were present in almost one third of girls' recurrent dreams, but that was more than ten times more frequent than for boys who have friendly interactions in fewer than 3 percent of their recurrent dreams. In studies of adults, women's bad dreams are more frequently centered on interpersonal conflicts, and women's dreams are twice as likely to contain friendly interactions as men's are.

The researchers speculate that recurrent dreaming is related to life stresses and that imagining these threats in a dream state simulates threatening events of real life, thereby enabling our mind to rehearse strategies to avoid and confront such threats. Because real-world threats to boys and girls, men and women, are somewhat different, so too are their threatening recurring dreams. For example, children are much more likely to experience threats from imaginary creatures and monsters in their dreams than adults are, but adults, through life experience, have learned that monsters do exist, but they are usually human beings.

## SLEEP PARALYSIS

> *I was awakened from a dream by a hooded figure standing there in the shadow of my bedroom doorway. I could see him clearly in his dark clothes and hood, although his face was obscured. He began to emerge from the shadows and slither slowly toward my bedside to attack me. Terror ripped through my body, setting my heart pounding furiously as a cold wave of panic drenched me. I feared my heart would burst from the explosive pressure expanding inside my chest. I couldn't breathe.* I am going to die of a heart attack even before the murderer gets me! *I thought. I struggled to move, but my body was completely frozen. I could not move a single muscle in my entire body or scream out to alert my husband sleeping beside me. I tried and tried to scream or at least moan, but I could not make any sound at all as I lay*

*paralyzed. I could hear my German Shepherd, Rusty, breathing heavily and fidgeting at the foot of our bed, but I could not rouse him to drive off the intruder. Unable to move a muscle, I poured all my strength into trying to roll my stiff body toward my husband to awaken him, but I could not. Now another man in dark clothing joined the intruder, and I could hear their muffled whispers. I was gasping for air, as if bound head to foot and gagged. Suddenly, my husband rolled over, touching me lightly, and I awoke instantly. I shot up in bed with my heart racing, drenched in sweat. But there was no one there—only my husband fast asleep at my side and Rusty snoring at the foot of my bed.*

That scenario is a compilation, created from multiple reports from people who have suffered this type of terrifying dream. They share in common the experience of absolute terror at being paralyzed while their bodies are asleep, fully awake mentally but unable to move a muscle. They feel as if they are under mortal threat. They are fully conscious and the experience is real, not dreamlike. Sleep paralysis has been described throughout history, and it is an unforgettable, terrifying experience for the 7 percent of the population who have experienced it at least once in their lifetimes.[3] While 28 percent of students and 32 percent of psychiatric patients report experiencing sleep paralysis (possibly because they are more forthcoming than the general public), the rate of sleep paralysis is thought to be much higher than reported, because the experience is so traumatic and, for many people, embarrassing. Some people may experience only one episode of sleep paralysis in a lifetime; others may suffer recurring episodes, each one as terrifying as the last. Some victims of repeated episodes of sleep paralysis can recognize what is happening to them and become consciously aware that they are not awake and paralyzed, but instead they are having an attack of sleep paralysis. Even as they struggle with all their effort to move, knowing that they cannot, they are fully conscious and understand that they are trapped awake and immobilized in a dream, but that in time they will be able to move again. The instant they are able to move a muscle—any muscle—they awaken. High blood pressure, lack of sleep, narcolepsy, sleep apnea, and alcohol use all increase the risk of suffering sleep paralysis. Also, people with anxiety disorders, including PTSD, are more likely to experience sleep paralysis.

## LUCID DREAMS

Being conscious and awake during a dream state is not the same thing as sleep paralysis. As indicated in the list on page 186, being able to fly is a common dream, but in the following scenario there is something very peculiar:

> *I was dreaming that I could fly. I soared above the heads of everyone with effortless breast strokes. I know that is impossible, to be swimming through the air. I was dreaming, but I was fully aware that it was a dream.*

(This is again a compilation of several reports of lucid dreaming.)

We experience an endless array of fantastic dreams, often vivid with color and sounds with complex character and elaborate scenes unfolding, but while we are conscious of the actions taking place in a dream, they are perceived as real. No matter how puzzling or enlightening, the experiences we have during a typical dream are perceived no differently from the experiences we have while awake. Lucid dreaming is different. This is a dream state in which a person becomes fully conscious during a dream, but they know that they are dreaming. Some people who have lucid dreams embrace the hallucination, enjoy the experience, and consciously direct the dream for maximal enjoyment. In a lucid dream state, a person can create their own virtual reality and cross freely over the line separating reality to participate in the fantasy.

## NARCOLEPSY

At the opposite extreme of being awake during sleep is instantly falling asleep. "I would be in the middle of a conversation, walking to the bathroom or watching TV and just fall over, completely dead weight, and unknown to the world for about three minutes," Amanda wrote on the Narcolepsy Network.[4] "One day I passed out and smacked my head on the concrete floor." At night, Amanda regularly had trouble sleeping. Often the episode of narcolepsy is triggered by emotions, such as laughing and joking. There are a number of experimental animal models, including fifteen breeds of dogs, with a genetic predisposition for narcolepsy, including Labs, poodles, and Dachshunds.[5]

### SLEEPLESSNESS

The absence of sleep, often used in torture, will cause madness, hallucinations, and, eventually, death. But consider Al Herpin (1862–1947). Herpin lived in Trenton, New Jersey, but his home lacked a bed. He claimed that he had no need for one because he never slept. Herpin appeared otherwise normal and he lived to be ninety-four years old. When you consider that we normally spend one third of our life asleep, this means that sleepless Herpin had the same amount of life experience as someone else (who sleeps) would have in a lifetime of one and a quarter centuries. Several other cases of people who do not sleep have been reported, but some of them are tragic. People with the rare hereditary condition of Fatal Familial Insomnia suddenly stop sleeping, and after months of sleepless nights and extreme exhaustion, they die.[6]

## THE ELECTRIC BRAIN AT NIGHT

In each of these bizarre cases, EEG analysis reveals what is happening inside the brain to generate these strange instances of sleep disturbances and bizarre dream states. But to see how, we first need to understand the stages of sleep and how the brain's electrical activity changes while we slumber.

Far from simply shutting down the brain and dozing off to sleep, brain activity transitioning from a conscious wakeful state to the unconscious state we call sleep is an extremely complex process. It radically

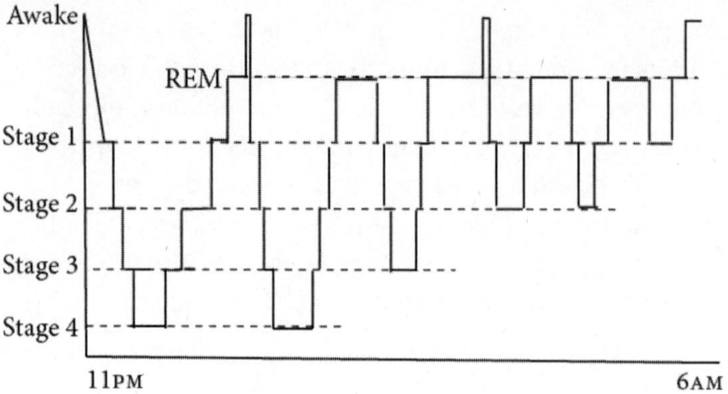


**FIGURE 24**: *The stages of sleep, progressing from wakefulness to slow-wave (or non-REM) sleep and then to rapid eye movement (REM) sleep when dreaming takes place. We go through several such cycles in a single night's sleep.*

alters every system of the body, and brainwaves are what orchestrate this coordinated shutdown and progression of the mind and body into and out of sleep on a regular twenty-four-hour cycle in sync with the rising and setting sun.

Research in the 1950s using EEG first revealed that the brain and body progress through several stages of sleep. Pull up the covers. Turn out the light, and over the next hour or so your brain will go through a progression of states while you are unconscious. As you close your eyes, your alpha waves quickly build strength, squashing the higher-frequency waves of electrical activity buzzing through your cortex just moments ago when you brushed your teeth and checked your alarm. With your alpha waves surging, you begin to enter that peaceful state—the alpha state. You feel relaxed and cozy. Focused, but tranquil, drifting as the curtain falls on the world's stage and your thoughts turn inward. As you doze off, your brainwaves begin to slow down. They become more synchronized and swell to great heights. This transition into sleep is called the "hypnagogic state." You are lightly asleep and entering slow-wave sleep, also called non-rapid eye movement (NREM) sleep. During slow-wave sleep your muscles relax and become flaccid, your heart rate slows, your blood pressure drops, and your body temperature cools. If your neighbor's dog barks and awakens you from your slow-wave sleep, you will not recall any active dream. The magic of dreams does not happen in slow-wave sleep; it occurs at later stages.

Now you drift back to sleep (if your slow-wave sleep was interrupted by your neighbor's pooch) and the evolving sleep process in your electric brain continues. Your brainwaves continue to slow down and synchronize, just as Dr. Brown described in the early stages of anesthesia. But then over the next half hour or so your brain begins to generate cortical brainwaves similar to those that occur in your brain when you are awake! At this stage of sleep, your brainwaves are active and desynchronized, and the waves are oscillating rapidly. But there is one profound difference from the awake state. Despite the awake-state EEG activity in your cerebral cortex during this stage of sleep, you are essentially a quadriplegic. All the muscles in your body are paralyzed (atonia), except for those that maintain your breathing and move your eyes beneath your closed eyelids. You are dreaming now. This busy eye movement, the telltale outward manifestation of the altered conscious brain state of vivid hallucination, is what inspired the name for this stage of sleep: rapid eye movement (REM) sleep. Your eyes dart about, tracking objects in your hallucinating brain. Your heart rate, blood pressure, and body temperature vary in reaction to your dream, because to you these hallucinations are real. The changes in vascular tone often cause men to have penile erections during the REM stage of sleep, without being sexually inspired, and women can experience clitoral engorgement. In the awake state, smooth muscles contract to restrict blood flow to the penis and clitoris, but during REM sleep these muscles relax and blood flows in, engorging and swelling these tissues.[7] If you are one of those people who say you never dream, if you are awakened during REM sleep, you will report a vivid dream unfolding in your mind. If you think you don't dream, you do, but you just forget them the next morning. (As we will see, forgetting is an important part of sleep.)

This first REM period of your nightly slumber usually lasts only ten to fifteen minutes. Just enough time for a short cerebral video clip. No wonder some dreams seem fragmentary. Then, a cycle of slow-wave sleep follows this REM state, and your brain alternates between REM and NREM (slow-wave) sleep throughout the night. Toward morning the slow-wave sleep becomes less deep, and the REM periods become longer. There is plenty of time now for elaborate dream scenes and plot twists in your predawn REM states. Soon your alarm will go off, but you habitually awake before it does. As you become groggily aware of the transition from sleep and dreaming to becoming partially awake, you enter a transition zone called "hypnopompic sleep." This is that fleeting phase of drifting in a suspended state of altered consciousness, as your unconscious and conscious mind duke it out for dominance. Your conscious brain, rising in strength, sputters high-frequency electrical activity throughout your cerebral cortex as it struggles against the unconscious dreaming mind trying to stay swinging in the hammock of cozy relaxation and hallucination. As your conscious and unconscious mind intermix in this fleeting moment, creative insights can often break through the stranglehold of the conscious mind, now in its weakened state, which normally commands and constrains our thoughts.

All vertebrates (animals with backbones) have NREM sleep, but REM sleep is something special seen only in mammals. Human infants spend sixteen hours a day sleeping, and, unlike in the adult brain, half of this time is in REM sleep. It is strange to imagine that the brain of an infant learning about the strange new world spends equal time in a conscious state of reality as it spends in an altered state of hallucination. (Maybe now I understand that quizzical expression babies give me.)

## EEGS, ELECTRICAL STIMULATION, AND SLEEP

In 1949, Giuseppe Moruzzi and Horace Winchell Magoun found that electrical stimulation of the midbrain reticular formation instantly desynchronized the slowly oscillating EEG and aroused sleeping animals.[8] Barry Sterman and Carmine Clemente later showed that electrical stimulation of the basal forebrain could produce slow-wave sleep in animals that were awake.[9] For present purposes it is less important to understand the neuroanatomy of how stimulating the brain at these spots could affect sleep. The important point is that together these studies combining electrical brain stimulation with EEG recording revealed that sleep is actively controlled by specific circuits in the brain. From this fact it is not difficult to understand how Al Herpin and others with

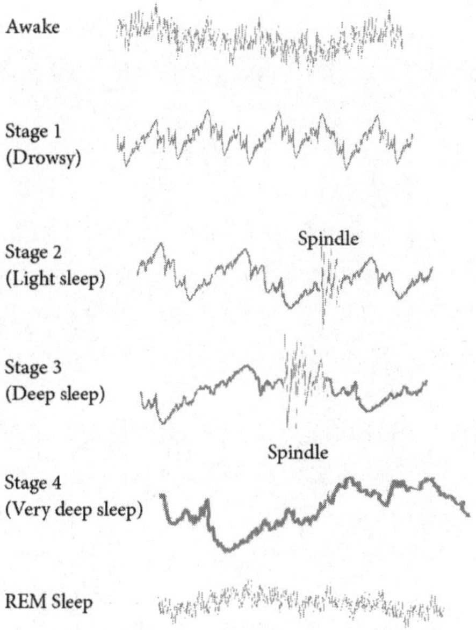


**FIGURE 25:** *Depiction of changes in EEG activity through the different stages of sleep. Brainwaves slow from wakefulness to deep sleep and exhibit distinct characteristics at each stage of sleep. Note the spindle activity in stages 2 and 3, and the similarity of the high-frequency EEG activity in REM sleep while dreaming to the high-frequency EEG activity in the awake state.*

dysfunctions in neurons that initiate sleep could have lost the ability to sleep.

However, consciousness isn't easily divided into awake or asleep. There is a progression, with the transitions in human brainwave activity during NREM sleep divided into four stages. In stage one sleep, when the eyes close, the EEG activity slows from the alpha-dominated brainwave state to the slower theta frequency (4 to 7 Hz).

Stage two sleep is characterized by development of sleep spindles. These are composed of quickly rising and falling transient oscillations at a frequency of 12 to 15 Hz (beta wave band). The name "spindle" refers to the shape of the sudden eruption in the waveform traced out on an EEG chart recording, which resembles a spindle, rising from sharp points at the leading and trailing ends. Like the sound of a wheel on an automobile with bad shocks suddenly jolted into wobbling after hitting a pothole, and then quickly settling down to its normal rumble, a spindle in an EEG recording is a transient event, a sudden eruption of an oscillation that is quickly damped down. We will return to these fascinating and important brainwaves shortly.

Stage three sleep is marked by spindling of very high voltage on a background of increasing activity of very slow waves (delta waves, 1 to 4 Hz). Your electric brain has slowed its oscillations down to the pace of a heartbeat.

In stage four sleep, high-voltage slow waves (1 to 3 Hz) dominate the EEG. You are dead to the world; because your brain and body are in the deepest stage of delta-wave sleep, it is most difficult to arouse you to a wakeful state from this stage. In parallel with the decrease in brainwave frequency and increase in voltage, muscle tone gradually relaxes. Release of growth and sex hormones from the pituitary increases in slow-wave sleep.

After descending through these four stages of NREM (also called slow-wave) sleep, the EEG cycle suddenly reverses itself and progresses backward in sequence, stepping up from stage four to stages three, two, and one sleep, increasing the frequency of its oscillations along the way. Then the EEG switches into its low-voltage, fast brainwave activity as seen in the awake brain—that is, REM sleep, which is when dreaming takes place. The eyes make tracking (saccadic) movements as the person hallucinates in a dream; however, unlike the awake state, muscle tone is strongly inhibited in REM sleep. If the body was not paralyzed, we would leap up to engage in whatever hallucinatory activity our brain is experiencing. This bodily shutdown of our muscles is obviously not working properly for people who sleepwalk or talk in their sleep. Sleep and brainwave activity are organized in this way in all mammals, including aquatic and flying species. Most mammals have about three or more of these cycles between REM and NREM sleep, but in an average night's sleep the human brain cycles through four to five REM periods, each one of about 90 to 100 minutes in duration. In recovery after sleep deprivation, EEG recordings show an increase in both slow-wave sleep and REM sleep.[10]

Cortical activity in sleep is regulated by neurons in the upper pons and midbrain, which send their outputs to activate the thalamus by releasing the neurotransmitter acetylcholine, as mentioned previously. The pons and midbrain control automatic brain functions without conscious thought, and the thalamus is a master switch box at the brain's core, relaying information to and from the cerebral cortex. When the thalamus is activated, it sends information to the cerebral cortex, activating it, and producing an alert awake pattern of brainwave activity. Although the acetylcholine synaptic inputs to the cortex produce a similar alert state of brainwave activity in wakefulness, in REM sleep, the brain is not responsive to external stimuli. The reason is that neurons in the brain stem, in the locus coeruleus region, where neurons use the neurotransmitter norepinephrine, and in the dorsal and medial raphe region, where neurons use the neurotransmitter serotonin, affect the brain's state of arousal. So do neurons in the hypothalamus that use histamine as a neurotransmitter. Norepinephrine and histamine neurotransmitters are most active in a wakeful state, but the firing rate of these brain stem neurons slows down during slow-wave sleep and it stops in REM sleep. (This is why the antihistamine Benadryl—diphenhydramine—makes you so sleepy. Other sleep-inducing drugs, like Ambien, which suppress arousal systems in the brain, can increase sleepwalking and other sleep disturbances.[11]) These brain stem neurons receive input from hypothalamic neurons, which control many basic body cycles, such as respiration, heart rate, reproduction, and circadian rhythms. An important set of neurons in the lateral hypothalamus use the neurotransmitter orexin, which excites arousal. Studies in humans with narcolepsy have detected abnormally low levels of orexin in their brains and spinal fluid, and genetic disruption of this neurotransmitter in animals is used as an experimental model of narcolepsy, as mentioned earlier.

The body's circadian clock, a cluster of neurons in the hypothalamus called the suprachiasmatic nucleus, contains genes that cycle like clockwork over a period of twenty-four hours. Inputs from the retina in the eye go to this brain region to set the clock, which is why we are advised to get some exposure to sunlight to overcome jet lag, and why people with blindness can sometimes experience sleep-cycle disruptions, called Non-24, because their sleep is not well matched to the time of day. The basal forebrain, with neurons that project directly to the wider cerebral cortex, is also important in controlling sleep/wake states, as mentioned in the brain stimulation experiments by Sterman and Clemente.

Shift workers, whose sleep cycles are out of sync with the normal circadian rhythm, experience a wide range of detrimental effects, including hormonal imbalance, increased inflammation, impaired glucose metabolism, obesity, metabolic syndrome, type 2 diabetes, gastrointestinal dysfunction, compromised immune function, cardiovascular disease, excessive sleepiness, mood and social disorders, and increased cancer risk.[12]

Clearly, the simple act of nodding off to sleep is anything but simple. No wonder sleep disorders are so common, and now the strange sleep disorders we examined at the opening of this chapter should be less mysterious.

## WHY SLEEP DISORDERS HAPPEN

People experiencing sleep paralysis are stuck in a purgatory of the confluence of REM sleep, with its accompanying paralysis, stage one NREM, and the partially awake hypnopompic state. EEG recordings show that sleep paralysis is a type of REM sleep but with a heightened state of arousal. When the mind is aroused, alpha wave activity is suppressed by increased higher-frequency beta-wave activity. The brain is stuck in an abnormal state where reality and hallucination are no longer separated. T. Takeuchi at the Department of Psychology, Waseda University, in Tokyo, Japan, and colleagues[13] observed that during sleep paralysis, alpha-wave brain patterns are frequently interrupted by beta waves, especially in subjects who reported visual hallucinations. Either something has alerted the dreaming mind and grabbed attention (like the terrifying hallucination of an intruder in their bedroom), or just the opposite: The shift from the sedated calm mental state of alpha-wave activity into higher-frequency beta-wave activity could enable brainwaves to open the thalamic gate to arousal and consciousness while still in a REM state of paralysis and hallucination. The thalamic gate can be kept closed by brainwaves in slow-wave sleep because when activity in large populations of neurons is yoked together in a slow, rhythmic oscillation, the discrete transmission of information between neurons in many different neural circuits is impeded. But here again is the chicken-and-the-egg dilemma of brainwaves—do these rhythmic oscillations of electrical activity in our cerebral cortex drive our cognitive functions or simply track them?

Sleep paralysis is thought to result from physiological dysfunction in the orderly transition from REM sleep to wakefulness in the cycles of sleep throughout the night and in awaking in the morning. Such individuals may have shorter REM and NREM sleep cycles and fragmented REM sleep. Specific neurons that use the neurotransmitter acetylcholine (mentioned earlier) heighten arousal, and they become hyperactivated in sleep paralysis, but at the same time signals to suppress activity of the sleep-inducing serotonergic neurons are too weak. The result is that the brain becomes aroused and consciously aware during REM sleep. In REM sleep, sensory input from the environment and muscle movement throughout the body are blocked, and the person is genuinely paralyzed. Becoming conscious during REM sleep in sleep paralysis is a similar horror of a person becoming fully conscious while paralyzed on the operating table by drugs that block contraction of muscles, or a person suffering Locked-In Syndrome from brain trauma or disease. Sensory input, like Rusty the guard dog snoring and sensing your husband sleeping beside you, gets through to the consciously aware mind and mixes with the hallucinations of dreaming.

The typical hallucinations experienced during sleep paralysis are indicative of the parts of the brain that are not functioning properly—that is, becoming activated when they should be asleep. The universal experience in sleep paralysis is the feeling of terror, typically that an intruder is coming for the person while they are immobilized in bed. This nightmare during sleep paralysis is thought to result from dysfunction in the multisensory processing area of the temporoparietal junction of the cerebral cortex. This part of the cerebral cortex coordinates activity from the thalamus (which regulates sensory input to the cerebral cortex) and the limbic system (where emotional and threat-detection functions are carried out), and it integrates information about the body's external and internal environment. This is the hub of our brain's security alert system. Damage to this brain region or electrical stimulation produces out-of-body experiences. Impairments in this cortical center are also associated with anxiety disorders, Alzheimer's disease, schizophrenia, and autism spectrum disorder. EEG recordings show that frontal lobe seizures often accompany sleep terrors and other sleep abnormalities.[14]

Turning from sleep paralysis to narcolepsy, brainwave activity also provides insight into how the brain is malfunctioning in controlling sleep. In people stricken with narcolepsy, a REM-like sleep EEG pattern (4 to 7.5 Hz) often accompanies the attack, in contrast to the normal 8 to 13 Hz alpha activity in the waking state. These people suddenly switch from an awake-state EEG into a dream-state EEG with no warning and without the usual sequence of sleep stages taking place in the brain.

At the opposite extreme are people who cannot sleep. Angelo Gemignani, at the University of Pisa, found that people with Fatal Familial Insomnia (who lose the ability to sleep and then die of exhaustion) are missing normal slow-wave EEG activity.[15] His recordings showed a marked reduction in slow-wave sleep oscillations and a reduction in spindle activity, especially in frontal areas in these individuals. These alterations are likely caused by neurodegeneration of thalamic neurons relaying information to the cortex. If these neurons are lost through disease, the thalamus fails to switch off input to the cerebral cortex to enable the person to sleep. The lights, so to speak, are always on.

## PLAY IT AGAIN, SAM (BACKWARD): TO LEARN, SLEEP ON IT

One of the reasons that we sleep is to dream, and one of the reasons we dream is to remember and forget.

First-stage sleep is characterized by theta waves, and stages three and four by delta waves. As mentioned earlier, peculiar bursts of brainwaves, called sleep spindles, are sudden eruptions of oscillatory activity that appear during stage two of light sleep. Also called sigma waves, the fast (13 to 15 Hz) and slow (11 to 13 Hz) frequency spindle activity increases as we fall deeper into sleep. Spindles are thought to be correlated with the transfer of information between the hippocampus (important for spatial memory) and the neocortex (important for integrating sensations, thoughts, and memories), and thus spindles are believed to be important in memory consolidation. Memory consolidation is the process of converting short-term memories into lasting long-term memories. Before proceeding with the new research on how sleep and sleep spindles are thought to be involved in memory, it will be helpful to take a brief aside and explain some key points about how memories are stored in our brain.

The fundamental difference between short-term memory and long-term memory is that short-term memory is sustained by temporary increases in the voltage that a synapse generates (called strengthening or potentiating the synapse). That synaptic connection, representing a vital link in neural circuitry that is temporarily holding a new experience in the mind, has its power boosted. So when we try to recall this recent experience, that circuitry is prominently represented, and thus easily recalled. Over a short time, however, the strength of the synaptic connections encoding that recent experience begins to fade, so the "chalkboard," so to speak, can be erased and a new message boldly written over it. Short-term memory is what we use to recall a person's name right after being introduced. This sort of memory is crucial, as is evident when short-term memory fails temporarily as we have a "senior moment" and forget why we went downstairs to get something, or when we temporarily misplace our car keys.

Long-term memory has a different purpose, and these more permanent mental records, like your name, address, what seven times six equals, are stored in the brain in a different manner. Obviously, these experiences were first held in short-term memory when we initially encountered them, but somehow they became permanently engraved in our memory. That transition is memory "consolidation," and that conversion requires a growth process that "hardwires" synaptic connections in the neural circuit holding this experience, so that these synapses are permanently strengthened. To make new proteins that cement these synaptic connections together permanently, appropriate genes in our DNA must be activated to read out the instructions for neurons to synthesize these needed synaptic proteins. Many labs conducting research in the last few decades, including mine, have identified the detailed cellular and molecular mechanisms of memory consolidation. The key distinction between short-term and long-term memory is that genes must be turned on and new proteins made for long-term memory but not for short-term memory.

We can block the conversion of short-term to long-term memory easily by using drugs or other measures that prevent activation of genes or the synthesis of proteins. The "neuralyzer" in the movie *Men in Black*, which erases the traumatic experience of witnessing an alien invasion, is not science fiction. We know how to do that. It should not be too shocking that scientists can create instant amnesia and erase any recent memory. We all know that after a bump on the head (concussion), anesthesia, or a "blackout" from alcohol intoxication, memory consolidation is blocked and recent experiences cannot be recalled because they were not encoded. What's more of a surprise is that our brain does this to us every night while we are asleep. These peculiar beta spindles are an important part of that "nightly neuralyzer" mechanism.

Interestingly, increased spindle activity is correlated with higher intelligence, and this correlation is especially strong in women.[16] The amount of spindle activity also increases at night after a person has learned something new, and tests prove that performance of the newly learned task correlates with the number of spindles experienced during sleep the night before. Spindles may represent the brain processing and integrating new information during light sleep, which is thought to be one reason why memory and cognitive abilities suffer after sleep deprivation.[17]

In regard to sleep and learning, unusual brain oscillations occurring in the hippocampus are sharp wave ripples. This spindle-like activity is seen as brief bursts of very-high-frequency oscillations of 200 Hz or higher, most prominent in the hippocampus. This high-speed burst of neuronal firing is extremely rapid: about ten times faster than the imperceptible frame rate of movies. During these high-frequency bursts, 50,000 to 100,000 neurons discharge in synchrony. Why would 100,000 neurons suddenly fire together simultaneously in response to no sensory input to the brain? Sharp wave ripples occur during slow-wave sleep as well as during quiet periods of wakefulness. Research in my laboratory on sharp wave ripples in the rat brain has revealed some intriguing findings.[18] We performed our studies on slices of rat brain hippocampus, kept alive in a carefully oxygenated and temperature-regulated chamber, and we used electrodes to stimulate hippocampal neurons and record the firing of individual neurons, as well as measure the local field potential oscillations (that is, brainwave activity), including any sharp wave ripple complexes. What Olena Bukalo, a postdoctoral fellow in my lab, and I discovered was something that violated all the standard thinking about how neurons normally operate.[19] This led us to discover a new form of synaptic plasticity that is very important in memory consolidation during sleep—specifically how our sleeping brain decides what events of the day to remember and which ones to forget.

The fundamental rule of learning is that "neurons that fire together, wire together." The bell and the food must be presented together for Pavlov's dogs to learn to associate the sound of the bell with food and begin to salivate in anticipation. To understand how this rule operates at a cellular level, recall some basic features of neurons and neural circuits.

A neuron can receive thousands to tens of thousands of synaptic inputs onto its rootlike dendrites. We refer to the recipient neuron, which receives synaptic input, as the "postsynaptic neuron" in a neural circuit. A complex neural network will have many neurons linked together this way, but each one can be termed either presynaptic or postsynaptic, because electrical impulses pass only one way through neurons—in the dendrites and out its axon to pass information on to the next neuron in the network. This fundamental law of neuroscience, conceived by Nobel Prize–winning neuroanatomist Ramón y Cajal at the turn of the twentieth century, is called the neuron doctrine. Each of these synapses onto the postsynaptic neuron releases a neurotransmitter when they fire, which changes the voltage in the recipient dendrite slightly. Synapses that excite neuronal firing depolarize (reduce the negative voltage) across the dendrite of the postsynaptic neuron slightly, and synapses that inhibit neuronal firing hyperpolarize (increase the negative voltage) across the dendrite slightly. Remember from an earlier discussion (see page 95–97) that the postsynaptic neuron will fire an electrical impulse (action potential) when its transmembrane voltage reaches a certain threshold voltage.

The thousands of synaptic inputs onto a neuron's dendrites are scintillating with activity, all firing at different times, reflecting activity in the many different neuronal circuits that feed into the recipient neuron, and each one acts independently to influence the overall transmembrane voltage of the postsynaptic neuron. The dendrite of the postsynaptic neuron simply takes a vote, summing the cumulative effect of all the signals it receives from its synapses, and the postsynaptic neuron makes an all-or-nothing decision—to fire or not to fire. If the collective input from the population of actively firing synapses on its dendritic tree move the transmembrane voltage to the trigger point, the postsynaptic neuron fires. An action potential is sparked at the junction between the cell body and the neuron's wire-like axon, and it shoots at high speed down the axon to fire its synapses onto the dendrite of the next neuron in the circuit.

At a cellular level, the "neurons that fire together, wire together" rule of learning means that when a synapse fires at the same time as the postsynaptic neuron, then a biochemical process is initiated that increases the strength of that synapse. That is, if this particular synapse depolarized a postsynaptic neuron by half a millivolt initially, then after it is strengthened, it will depolarize the postsynaptic neuron by approximately twice as much—a millivolt. Now that synapse has a more dominant influence on whether the postsynaptic neuron fires an action potential. This effect increases the power of that synapse to fire the postsynaptic neuron, relative to weaker synapses from other circuits that are also connected to that same postsynaptic neuron. That strengthened synapse, in effect, talks louder, and so the group of neurons that it represents, with their circuits feeding into it, have more influence on the actions of the postsynaptic neuron.

I can see this happen in slices of rat hippocampus when I use a microelectrode to record the transmembrane voltage and firing of a postsynaptic neuron in response to delivering electrical stimulation to fire neurons connected to the postsynaptic neuron. I watch a glowing line sweep across the oscilloscope and on computer screens, plotting how much voltage the neuron is producing across its cell membrane. When I fire the synaptic input to this neuron, I see a sudden bump in the trace, looking something like a heartbeat on a hospital monitor, indicating that the synapse has fired and momentarily poked the postsynaptic neuron closer to the threshold for it to fire an action potential. The size of that voltage bump caused by the synapse firing is the "postsynaptic potential," and I can easily measure its amplitude (voltage).

Now, if I fire that synapse simultaneously when the postsynaptic neuron fires an action potential, the size of that postsynaptic potential is suddenly much larger when I test-fire it again. Neurons that fire together, wire together. If my timing is off and I fire the synapse just a few milliseconds before or after the postsynaptic neuron fires an impulse, the synapse (postsynaptic potential) is not strengthened. Remarkably, synapses strengthened in this way maintain their increased voltage kick for as long as I can keep recording (for weeks when this experiment is performed in freely behaving animals with electrodes implanted in their brain). This long-term increase in synaptic strength is called "long-term potentiation," or LTP. LTP is induced by the firing of a synapse simultaneously with the postsynaptic neuron firing an action potential.

When the phenomenon of LTP was first described by Tim Bliss and Terje Lømo in 1973, something very interesting was discovered. To make the postsynaptic neuron fire an action potential, intense, high-frequency electrical stimulation was needed to recruit enough synaptic inputs to fire simultaneously to depolarize the postsynaptic neuron enough to reach the threshold to produce an action potential. When that happened, all of those simultaneously activated synapses became strengthened, but such intense, high-frequency firing is artificial. Because hippocampal neurons do not fire in this way, this artificial stimulus tended to diminish the importance of the finding. However, the researchers found that if they instead delivered milder stimulation in a pattern of a few action potentials repeated in a way that is normally seen in the hippocampus, they could induce LTP without the artificial high-frequency stimulation. The effective stimulus pattern for inducing LTP turned out to be the theta frequency—short bursts of a few action potentials, repeated at about 5 Hz.[20]

This in itself demonstrates the fundamental way in which brainwaves can regulate the transmission of information through neural circuits. Rhythmic firing of appropriate frequencies, matched to what might be called the resonant frequency of a neural circuit, propagate readily, but if the rhythm is not the appropriate frequency, transmission of information through the circuit will be impaired or fail. (This is, as discussed previously, one of the features of all waves in nature—the efficient compounding of energy in synchronous oscillation, as when rhythmic clapping rocks a stadium to its foundation, but uncoordinated clapping with the same energy generates only noise.) Different rhythms of neural firing will be better matched to the resonant properties of different networks, just as an opera singer can shatter crystal stemware by singing the proper note (sound frequency) that is matched precisely to the physical characteristics of that wineglass that let it vibrate robustly at this frequency. As a result, brainwaves of different frequencies join different neural populations together into functional circuits. Theta rhythms work particularly well to depolarize the postsynaptic neurons in the hippocampus because, as was discussed earlier in the debate about gamma waves (see pages 162–168), the synchronized rhythmic firing of synapses at the theta frequency locks more synapses together into synchronized firing at the optimal rate, thereby very effectively impacting the postsynaptic neuron voltage.

Years after the discovery of LTP, another form of synaptic plasticity was identified by Mark Bear, then at Brown University, and his graduate student Serena Dudek and others, called long-term depression (LTD). It turns out that the synapses that do not fire at precisely the right time (that is, the ones that fire out of sync when the postsynaptic neuron fires an action potential) become weakened. This makes good sense. Selective suppression of synapses that fire out of step with the postsynaptic neuron will improve the overall performance of the neuronal circuit in the same way that hushing a clarinet player does when she is playing out of sync with the rest of the symphony. The biochemical mechanism that weakens synaptic strength in LTD is also known in detail, but the big picture significance of LTD is appreciating the gob-smacking moment for neuroscientists realizing that there was a cellular mechanism for forgetting—weakening and ultimately breaking functional synaptic connections. So fixated were scientists in their passionate search to discover the cellular mechanism of encoding memories by strengthening synapses in some way, for decades neuroscientists completely overlooked the possibility that synapses could be actively and selectively weakened.

When Serena Dudek was a postdoctoral fellow in my lab, we explored the mechanisms that direct the changes in synaptic strength to take place in only the correct manner (when the synapse fires simultaneously with the majority of synapses on the dendritic tree).[21] This synapse specificity is vital for learning, because discrete information is encoded in discrete neural circuits that feed into a postsynaptic neuron through the specific synapse tied to the appropriate neural circuitry.[22] The important concept for both LTP and LTD is that strengthening or weakening synapses, according to their simultaneous firing with the postsynaptic neuron firing, is synapse-specific, and that the change in synaptic strength is dependent on neurotransmitter signaling, notably the glutamate neurotransmitter receptor called the NMDA receptor.

You can imagine our surprise, then, when Bukalo's and my experiments in the slices of rat hippocampus uncovered a mechanism of synaptic plasticity that completely violated this rule. We were studying how postsynaptic action potential firing signals to the nucleus of the neuron to alter readout of genes because proteins that strengthen synaptic connections permanently are made by genetic instructions. Somehow, the right genes in the nucleus must become activated to make these proteins and they must be shipped to (or only act specifically upon) the appropriate synapses among the thousands in that neuron. However, we found a form of LTD that was not synapse specific, not dependent on the NMDA receptor, and, in fact, was not driven by synaptic firing at all. Instead it was driven by the postsynaptic neuron firing action potentials in a very peculiar way during sharp-wave ripples, with the result that the strength of all the synapses on the dendrite of the postsynaptic neuron was altered. We called this new form of plasticity action-potential-induced LTD, or AP-LTD. To understand why a form of plasticity that violates the synapse-specific learning rule, which occurs during sharp-wave ripples, makes any sense, we need to consider the importance of forgetting, and how sleep is a time when the brain is actively engaged in resetting and forgetting.

## HOW AND WHY WE FORGET . . . AND REMEMBER

By far most of the experiences we have are quickly forgotten. Who needs to remember that parking spot they had at Walmart that Sunday afternoon two years ago? But how does the brain decide which experiences will be important for the future and need to be retained, and which ones can be discarded? That decision, like cleaning out the barrage of incoming messages in your email inbox, requires evaluating each new experience, considering its relevance compared to other information stored in the brain, and judging the potential importance this event may have in the future. Active and precise forgetting is just as important for learning and memory as is active and precise remembering. As we learned from the research of Benchenane in Paris (see pages 123–126), EEG recordings in hippocampal neurons in rats that had learned to run a maze during the day revealed that, as the rat sleeps, it relives in its brain the experience of running the maze. The same sequence of neurons in the hippocampus that were activated while the rat ran through the maze replay their firing activity in the sleeping brain. The stronger the nocturnal replay, the better the rat's performance in running the maze the next day. If the rat's sleep is disrupted, the poorer the rat's memory is in running the maze the next day. Similar studies have shown that much the same process happens in humans. Together, this body of research has determined that as we sleep our brain replays the experiences of the day, associates this new information with existing memories, consolidates those important synaptic connections (memories) into long-term storage, and lets the others slip away. Strategic forgetting is vital.

The very rapid oscillatory pattern of sharp-wave ripples in the hippocampus happens during quiet periods of wakefulness (momentarily "zoning out") and during slow-wave sleep. Remember that during slow-wave sleep we are not engaged in dreaming in the way we are during REM sleep when our eyes are darting about and our brainwaves are jazzed and rapidly firing similar to how they do when we are awake. In slow-wave sleep, we are cut off from the outside world and our brain is idle, with its brainwaves slowed down to their very slowest rate of oscillation, but interrupted periodically by curious transient higher-frequency bursts. These are times when the brain stops taking in information from the outside world and self-generates this peculiar pattern of firing that is specific to the hippocampus and cortical areas associated with it. These very-large-amplitude, short bursts of extremely rapid oscillations do not report incoming information to the brain. They must shuttle information internally within the brain, like the process of autosaving in storing a document during word processing. A body of research has shown that that's exactly what they are doing, and that this is happening to consolidate memories.

EEG recordings of the brain in resting state, and similar data using implanted electrodes, reveal that much of the neuronal activity self-generated by the brain is strange. In contrast to brain activity driven by sensory input, much of the intrinsic activity in the brain is oscillatory, and some of this neuronal firing operates completely outside the textbook view of how neurons operate. Intrinsically generated neuronal firing can involve nonclassical modes of firing that make no sense from the standpoint of viewing the brain as an input-output electronic device. (That analogy works fine during the day when our brain is responding to sensory input, but not at night when we are asleep. Obviously, our brain is still vigorously active electrically when there is no sensory input, but during these periods, activity in our neural networks is driven by the brain itself.) The coherent high-frequency oscillations of electrical activity during sharp-wave ripples take place in thousands of hippocampal neurons simultaneously firing together; thus, by the rule of "neurons that fire together, wire together," the oscillations functionally connect these thousands of neurons into a transient functional ensemble. This is analogous to certain members of a choir momentarily singing together, while people on either side of them remain silent.

The mind-boggling finding about these neurons participating in the sharp-wave ripple firing is that the synchronized action potential firing is not caused by synaptic input: These neurons are firing backward! The action potentials are triggered in the distal region of axons—not in dendrites—and it zings backward (antidromically) into the cell body and invades the dendrite. This odd behavior, called "antidromic firing," violates the neuron doctrine, the fundamental principle by which neurons transmit and integrate information—from dendrite to axon. This antidromic firing had been seen and described by others, but why neurons would do so was a mystery. Clearly, a neuron firing backward during sharp-wave ripples is not sending information through the neural circuit like its adjacent neighboring neurons, which are firing normally and not participating in sharp-wave ripple firing. What could neurons firing backward together at high frequencies be doing?

Returning again to the fundamental learning rule of "neurons that fire together, wire together," an obvious problem becomes evident. Our neurons are firing together all day. The barrage of input a postsynaptic neuron receives as we go about our day would strengthen synapses until eventually all of them would reach the maximal strength they can produce. At this point, neural circuits would become saturated and could no longer encode or convey new information. But while considering the learning rule, what would happen when a neuron fired impulses when there was no synaptic input?

What we found in my lab is that when a neuron fires antidromically during sharp-wave ripples, all of the synapses on its dendritic tree become weakened. This happens because these synapses are not firing when they should be, according to the theory of synaptic plasticity; that is, they are not firing when the postsynaptic neuron is firing impulses. It is as if the conductor of an orchestra snaps his baton on the downbeat to launch the musicians into playing, and all the musicians sit there in silence. Do that a few times and those musicians sitting in silence will be eliminated. We found this by using drugs to block synaptic activity in the hippocampal brain slice, then we positioned a stimulating electrode in the bundle of axons that normally carry action potentials out from hippocampal neurons (the alveus), so that we could make the neurons fire backward (antidromically) as they do in sharp-wave ripples during slow-wave sleep. When we washed out the drugs to recheck the strength of the synaptic inputs to these neurons through their dendrites, all of the synapses had been weakened. As a consequence, synapses that were weakest before antidromic firing would be lost—reduced from their initially feeble strength to nothing. Stronger synapses became weakened, too, but they were not completely eliminated—their synaptic strength had only been adjusted downward. The process that reduces the strength of all synapses on the dendrite prunes inputs to the postsynaptic neuron. It trims away the weakest ones to refine the shape of neural circuitry feeding into it and cuts back the strength of the robust synapses a bit. Like pruning a bush, this dendrite-wide clipping and pruning of synapses increases the precision of neural networks.

For example, recall that neurons in your hippocampus called place cells fire when you move to a particular location in your environment. At first, the firing of a place cell neuron has a rather broad "tuning curve," meaning that its GPS-like activity has less precision until experience improves its accuracy. Imagine you check into a hotel, take your key card to your assigned room number 201, and insert it into the door lock. As that green light flashes on the door lock, a place cell neuron in your hippocampus fires, signaling that location in your environment.

The next day when you walk up to that door, that same place cell will fire again. However, that neuron also fires when you are in the general vicinity of your door, because it has not yet been tuned to recognize the exact spot where your key card works. You might even walk up to the incorrect door on the opposite side of the hall, for example, and be surprised when your key card fails. That happens because this cell's unrefined GPS behavior is also activated when you are in proximity to the neighboring door. Eventually, with experience, this place cell neuron increases its precision and does not fire until you are right at door 201. Now when you step off the elevator and walk down the hall, you don't hesitate or miscalculate; you walk straight to your door and open the lock.

At a cellular level, what this means is that this neuron is stimulated to fire because a network of neurons feeding into it are activated by your location in your environment. This is a new location and it is important, so your brain identifies this experience as something that needs to be recalled for the future. The first time you checked in, some of the inputs to this neuron were more highly tuned to adjacent spots in the hallway, but their sensitivity to location overlapped somewhat with the position of door 201. You need them, for example, if you need to learn and remember that the fire extinguisher is just across the hall from door 201. As you gain experience, the strength of synapses that fire together increases, which builds this memory circuit for the location of your hotel room door, but when these neurons fire antidromically during sharp-wave ripples, the strength of all the synaptic neurons to this functionally connected population of neurons gets weakened. The result is that the strongest connections prevail, forming a more specific memory circuit, while the weakest ones get removed to join in firing cooperatively with another population of neurons that hold slightly different information (where the fire extinguisher is, for example). In this way, selective forgetting (pruning away weak synapses) provided by strange brainwaves while we sleep sharpens memory.

We place so much emphasis on factual, academic learning and memory that we tend to overlook the amazing and very complex learning that effortlessly goes on in the brain all the time. It is not until we see someone struggling with a cognitive dysfunction, such as Alzheimer's disease, in which this automatic learning and recall circuitry has been degraded by disease, do we glimpse how attuned our brains are to learning and using information on a daily basis. An elderly man with dementia may very well walk up to the fire extinguisher and try to insert his room key.

It is thought that downscaling synaptic strength during slow-wave sleep is necessary to prevent saturation of synapses that become potentiated during wakefulness,[23] helping assimilate new memories into a schema (a scene in memory) by refining and sharpening of previously acquired memories.

Our results show that whether the action potential is generated antidromically or through synapses (orthodromically) is a critical factor in synaptic plasticity that had not been previously considered. The fact that sharp-wave ripples occur during sleep and quiet periods of wakefulness makes sense too. The reduced sensory input during such periods, when the brain is not processing external input, would be favorable to replay neuronal assemblies induced intrinsically and through antidromic firing, thus contributing to memory consolidation.

The synaptic plasticity induced by antidromic action potentials may have practical significance for human therapy. Later we will consider how several modes of electrical brain stimulation can help people with Parkinson's disease, obsessive-compulsive disorder (OCD), depression, and addiction.[24] Parkinson's disease (and the others) can alter neural circuit properties, but how deep brain stimulation helps these patients is unclear. When an electrode inserted in the brain is stimulated, it will, just as we did in our experiments, activate both synaptic firing and antidromic firing. Possibly synaptic depression induced by antidromic firing may weaken synapses from dysfunctional networks, prune them away, and restore the network to a more functional state.

With this understanding of the electric brain obtained from the latest research in leading laboratories around the world, part III will explore how this new information can be used by you personally, and how the advancements in brainwave research will change the world. Whether brainwaves are orchestrating neural information processing in complex ways or they are simply a by-product of millions of neurons firing bursts of electrical energy is irrelevant when it comes to the power of using brainwaves to read and control the brain. The stunning applications include the ability to read your mind, predict your actions, know your individual intellectual strengths and weaknesses, foretell your potential for learning specific types of information, divine your personality, diagnose psychological and neurological disorders, interface your brain to a computer, transmit thoughts brain to brain, and more. Brain stimulation and other methods to manipulate your electric brain can enable you to change your brain. Harnessing brainwave power gives human beings capabilities beyond anything that our species has achieved throughout human history: the ability to control behavior by direct intervention in brain circuitry, and the ability to control minds with machines and machines with our minds.

PART III

# Harnessing Brainwave Power

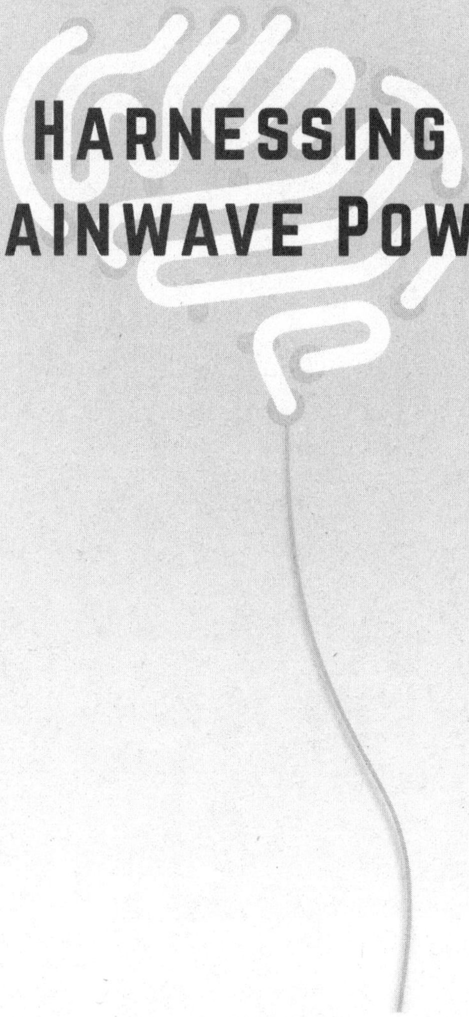

## CHAPTER 8

# Mind Control: Brain-Computer Interface

Breakthrough research on how brainwaves arise, and how they encode and orchestrate information processing in the brain, is leading to revolutionary technology that couples the brain directly to computers. This new technology, called brain-computer interface (BCI), enables the mind to link directly with computers and control an astonishing array of devices, from wheelchairs to military drones, and enables signals from computers to control a person's brain.

## RADIO WAVES TO BRAINWAVES

José stood in the blistering-hot bullring as a raging bull, bred for aggression, charged with its sharp horns lowered to kill. José was not a trained matador. He carried no sword. Cloaked in dust erupting from its thunderous hooves, the charging bull screeched to a halt within inches of brutally goring him. José Manuel Rodriguez Delgado, a Spanish neuroscientist, had stopped the charging bull by pressing a switch on a small radio transmitter that he concealed in his hand behind the red cape. The radio signals energized electrodes he had surgically implanted into the brain of the bull. Vividly demonstrating that his miniature electrodes could do what swords might not, Delgado startled the world in the summer of 1964 with the reality of mind control by directly manipulating brain circuits and thus behavior. With his electrodes, Delgado squelched

**FIGURE 26**: *José Manuel Rodriguez Delgado, a Spanish neuroscientist, stops a charging bull in 1964 by pressing a switch on a small radio transmitter that energized electrodes he had surgically implanted into the bull's brain. Delgado's research was motivated to understand and control normal and abnormal human behavior through brain stimulation.*

the bull's deadly aggressive behavior, but this was simply a dramatic presentation. Delgado's real interest was in controlling humans.

Delgado attracted a cult following among science-fiction enthusiasts, and he alarmed some members of the public by the frightening prospect of "mind control" through radio-activated brain stimulation. Mind control is a lingering fear among many. It feeds paranoia among some who suffer from mental illnesses that cause auditory hallucinations or irresistible compulsions, which they attribute to a microchip implanted in their brain by the CIA or other nefarious government agencies. But from the time Delgado first began to experiment with controlling animal (and human) behavior by directly stimulating neural circuits in the brain through implanted electrodes, humanity has been forced to confront fundamental questions that are at the core of what it is to be human, and to trace the critical but tenuous lines of separation and interdependency between the individual and society. From

Delgado's perspective, society creates order by controlling individuals, and now scientists had their hands directly on the levers. Undercutting free will—the mind, even—and directly manipulating people's brains meant the last and most sacred sanctuary had been breached. After the stunning advances of Delgado and other researchers in the mid-twentieth century in brain stimulation research to control behavior, this line of research declined for decades, largely out of public concern that such investigations were opening a Pandora's Box that was best left sealed. Slowly, however, research in this area has become active again, primarily as a result of the discovery that deep brain stimulation could have profound therapeutic benefits for developing artificial limbs and other prosthetic devices that must be connected to neural circuits, as well as for people with neurological disorders such as Parkinson's disease. Now the lid is cracked open; and the future—the near future—will reveal whether we've lifted the lid on Pandora's Box or a treasure chest.

José Manuel Rodriguez Delgado was born in Ronda, Spain, in 1915. He began his studies for a medical degree at the University of Madrid, but after the Spanish Civil War (1936–1939) broke out, he joined the Republican side[1]—the same side that writers Ernest Hemingway and George Orwell supported—serving as a medical corpsman. George Orwell fought for seven months in a rebel militia and published his memoir *Homage to Catalonia*. Hemingway's reporting on the war led to his great novel *For Whom the Bell Tolls*.

By nature, civil wars, which rip a society apart, are uniquely tragic and brutal, but the Spanish Civil War was a boiling cauldron infused with the world's most vicious forces, the Nazis and Fascists on the Nationalist side, and the Soviet Union and Anarchists on the Republican side. Marked by massive civilian deaths, mass atrocities, systematic killings, and mob violence committed by both sides,[2] the Spanish Civil War took the lives of nearly one million.[3] Ultimately these three men, Delgado, Hemingway, and Orwell, found themselves on the losing side, defeated by the Nazi- and Fascist-backed Nationalists, who were led by military dictator Francisco Franco.

Delgado was held in a concentration camp for five months after the war ended.[4] The details of his internment are sketchy, but he was lucky to have survived. When the Civil War ended in 1939, about 500,000 Spanish Republicans escaped to France, one of the few countries accepting

Spanish refugees. The refugees were placed into internment camps, but when the Germans defeated France in 1940, Nazi authorities condemned the Spanish Republicans to forced labor camps and deported more than 30,000 to Germany, where about half of them were imprisoned in concentration camps. Some 7,000 of them became prisoners in Mauthausen concentration camp in Austria, where more than half of the Spanish refugees perished.[5]

Mauthausen was a massive concentration camp, one of the first to be established by the Nazis and the last to be liberated by the Allies. Noted for extreme cruelty and sadistic methods of extermination, primarily through forced labor (*Knochenmühle*, meaning bone-grinder) and starvation, Mauthausen was intended to be the toughest of all concentration camps. Its special purpose was extermination of perceived political enemies and the intelligentsia, the more highly educated members of society such as scientists, artists, teachers, and university professors. An engineer who survived the Mauthausen concentration camp, Simon Wiesenthal became world renowned for devoting the rest of his life to hunting Nazi war criminals and bringing them to justice.[6] Wiesenthal had been a Jewish resident of Lwów, (now Lviv, Ukraine), where Professor Adolf Beck had devoted his life at the university to studying electrical activity in the brain (see pages 40–50). The Nazis first imprisoned Wiesenthal at the same former factory in the northwestern suburbs of Lwów that was converted into the Janowska concentration camp, which Adolf Beck avoided by taking his own life.

After completing his MD degree and then receiving a PhD degree at the Ramón y Cajal Institute in Madrid, Delgado won a scholarship to Yale University in 1950 to study under John Fulton. Fulton's pioneering work on prefrontal lobotomy in chimpanzees inspired the Portuguese psychiatrist Egas Moniz to perform the operation on patients with schizophrenia. Moniz received the Nobel Prize in 1949 for this revolutionary surgical approach to treating mental illness, but prefrontal lobotomy eventually became discredited due to its indiscriminate application to manage unruly mental patients and even children with behavioral problems. Prefrontal lobotomy was also abandoned not only because of the debilitating side effects, but because drugs were developed to manage schizophrenia and other mental illnesses, without the risks that brain surgery entails. It was in this environment that Delgado began to conceive of a revolutionary new approach to controlling mental illness—by implanting electrodes into the brain to activate or inhibit the neural circuits underlying behavior, emotion, and mental illnesses.

Mired in controversy brewing after his dramatic demonstration in the Madrid bullring and his other research on mind control by electrical stimulation, which he conducted as a professor at Yale University in 1974, Delgado decided to accept an offer to become chair of physiological science at a new medical school in Madrid, which is now the Ramón y Cajal Hospital. The Cajal Institute (near the Cajal Hospital he joined in 1974) is where Delgado had earned his PhD. A red brick building built in the 1980s, it is sheltered from the busy street by an iron fence painted the color of rust, which frames a courtyard lined by trees. Inside the building a helical staircase ascends from the small lobby up three floors of laboratories, each one concealed behind green doors with a round porthole in them, which gives one the feeling of being on a ship. Here I met with a scientist who was visiting the Institute to attend an international symposium of neuroscience where I was presenting my research on brain plasticity. He is now retired, but I learned that he was a former colleague of Delgado's at the Cajal Hospital, having worked in a laboratory in the same department with him.

"What was he like?" I asked.

"He was crazy," Delgado's former colleague instantly replied.

"Crazy? What do you mean, 'crazy'?" I prodded.

"Read his book. He's crazy," the neuroscientist replied.

Delgado published 500 scientific papers and six books. I presume the colleague was referring to Delgado's book *Physical Control of the Mind: Towards a Psychocivilized Society*, published in 1969.[7] The provocative book expounds on the transformational potential of brain stimulation to control the human mind and bring human beings into a new era where mental illness and the horrors of human nature are at last brought under control by this new technology. Having lived through the horrors of the Spanish Civil War, Delgado believed that brain stimulation could be humanity's salvation—his electrodes would bring, as he wrote in this book, "a future psychocivilized human being; a less cruel, happier, and better man."[8] In Delgado's eyes, brain science could accomplish what religion, regulation, and punishment, which have been used to keep human nature in check for millennia, have failed to do.

"By 'crazy' do you mean he was ahead of his time, or that he was really crazy?" I asked.

"He was crazy! Crazy in the way he believed the human brain worked and that he could control it with his electrodes. He did his brain stimulation on people too [not just on animals]! Much of his work could not be repeated. He did not even know where his electrodes were placed in the brain. They were long and very flexible—he had no idea. His work was sloppy," the scientist said, with disdain for Delgado's motive, his methods, and his nonscientific pandering to the popular press.

"Where did he place the electrodes in the bull's brain?" I asked. "If they had been placed off target, in the striatum, for example, he might have simply paralyzed the animal with his electric shock."

"Or in the reticular formation!" he said. (The reticular formation controls arousal and consciousness. Zapping this part of the brain wouldn't be any more elegant than knocking an animal unconscious with a blow to the skull.) "We don't know where the electrodes were—he never knew. This was only one bull. The experiment was not repeated. We don't have any idea how many attempts he made before this one."

"Maybe he was ahead of his time too," he admitted. "He did accomplish a great deal bringing brain science—electrophysiology—to Madrid. He built a very strong department. But he was crazy."

I was unable to locate Delgado's notebooks or experimental apparatus during my short stay in Madrid, but I do end up finding one of his former PhD students, Dr. Oscar Herreras Espinosa (see page 158), who is now a faculty member at the Cajal Institute.[9] Herreras Espinosa, you will recall, shared his recordings of neuronal oscillations in the hippocampus with me, working to understand what causes neuronal oscillations and deliberating over whether they are critical to how the brain processes information, or are instead by-products of neural circuits at work. Herreras Espinosa joined Delgado's lab in 1981 at the age of twenty-one. His thesis research, supervised by Delgado, concerned neurotransmitters involved in the theta rhythm and epilepsy, and Herreras Espinosa used special electrodes to record neural activity, to stimulate neuronal firing, and to sample the extracellular fluid around neurons through microscopic cannulas, in search for neurotransmitters released by synapses when neurons fire. This chemical signaling between neurons, and the imbalance in neurotransmitters exciting and inhibiting neuronal firing, is associated with many neurological and psychological dysfunctions, and thus is a potential avenue for treatment by drugs manipulating levels of specific neurotransmitters. (SSRIs, for example, augment signaling by the neurotransmitter serotonin.) Fishing through some drawers in his office, Herreras Espinosa finds a dusty glass tube the size of a runner's baton, stuffed with bunches of these electrodes from his days in Delgado's lab. Smiling with delight, as one does when discovering an old snapshot in a shoebox of faded photographs, he hands them to me. They look something like a sheaf of clear fiber-optic cables, but the newness has worn off with the corrosion of time.

"What was Delgado like?" I ask.

"Well, I had really very little contact, because he was running the entire department. He used to come every now and then and say, 'Well, I have this idea, please go ahead with it.'

"He was a very curious person. Some people loved him and others hated him. I think he had great ideas," he says.

"I saw him many times talking to the press, the media. It was amazing! He had such a powerful . . . so convincing—you know, he managed the press very, very, well. He was a showman" but a very brilliant scientist, he says. Delgado's perceived "showmanship" drew criticism and resentment from some academics, because they felt the scientific method works by taking the personality out of the picture and replacing it with experimental data. This pitfall of showmanship can happen anytime a notable scientist tries to communicate directly with the public. The phenomenon is now referred to among scientists as the "Carl Sagan" effect, after the acclaimed astronomer who became a celebrity—it seems you cannot be both.

Herreras Espinosa clearly retains deep respect for Delgado. "When I look at those days, and I see what we are doing now—we now have animals that are chronically implanted with electrodes, so we are more or less doing what he was doing fifty years ago." (A "chronically implanted" electrode is one that is implanted in the brain of freely behaving animals for long-term recording, as opposed to the type of recording I described earlier in which electrodes are inserted into the brain of an animal under anesthesia with its head immobilized for stability, and then removed.)

"He was ahead of his time?" I ask.

"I'm not sure," he says, surprisingly. "I think we are late. That was the right time to do those things," he adds, implying that modern scientists have dropped the ball. "The technique was already good for those experiments, and it was a very nice approach to modulate behavior. It was kind of stopped, or slowed down, and it's only now—since no more than the past ten years, that behaviorists are coming into electrophysiology—that these things are coming up again."

I comment that Delgado's work was frightening to the public, and he agrees. "Deep brain stimulation for refractory epilepsy or Parkinson's in humans has been out for twenty years, but you talk to people, and they do not know it. When I talk to someone, common people, they say, 'Well, I don't want anyone to put anything inside of me!' This is a barrier that we still have to fight." He says he works with neurosurgeons on migraine and spreading depression, but "When I tell them, 'Why don't we put an electrode in and do this or that,' and they say, 'No, no, no . . . you cannot put anything inside the patient.' Well, they are doing that already for people who are stricken with epilepsy, but they say, 'No, this is not socially accepted.'" (Spreading depression is a loss of electrical charge in brain cells that spreads as a wave across the cerebral cortex, seen in EEG as a cessation of brainwave activity sweeping over the brain. Spreading depression is associated with pain, migraine, and other types of brain injury and disease.)[10]

Science is a luxury that can only be practiced in societies after all the basic needs of life have been obtained, because scientific research requires substantial funding and public support. For this reason, science does not proceed at the pace of scientific innovation; it proceeds at a pace, and in the specific directions, that is funded by the public or business. Research can be stalled or halted by regulations. The ban against stem cell research by the United States government during the George W. Bush administration in 2001 is a good example,[11] but one could cite many recent examples under the Trump administration: Terminating research on climate change is perhaps the most widely recognized instance of government control of scientific research.

Reflecting on Delgado's research conducted well over a half century ago, Herreras Espinosa says, "When I look at those years, I think, we have lost a lot of time. They had the technique and they could have begun to study control of behavior or just the neural basis of behavior many, many years ago." With his gray hair pulled into a long ponytail,

Herreras Espinosa somewhat resembles an aging hippie, with a radical's thoughtful, independent perspective.

When I ask if Delgado was sloppy, because he didn't even know where his electrodes were placed, Herreras Espinosa answers, "Well, we don't know where the electrodes are placed even now." The reason, he explains, is that we have such a superficial knowledge of the brain's detailed neural circuitry. He gives an example from his own work mapping circuits in the hippocampus, which deeply troubles him. If the neuron is silent when his probing electrode passes by it, "you don't even know it is there." Even with our most sophisticated modern methods, we can't be too confident that we have traced out a complex neural circuit properly, because there is no assurance that we have detected all the connections. "In Delgado's time, no one knew precisely where they were recording from."

I ask about the future—for example, Mark Zuckerberg and Elon Musk pursuing brain-computer interfaces to communicate over the internet without typing or to operate machines by brainwaves—and Herreras Espinosa replies, "I don't like to think about the future, because I think I am doing the future." Regarding such science fiction-like applications connecting mind and machine through brainwaves or implanted electrodes, he says, "My view is that this is unstoppable, and we had better get along with it."

In his book published in 1969, Delgado wrote, "Physical control of many brain functions is a demonstrated fact, but the possibilities and limits of this control are still little known."[12]

## DELGADO'S RESEARCH

Before examining the work of scientists who are "doing the future now" on brain-computer interface research, let's reveal what Delgado and his colleagues were doing and thinking decades ago.

Delgado wrote, "There are basic mechanisms in the brain responsible for all mental activities, including perceptions, emotions, abstract thought, social relations, and the most refined artistic creations. These mechanisms may be detected, analyzed, influenced, and sometimes substituted for by means of physical and chemical technology . . . Predictable behavioral and mental responses may be induced by direct manipulation of the brain."[13]

As early as 1955 and through the 1960s, Delgado showed that in colonies of cats and monkeys, aggression, dominance, mating, and other social interactions could be evoked, modified, or inhibited by radio-controlled stimulation of electrodes in specific cerebral areas.[14] Delgado's work built on the pioneering work of Walter Hess conducted in the 1920s who probed the brains of cats with a stimulating electrode and found a spot deep in the brain, now called the hypothalamic attack area, which when stimulated would cause the animal to launch into a violent attack on another animal in its cage.[15] In 1954, neurosurgeon Wilder Penfield had found that stimulating specific spots in the brains of his patients using electrodes that had been implanted during surgery could block the thinking process, inhibit speech or movement, provoke talking, and evoke pleasure, laughter, friendliness, hostility, fear, hallucinations, and vivid memories.[16]

Summarizing his findings, Delgado reported, "Cats, monkeys, or human beings can be induced to flex a limb, to reject food, or to feel emotional excitement under the influence of electrical impulses reaching the depths of their brains through radio waves purposefully sent by an investigator [scientist]."[17]

"The diameter of the pupil can be electrically controlled as if it were a diaphragm of a photographic camera," Delgado explained. He found that electrodes implanted in the left parietal cortex of his patients evoked flexion of the right hand when stimulated. When his patient was warned of the oncoming stimulation and was asked to keep his fingers extended, the radio-controlled signals overruled the man's own intentions. "I guess, Doctor, that your electricity is stronger than my will," the man said, seeing his own hand operate autonomously.

"Electrical stimulation of the rostral part of the internal capsule [part of the brain controlling movement] produced head turning and slow displacement of the body to either side with well-oriented and apparently normal sequence, as if the patient were looking for something." When Delgado asked the patient, whom he had controlled by radio signals, why he had turned his head, the patient gave clear explanations for his actions: "I am looking for my slippers." "I heard a noise." "I was looking under the bed." Not only had Delgado's electrodes commanded the person's body, but the mind had concocted a rational explanation for the movement that was not in the subject's control.

This radio-controlled brain stimulation was so simple, a monkey could do it. Delgado trained a monkey to press a lever inside its cage to trigger radio stimulation of another monkey's brain. Pressing the lever induced the monkey implanted with electrodes to run with excellent coordination around the cage—better entertainment, no doubt, than a radio-controlled car is to a child.

Electrodes that were controlled by radio signals were implanted into the brains of children and adults who were suffering from epileptic seizures and other behavioral disturbances. Delgado called his device a "stimoceiver" and showed that brain-to-computer-to-brain feedback (that is, recording EEG activity in the brain and then stimulating the brain when epileptic activity was detected) could squelch abnormal seizure activity in chimpanzees and human patients.

Delgado repeated the early experiments on aggression that Hess had performed on cats, but he used radio-controlled electrodes implanted in the brains of monkeys (a much better model of the human brain) and he obtained the same results. Stimulation of this and related regions provoked rage and a violent attack on another monkey or inanimate object in the cage. These same brain regions are also present in humans, and stimulating them in the human brain evoked similar rage responses. Research on aggression triggered in animals always draws the criticism that perhaps shocking the brain in this way simply hurts or irritates the animal, provoking the rage in response. However, when done with humans, they can describe what is happening. "I don't know what came over me," one patient said. "I felt like an animal."[18]

Summarizing this line of research, Delgado observed that "while individual and collective acts of violence may seem rather distant from the electrical discharges of neurons, we should remember that personality is not in the environment but in the nervous tissue."[19] No doubt recalling his experience during the war and imprisonment in a concentration camp, Delgado wrote, "Violence, including its extreme manifestation of war, is determined by a variety of economic and ideological factors; but we must realize that the elite who make the decisions, and even the individual who obeys orders and holds a rifle, require for their behavioral performance the existence of a series of intracerebral electrical signals . . ." These violent acts could be inhibited by electrical stimulation, he said, but then added, "Human relations are not going to be governed by electrodes, but they could be better understood if we considered not only environmental factors, but also the intracerebral mechanisms responsible for their reception and elaboration."

Not only are neurological conditions like epilepsy amenable to treatment by radio-controlled electrical stimulation of electrodes implanted in the brain, psychological illness can also be studied and addressed therapeutically in this way. Delgado describes a female patient who was crippled by overwhelming anxiety attacks that had led her to make several suicide attempts and suffer from chronic depression. Drugs and psychotherapy had failed her. "Stimulation of the dorsolateral nucleus of the thalamus evoked precisely the same type of [anxiety]."[20] Moreover, the amount of anxiety the woman experienced could be controlled by the intensity of the stimulus applied. "It was possible to find the electrical threshold for a mild anxiety or to increase it to higher levels simply by turning the dial of the stimulator. One could sit with one's hand on the knob and control the level of her anxiety." Similar experiments on another patient caused her to swivel her head side to side with an expression of fear, searching behind her for danger. She reported that she felt that a threat was approaching and something horrible was going to happen to her.

Delgado explored the brains of people with schizophrenia as well as the criminally insane who committed uncontrollable acts of violence. For example, a "charming and attractive" twenty-year-old girl who "on more than a dozen occasions [snapping in rage] resulted in an assault on another person, such as inserting a knife into a stranger's myocardium or a pair of scissors into the pleural cavity of a nurse."[21] Studying such patients with normal EEG recording was practically impossible, because they would not tolerate sitting still with electrodes on their heads, so for this patient he implanted a stimoceiver in the young woman's brain so he could see what was going on inside her head as she went about her normal activities. He recorded neural activity from her amygdala and hippocampus, and observed abnormal brainwaves (spontaneous sharp waves) when she suddenly broke away from what she was doing and began to walk about aimlessly. During these episodes, her speech was inhibited for several minutes. "Transitory emotional excitement was related with an increase in the number and duration of 16-cycles-per-second bursts [of brainwave activity]."

When Delgado stimulated one of the electrodes in the woman's right amygdala with 1.2 milliamperes of current, she stopped playing the guitar and singing, threw it away in a fit of rage, and launched an attack against the wall.[22] After which "she gradually quieted down and resumed her usual cheerful behavior. This effect was repeated on two different days."[23] H. E. King described a similar result in another patient, a woman suffering from depression and alienation.[24] Suddenly she said in a flat tone of voice with a blank facial expression, "I feel like I want to get up from this chair! Please don't let me do it! Don't do this to me. I don't want to be mean . . . I want to hit something. I want to get something and just tear it up."[25]

But the telemetric approach of controlling electrical activity in a person's brain could also provoke peaceful responses and vivid memories, and alleviate chronic pain. "Intractable pain, which cannot be alleviated by the usual analgesic drugs, and their unbearable suffering could be blocked by direct intervention in brain structures where sensations reach the perceptual level of consciousness," Delgado wrote.[26]

In animals with electrodes implanted in parts of the brain to provoke pleasure and a sense of reward, the creatures would persistently and sometimes frantically press the lever up to as many as 5,000 times per hour. Pressing the lever stimulated the same reward circuit in the brain activated by cocaine and heroin, but electrical stimulation drove the brain's pleasure circuits to the maximum. Stimulating the septal region of patients' brains made them feel sudden sexual pleasure building to an orgasm. "She started giggling and making funny comments, stating that she enjoyed the sensation very much. Repetition of these stimulations made the patient more communicative and flirtatious, and she ended by openly expressing her desire to marry the therapist."[27] Working his radio-controlled stimoceiver a half century before Wi-Fi and the computer chip, Delgado was playing the human brain and controlling behavior like a video gamer commanding an avatar. Technology of mind control is vastly more sophisticated and powerful today, but in transcending the border between mind and machine, the ethical and philosophical questions confronting us have not changed.

## CURRENT BRAIN STIMULATION RESEARCH

Much current research is done in mice, but the mouse brain is a feeble substitute for the human brain in many respects, especially for higher-level cognition and neuropsychiatric disorders. (Imagine if anyone, through injury or disease, had brain function reduced to the level of the most intelligent mouse—there is no comparison.) Our closest biological relatives, nonhuman primates, have a more complex brain and cognitive function that make it the best possible model of the human brain. However, research on chimpanzees, which was the mainstay of this type of research in the 1950s and 1960s, is now outlawed in the United States and most of Europe,[28] and brain stimulation research on human patients is highly regulated today.

### CONNECTING THE BRAIN TO PROSTHETIC DEVICES

The cochlear implant, which restores hearing to many deaf people, is a great brain-computer interface success story. A microphone picks up sound and stimulates an array of electrodes implanted in the cochlea (located in the inner ear) to stimulate the auditory nerve. While the implant provides nowhere near the fidelity of normal hearing, cochlear implants are life-changing because they enable profoundly deaf people to hear sound and understand speech. Similar devices have been used experimentally to help those who are blind, by feeding signals from a video camera to an array of electrodes implanted in the retina or visual cortex. The visual sensation evoked is extremely crude, but sensation of changes in light can be helpful to those who are profoundly blind, and the hurdle, although a major one, is in increasing the number of electrodes that can receive the signals. But this is only the beginning of what BCI researchers are working to achieve with neuroprosthetic devices in the future.

#### *UNLOCKED*

She slumps in her wheelchair as air is forced in and out of her paralyzed lungs through a tube in her neck. Neuroscientist Erik Aarnoutse of Utrecht University and his colleagues gather around the woman, Hanneke, in the living room of her comfortable home. Observing her intently, Dr. Aarnoutse, with a Viking physique and shoulder-length blond hair, towers above her with his arms crossed to avoid influencing her responses through his body language. "She is very intelligent," Aarnoutse tells me, "and driven—highly motivated to communicate." But she cannot speak. The fifty-nine-year-old blonde woman is trapped inside her own body, completely paralyzed by Lou Gehrig's disease (or ALS) that struck her in 2008 and attacked the neurons that make her

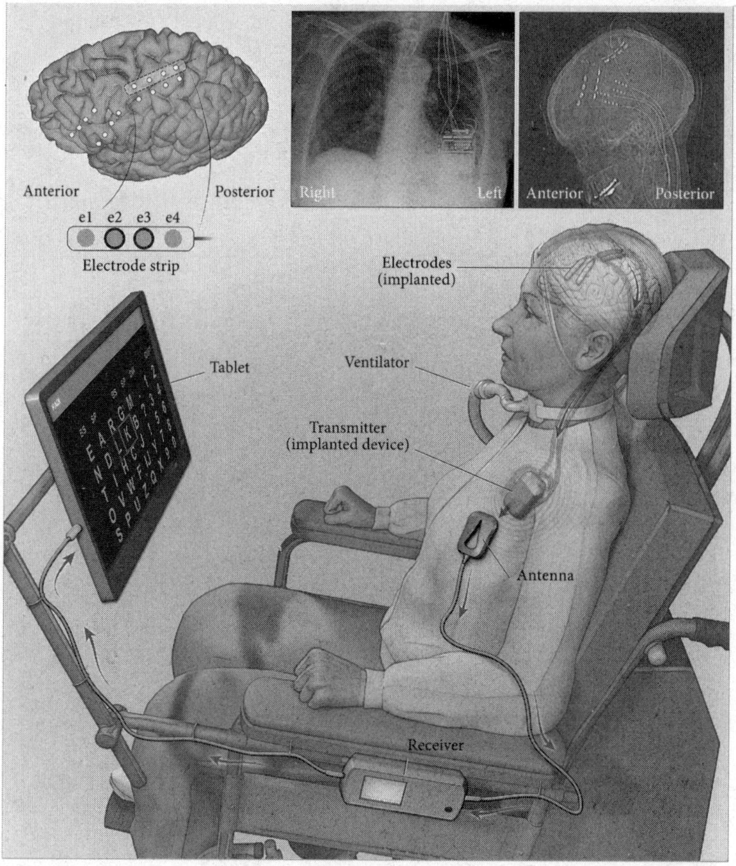


**FIGURE 27**: *Brain-computer interface, enabling a person paralyzed by Locked-In Syndrome to communicate. ECoG electrodes sense changes in brainwaves when the person imagines selecting the appropriate letter to type a message.*

muscles move. She can only operate the tiny muscles that shift her blue eyes to gaze out from her frozen body on a world she can no longer interact with, a condition known as Locked-In Syndrome.

A technological achievement, announced at the Society for Neuroscience annual meeting in November 2016, and simultaneously published in the *New England Journal of Medicine*, has helped break the binds that lock her in, and has done so by deciphering her brainwaves.[29] Using her mind alone she wills her thoughts to appear on a computer screen, spelling out words one letter every fifty-six seconds to share her thoughts. Slowly she types out one of her first thoughts, one letter at a time to the eager scientists gathered around her: "T . . . R . . . Y . . . I . . . N . . . G . . ." she pecks out with her mind until she has completed her sentence, ". . . to communicate like this is like tacking a sailboat," she tells them, evoking a metaphor from a time in her life when she must have struggled to beat a sailboat in triumph against a headwind.

Prior to this breakthrough in brainwave communication, the only way the woman could communicate was through a device that used an

**FIGURE 28:** *A patient with Locked-In Syndrome named Hanneke, in her living room, using the brain-computer interface (BCI) device to communicate by typing out messages.*

infrared camera to track her eyes, but the device is awkward to put into position and operate for a person who cannot move, and it does not function well in many situations, such as outside in the bright sun.

Her mind is clear, and the part of her brain that controls her bodily movements operates perfectly, but the signals from her brain never reach her muscles because the motor neurons that relay signals from her brain to her muscles have been attacked by ALS, which also afflicted renowned physicist Stephen Hawking. But Hawking had a limited ability to move his cheek, and he used those feeble nudges to operate a computer to translate text into computer-synthesized speech.

Devices that couple neural activity to computers have been used experimentally to help patients with a range of neurological disorders, including Locked-In Syndrome, but have met with limited success. In pioneering work in 1998, neurologist Philip Kennedy, of Neural Signals Inc., implanted an array of electrodes into the brain of a patient who was paralyzed by a stroke, and, in 2015, a team of researchers led by neuroscientist Leigh Hochberg of Brown University implanted an array of ninety-six electrodes into the cerebral cortex of a fifty-eight-year-old woman with Locked-In Syndrome. In these experiments, patients were able to select words displayed on a computer screen by mental concentration alone. But implanting electrodes into the brain carries inherent risks. Frustrated by regulations restricting brain implant experiments on humans because of the significant health risks, Kennedy had electrodes implanted into his own brain.[30] He succeeded in obtaining vital experimental data for his research, but he experienced complications after the surgery that left him unable to speak for a time, and ultimately he had the electrodes removed because of an infection. "Anytime there is a wire penetrating the skin there is risk of infection," says Andrew Schwartz, an expert on brain-computer interfaces at the University of Pennsylvania who is developing prosthetic devices controlled by electrodes implanted in the cerebral cortex. Furthermore, previous attempts at brain-computer interfaces could only be performed in the laboratory because of the bulky instrumentation required.

To overcome these obstacles, the researchers at Utrecht University developed a minimally invasive brain implant that communicates wirelessly to a computer notebook that patients can use to communicate at home and outside. Surgeons lifted a small flap of scalp and drilled two finger-size holes through the woman's skull. Then they slipped a thin plastic strip, which looks something like cellophane tape with four tiny dots on it, through the holes to rest on the surface of her brain. The four spots are miniature electrodes that do not penetrate brain tissue, but because they are beneath the skull, they make good electrical contact with the brain to record brainwaves in high fidelity (electrocorticography, or ECoG). Because the surgery does not enter the brain, it carries less risk. In fact, drilling holes in the skull, called trepanation, is one of the oldest of all surgeries, dating back to Neolithic times when the surgical instruments were chipped from stones. The surgeons then threaded tiny wires from the electrode under the skin to a small electronic control device that was implanted beneath the woman's skin in her chest region, to communicate by a radio transmitter to an ordinary tablet computer.

Surgeons placed the electrodes over the part of the brain (motor cortex) that becomes activated when the woman imagines closing her fingers. By analyzing the brainwaves, the researchers observed a simple but reliable pattern. Every time she imagined pinching her fingers, the power of certain frequencies of brainwaves abruptly changed. Like ocean waves stirred by a sudden gust of wind, the low-frequency brainwaves oscillating at about 20 Hz (beta waves) abruptly ceased as higher-frequency brainwaves oscillating at about 80 Hz (gamma waves) whipped up. By simply measuring the ratio of gamma- to beta-wave power in ongoing brainwaves sweeping through her motor cortex, the computer could detect when the woman was imagining closing her fingers. The woman quickly learned to operate a cursor on a video game, mastering that task only two days after the surgery. Next, the scientists presented her with the alphabet arrayed in rows and columns on a computer tablet. As the display swept over individual letters in sequence, the woman imagined selecting the appropriate letter as if she were clicking a mouse.

Some experts remain concerned that any surgery on fragile ALS patients to help them communicate is unjustified. "Implantations like the one reported here may carry an unknown risk for advanced ALS patients," warns Niels Birbaumer, an expert in brain-computer interfaces at Tübingen University in Germany who was not involved in the study. But according to Schwartz, many people with Locked-In Syndrome choose not to use ventilators to breathe when their disease reaches an advanced stage because they cannot communicate, and they feel they are a burden on their loved ones. Studies have found that, like Hawking, locked-in patients can lead meaningful and productive lives if they can communicate in some way. "We need to do anything we can to help these people," Schwartz says. "We are talking about life and death."

More than a year after the device was implanted, Hanneke lives at home with her husband and one of her children, and she has gotten much faster at typing out her thoughts. Also, the device works well outdoors in the sunshine where her eye tracker fails. "She's happy," Aarnoutse says. "The ability to communicate has given her more freedom and made her more independent. Recently she used the device [to go] on holiday with her husband."

The pace of this research is accelerating rapidly. In 2019, I heard Krishna Shenoy,[31] a BCI researcher at Stanford University, present his research on three ALS patients using a point-and-click typing system to communicate. This system is operated by an array of microelectrodes implanted into the motor cortex, and these, unlike ECoG, can detect the firing of individual neurons. Using this BCI device, the paralyzed patients are reportedly able to type at a rapid rate of 24 to 32 characters per minute without moving a muscle!

## ARTIFICIAL EYES, LIMBS, AND EARS

Research on prosthetic devices interfacing directly with the nervous system began to accelerate in the 1970s when electrodes implanted in the brains of laboratory animals (rats and nonhuman primates) were used to determine how neural circuits in the cerebral cortex operate (for example, when an animal moves its limbs). Then, by delivering electrical pulses to neural circuits in the right pattern to mimic the signals that the neurons normally would have used to energize the limbs, scientists began to command the brain to make the monkey's hand grasp a cup. With this information and technology mastered, it then became possible to use a computer to analyze the neural spiking occurring in regions of the brain that control a person's bodily movements, as when someone with spinal cord damage imagines trying to grasp a cup. The researchers simply bypassed the injury to deliver the appropriate electrical stimulation to the muscles directly. In this way the paralyzed limb can be controlled by the person's thoughts. Similarly, signals from a video camera can be used to stimulate the visual cortex to give a sense of vision to the blind. Cochlear implants work by picking up sound through a microphone and stimulating auditory nerve fibers in the cochlea of the inner ear to restore hearing.

Animal studies show the great potential of BCI interfacing with the brain in this manner. Electrodes detecting cortical activity in the motor cortex of macaque monkeys can be used to manipulate a robotic arm so they can feed themselves, for example.[32] The drawback to this approach for use in humans is the need to implant electrodes into a person's brain, and the slow corrosion and displacement of electrodes over time that ultimately renders them ineffective. However, new research on brainwaves that are generated during cognitive activity has advanced so that implanting electrodes in the brain is no longer necessary—in some cases. A person can simply wear a cap wired with electrodes that detects their brainwaves. A computer analyzes the signals to recognize what the person's brain would like to do, and in turn operates a robotic prosthetic arm or other device.

Such research is being conducted in many labs, such as in the laboratory of Duke University neuroscientist Miguel Nicolelis. I first met Nicolelis at a neuroscience conference in Germany, where he presented some of his remarkable and as-yet-unpublished research, which transforms a paralyzed man into something of an iron man. During the 2014 World Cup Soccer tournament, Nicolelis, who is a Brazilian-born soccer lover, gave a dramatic public demonstration of this futuristic technology. He fitted a paraplegic man with a robotic exoskeleton that was controlled by the man's own EEG, and, as thousands of soccer fans watched, the young paralyzed man delivered a ceremonial first kick to open the games.[33]

Some have criticized the achievement because the kick was quite feeble, and because other robotic prosthetic devices that are controlled by muscles that are still functional in a paraplegic's body can control a prosthetic device much better.[34] Indeed a paraplegic colleague of mine, between sessions at a meeting on neural regeneration in California, demonstrated to me how he can move his legs and walk by pushing buttons on an experimental device he made that shocks his leg muscles. Sitting in his wheelchair, he pushed one button that stimulated electrodes in his thigh muscle. His inert lower leg shot up forcefully. But if the soccer kick was indeed triggered by the paralyzed man's EEG (and not, say, by the much larger electrical signals generated by muscle contractions in the head and face that are the bane of technicians who do EEG recordings), Nicolelis has made a significant achievement.

Technological improvements in EEG recording and in understanding what brainwaves mean in detail will bring improvements in the ability to control prosthetic devices by EEG. Keep in mind that the brainwave activity that reaches the scalp is the result of millions of neurons firing, generating electrical signals that interact and propagate over large expanses of the brain. So the challenge in controlling prosthetic devices and computers by EEG is the "crowd-in-the-stadium versus the spectator" problem. By listening from the parking lot to the roar of the crowd in a stadium, some basic information about what is going on inside can be gleaned, but the individual conversations of the thousands of spectators cannot be extracted. You cannot carry on a conversation with a stadium. Fine control of motor and sensory function requires monitoring and controlling activity in individual neurons operating in complex circuits—thousands of them—instantly analyzing the complex traffic of information, understanding it, and selectively delivering the appropriate firing patterns to hundreds or thousands of individual neurons.

Although implantable electrodes carry risks and the current devices are much cruder than required for the desired tasks, there have been great technical advances in fabricating more biocompatible multi-electrode devices that can be implanted in the brain for long periods of time to interface with thousands of neurons at a time. Justin Sanchez, director of the Biological Technologies Office of the Defense Advanced Research Projects Agency (DARPA), is funding research in many labs to develop brain-computer interface devices to control prosthetic limbs for people with brain injury from stroke, trauma, or disease. The technology can allow people with disabilities to interface with a computer productively in other ways—send an email, for example—but DARPA is funding research to take this technology a step further. This research is developing methods of feeding information from sensors on the prosthetic limb back into the brain to provide sensory feedback. It is very difficult to grasp a glass of water with a robotic hand when you don't have any idea how hard you are grasping it or whether the glass is slipping. "Closing the circuit to take signals from the arm and send them back to the brain is really important to carrying out daily activities," Sanchez says.[35]

When President Obama visited Andrew Schwartz's laboratory at the University of Pittsburgh in 2016, a paralyzed man in a wheelchair maneuvered a robotic arm to shake the hand of the president of the United States.[36] By using only his thoughts, which were picked up from his brain and analyzed by a computer, the paralyzed man achieved his impossible dream to interact again with the physical world. As the then twenty-eight-year-old Nathan Copeland imagined shaking the president's hand, he saw his robotic arm do it. What's more, as he shook the president's hand with his robotic arm, he felt the precious lost sense of touch return. Neurosurgeon Elizabeth Tyler-Kabara, associate professor in the Department of Neurological Surgery at the University of Pittsburgh, implanted four tiny microelectrode arrays, each about half the size of a shirt button, in Nathan's motor and sensory cortex. The mechanical hand is equipped with sensory feedback that activates the electrodes implanted in the sensory cortex of Nathan's brain, so he felt the president's grip on the robotic hand as if it were in his sensationless, paralyzed limb.[37]

"I can feel just about every finger—it's a really weird sensation," described Copeland, who suffered a spinal cord injury in a car accident. "Sometimes it feels electrical and sometimes it's pressure, but for the most part, I can tell most of the fingers with definite precision. It feels like my fingers are getting touched or pushed."[38]

## *PIONEERS INTO THE FUTURE: PEOPLE WITH COMPUTER CHIPS IN THEIR BRAINS*

Nathan Copeland, who shook President Obama's hand with his BCI-driven robotic arm, has a wry sense of humor. He loves video games, and, with his ear piercings and bushy black beard, he strikes me as a techy. "I was eighteen when this happened to me," he says. "It sucked a lot. And it will suck for every person that it happens to."

But imagining the future, which he was helping to make a reality, Nathan predicts a time when BCI will transform the lives of people with disabilities from spinal cord injury and other traumatic injuries. "Someday it will be like *Star Wars*. You wake up—you know, like, 'A terrible thing happened to you. You were paralyzed. You couldn't move your hand, but we already fixed it . . . Your insurance covered it. It's all good.'"

I had the pleasure of meeting Nathan and several other pioneers who have experimental devices implanted in their brains to control prosthetic devices. They were invited by Florian Solzbacher, a BCI researcher at the University of Utah, to share their experiences at the Society for Neuroscience meeting in San Diego in November 2018. These remarkable people came from around the country in their motorized wheelchairs, flanked by medical attendants, to give working neuroscientists a patient's perspective on BCI. Clearly their passion is to bring wider understanding of spinal cord injury and to inspire scientists to develop better devices to enable people with disabilities to regain some lost function by connecting computers directly to electrical circuits in the brain.

"I'm thirty-two," Nathan says, lying under blankets in a reclined position in his four-wheeled motorized wheelchair. "When I was eighteen, I was in a car accident. I broke my neck and now I'm a C5 quadriplegic."

C5 is the fifth of seven vertebrae in the neck, called cervical vertebrae, which are numbered in order from C1, the neck bone supporting the skull, down the neck to the vertebra at the shoulders. Vertebrae in the chest and lower back are the 12 thoracic, 5 lumbar, and 5 sacral vertebrae, respectively. The higher up an injury is on the spinal cord, the larger the region of the body that will suffer paralysis and loss of sensation. A neck broken at the C1 or C2 region, as the American actor Christopher Reeve suffered when he was thrown from a horse, causes severe life-risking quadriplegia, typically requiring a respirator to sustain life. An injury in the C6 or C7 vertebrae results in paraplegia, leaving the patient with some control and sensation in arms, but paralysis of the trunk and legs.

"I have four arrays implanted in my cortex right now. I'm the first human to have two implanted in the sensory cortex," he says with a well-earned sense of pride.

Nancy Smith, a middle-aged woman with blonde hair, shares her story. "Ten years ago, I was in a head-on collision with an eighteen-wheeler. I was on a Girl Scouts trip with my daughter's Girl Scouts troop." Nancy was airlifted to a hospital in grave condition. "I suffered a broken nose, collarbone, complete fracture of my femur on my left leg, several broken ribs, internally I was bleeding, and the worst was the spinal cord injury—C3, 4, and 5 are fused." (Vertebral fusion is a surgical procedure to stabilize a broken spine.) "It left me unable to move."

"I did not fully understand what everything meant—how my life was now changed. I was having symptoms of PTSD from the trauma." Nancy shares with evident embarrassment how she suffered anxiety and depression from having her life shattered in an instant. "Having been a very active person as a mother, also a high school teacher, I was not used to *not* being able to do anything for myself at all."

Explaining her situation in the manner of a skilled teacher bringing understanding to her students, Nancy lays out her story, not as a plea for sympathy, but as a life lesson. "I lost everything. I had been teaching for twenty-two years. I really enjoyed being around adolescents, helping them through their most difficult time in their life, and I miss them horribly. I was not ready to retire, but there was no way I could go back to work. I can't use my hands [or legs]. I certainly wouldn't be able to man a classroom of young people."

The injury was horribly traumatic and excruciatingly painful, but, for Nancy, disability was crushing. "The trauma of going through an accident like that and then having to have someone to do *every . . . single . . . thing* for you. From scratching your nose to washing your hair—to, uhh, bowel and bladder. That's *extremely* humbling. A lot of tears got shed, because I was embarrassed."

She shares the desperation she felt for months after the accident and rehabilitation, suffering such dark moods that her doctor and husband feared what she might do. "The depression and anxiety of no longer being in charge," she explains. "You could not get me past that. My husband wouldn't leave me at home alone with the back door open because we had a swimming pool . . . It got to that point."

Suddenly she switches from an emotional tone of suicidal despair to outraged anger, saying, "Why did I work so hard to get these degrees and get this knowledge and be successful in what I was doing, just to have it all taken away? That wasn't fair."

The accident took away her abilities, her career, and her independence, but it also took away something she had not anticipated. "One of the very difficult things is that you lose a lot of friends." She recalls how friends would bring enchilada casserole over at first as a way to help and express sympathy, but then they stopped coming around. "I don't know why. They can't handle seeing me like this? They'll come around one or two times and then they are gone. That *hurts*. That hurts a lot."

But Nancy and the others who have volunteered to be experimental subjects in BCI research are courageous and determined fighters. "We knew we had to keep going. Our daughter was fourteen when it happened. Good grief! The worst age *ever* for something to happen to a daughter's mom. She was very active in Girl Scouts and water polo. Academically she was doing well. We didn't want to lose that. I was scared to *death* that I was going to lose that. I was going to lose *her*."

With the fierce determination of a momma bear, Nancy—no matter how dire her own situation may be—was going to protect her child. "It is difficult to parent when you are lying flat on your back down the hall. That doesn't work very well." She describes how she insisted on attending her daughter's water polo games and Girl Scouts activities. "I had to be there for her. We had to carry on."

Ian Burkhart, another of the participants at the meeting, was born in Columbus, Ohio, in 1991. Athletic and adventurous, Ian was a Boy Scout and lacrosse goalie, and he had just started college at Ohio University in 2009. While on summer vacation at the Outer Banks in North Carolina, he dove into the surf at the beach, something most of us have done numerous times, but he struck a hidden sandbar and broke his neck, rendering him a quadriplegic. The doctors told Burkhart that after his injury he would be able to shrug his shoulders a bit, but he would not be able to move or feel anything below his neck for the rest of his life.

As he speaks, I spot what looks like a black plastic bottle cap stuck to the top of his head above his left ear, the connection post bolted to his skull to connect the electrodes implanted in his cerebral cortex to the computer. The electrodes in Ian's brain, like those in Nancy's and Nathan's, are not the voodoo-doll, pincushion-type electrodes of Delgado's day. The human cerebral cortex is only 3 millimeters thick, so these electrodes, called the Utah Array because they originated at the University of Utah, look like a small patch of Velcro.[39] The bristles on the patch are actually 128 microfabricated hairlike microelectrodes, only 1 millimeter long.

Strapped around Ian's paralyzed forearm is something that looks like a Roman gladiator arm cuff. This is where electrodes stimulate the nerves and muscles through his skin to move his hand in response to his mental effort that is picked up by the brain implant and fed into a computer inside the laboratory. The computer interprets his intentions—the thought process is no different from you thinking of grasping a coffee cup, for example—and then sends the proper electrical commands to the electrodes in his forearm to clench the correct muscles to carry out

**FIGURE 29:** *Microelectrode "Utah Array" for implantation into the cerebral cortex for BCI and used to control prosthetic limbs. The array provides 128 channels for recording neural activity through microfabricated electrodes that are about a millimeter long.*

the desired hand movements. However, Ian has to deliberately concentrate on what we all do subconsciously.

"The first time I was able to move my hand it was a big shock. It was something that I hadn't moved in about three and a half years. Now it's something so fluid, it is kind of like it was before I had my injury. I just think about what I want to do and then I can do it. The system is essentially replacing my spinal cord with an artificial spinal cord," Ian says.

In a press release accompanying publication of the scientific paper announcing this achievement in 2016, Chad Bouton, a member of the team of researchers at the Feinstein Institutes for Medical Research, said, "This is the first time a completely paralyzed person has regained movement just by using their own thoughts."[40] This technology could not only help people with spinal cord injury, it could also be used to enable people with paralysis from a stroke or traumatic brain injury to regain control of their inert limbs.

Nancy's brain implant is not connected to a device that stimulates her own muscles; it is connected to a computer, which she is learning to control by her thoughts. On the computer screen an animated robotic hand responds to her mental commands. Nancy describes her elation when she was able to move the correct fingers on the robotic arm by her mental effort. It was the first time she could move something to interact with the world since her spinal cord was severed in her car accident. The task was to touch the correct dots on a keypad with the correct finger of the animated robotic arm on the computer screen. "I was able to touch the dots, depending on which finger I needed. We managed to get this done the very first day that I tried it!" The animated robotic arm could just as easily be a real robotic arm, like the one Nathan used to shake President Obama's hand, or it could be a computer keyboard on the screen to enable her to type words or commands to operate other devices that will help her regain independence.

"I played piano before my injury," Nancy says with a smile. So, to her delight, the scientists programmed an electronic keyboard on the computer monitor connected to her brain through the implant, and Nancy began to play *Twinkle, Twinkle, Little Star,* a musical performance conducted by her mind.

"I could play the piano by thinking about it, as if my hand were actually on the keyboard! It is pretty amazing to me." She laughs as she tells us that the technician who had programmed the keyboard had no musical training, and he did not realize where middle C was. When Nancy played what should have been the middle C key with her imagination, the wrong note sounded. She laughed, realizing the error, and the technicians quickly reprogrammed the keys properly so that the middle C key was beneath her imagined right thumb poised in her mind on the piano keys.

Keven Walgamott of Salt Lake, Utah, gingerly picks up an egg from an egg crate and delicately transfers it into another empty crate. There is nothing extraordinary about such a simple task that anyone might do in their kitchen without a thought, except that in this case, the fingers embracing the egg are plastic digits on a humanlike robotic hand.

Keven is a double amputee, having lost one hand and one foot in an electrical accident. He is operating the prosthetic hand as if it were his own, except that the sensors and motors in the artificial hand connect with his brain through electrodes in nerves and muscles in Keven's stump. Connecting into the peripheral nervous system, rather than through electrodes inserted directly into the brain, has the advantage of using the powerful computation that the human brain performs to interact physically with the world, by streaming information through appropriate neural pathways in our limbs. If you have ever had your stomach drop watching a child fumble and break a raw egg while handling it, the high demands of this task with an artificial arm are obvious. It would be nearly impossible for Keven to grasp the oval shell without breaking it, if not for the fact that this prosthetic device also has sensors in it that stimulate the sensory nerves feeding into Keven's brain, restoring the remarkable sense of touch. Information in our nerves is conveyed to the brain by action potentials. Whether those action potentials are triggered naturally or by electrical stimulation makes no difference to the brain or to the sensation of touch one feels.

"It almost brought tears to my eyes the first time I felt that," Keven says as he watches a video of himself operating the prosthetic limb and, in another scene, controlling a virtual robotic hand on the computer screen in training before being fitted with the physical device. "It was the first time I had the feeling of shaking my wife's hand since my amputation."

After fourteen months of experimental research, doctors removed the experimental device. "They've taken all the implants out, unfortunately, so I'm looking forward to the next phase when they have wireless and I can take it outside the lab."

I hope that the "gee whiz" dazzle of this amazing technology does not overwhelm a deeper appreciation of what is going on here. These are people. Truly courageous and selfless people, giving of themselves and taking real risks in the hope of helping others. Ian recalls grappling with the momentous decision he faced, as did the others, as to whether or not to have a device implanted in his brain for neuroscientists to use in their experiments. "Do I want to subject myself to not one, but at least two neurosurgeries?" Ian recalls, deliberating.

"It is a lot to ask of someone, because it is very invasive—especially when people are sticking stuff inside of your head," Nathan adds.

"I had very bad headaches," Ian says, recalling his recovery after surgery. "I mean, it's brain surgery. It is a big risk."

"You have to be prepared for tough recoveries," Nathan agrees, but then adds what many might overlook—the effect the irreversible decision has on loved ones. "Some of your family members may not agree with your decision." Nathan's mother, deeply concerned for his welfare, was firmly against her son risking further misery by becoming a human guinea pig in a bunch of science experiments. "My mother would not talk with me for many months after I said I'm doing this." Nathan went through the surgeries on his brain without his mother at his hospital bedside.

I mentioned that Nathan was the first person to have four implants in his brain, but I did not explain why. "My recovery was really rough after my first surgery, which took twelve hours."

After such a long surgery, recovery can be difficult. "After being under anesthesia that long you don't feel that great. I was not having a good time. I think one of the first things I said to my grandma—she is the one who took me because my mom was not talking to me at that time—I was, like, maybe Mom was right, 'cuz it hurt a lot."

Nathan had great trouble sleeping, and he experienced other serious complications. "I had aphasia. So right out of the blue, every now and again, I just couldn't say the right words. That was super frustrating and gave me anxiety."

After undergoing speech therapy, Nathan eventually overcame his aphasia. "He was a great guy—super good at his job," Nathan says of the speech therapist.

In addition to the technical hurdles of brain surgery, a larger issue neurosurgeons confront is that everyone's brain is a bit different. The same anatomical spot in one person's brain does not necessarily do the same thing that same spot does in someone else's brain. This is why neurosurgeons probe brain functions with electrodes while their patients are awake to map out functions in the person's cerebral cortex before they remove brain tissue to treat epilepsy and other disorders. In Nathan's case, his implant surgery was a failure.

"The positioning of the first implant was not right, so it did not work well. They gave me some options: 'You can work with it for now. We can see what that array can get . . .'"

Alternatively, "We can quit."

And a third option: "Or . . . you know . . . we can redo it."

Listening to the conversation at his bedside, Nathan's best friend, who had brought him to the hospital, told his friend on the way home, "I knew what you were going to say before you said it. 'Cuz you've been my best friend since high school."

There was not a moment's hesitation. "Yeah, we're redoing it," Nathan said, still struggling to speak with his aphasia at the time.

Unable to move a muscle, Nathan, like a courageous marine on a mission, charged back into the danger, regardless of risk, to succeed. "I said to myself, somebody needs to. I can. Why not me? Someone has to start somewhere."

All the others had the same motivation and courage. "Going into the study I didn't really think that I, myself, could benefit from it, but after the study, for younger people it could be life changing," Keven says.

A former Eagle Scout, Ian echoes the same selflessness. "I really saw it as an obligation to other people in a similar situation as me, but who may not have an opportunity to participate in a trial like this, and paving the way forward for future generations to help them."

As in all surgeries, patients must give informed consent, and doctors and scientists must thoroughly explain to the patient or experimental subject all the potential risks in great detail. "I agreed way faster than

I should've," Nathan says. His commitment was so strong, he says, "I didn't even read the packet they sent me. It wouldn't have mattered what that paper said. It's not going to make me better. Hopefully it can make other people better."

Dismissing the risks with the rodeo cowboy humor of a wild bull rider strapping his gloved hand to the beast, Nathan says, "The risks . . . I mean, you can die from Tylenol. Everything has a risk."

Nathan's second ride in the ring went famously. "Thankfully the second surgery was much quicker, because there was already a hole in my head. I still got aphasia sometimes, but I learned how to deal with it," he says, brushing off the side effects like dust slapped off blue jeans with a beat-up Stetson in the hands of a man standing victorious in the rodeo ring.

As you well appreciate by now, no matter what you may read in sensational articles, neuroscientists do not yet understand how thoughts, emotions, and intentions are coded in the pattern of neural impulses zipping through neural circuits and sweeping through brain tissue as oscillating brainwaves. The neural code is still a mystery, but computers using advanced machine learning can begin to recognize patterns of electrical activity in the brain that are associated with a specific sensory or motor function, and use that insight as a reliable signal to trigger prosthetic devices to perform useful functions. This is a complex process, far from being able to decode neural impulses as one would read computer code.

Even more powerful than the advanced computer computation in machine learning, the human brain learns to control the artificial device. After all, machine learning, as impressive as it has become, is only a shadow of the learning a living human brain can perform. This learning is largely an unconscious trial-and-error process, just as I experienced when I finally controlled the power of my own brainwaves using neurofeedback (see page 152). My brain did it, but I can't say how. Neurofeedback training is how all BCI patients learn to interface their brain with a computer.

It is a lot of work for the brain to learn to operate an artificial implant connected to a computer to execute a desired function. "I'm exhausted, and I have to be put back in bed immediately," Nancy says about the extreme effort involved in the neurofeedback sessions in which her brain learns to communicate with the hardware. "I have to really concentrate on what I'm doing."

Keven agrees. "Very exhausting! It was amazing how tiring it was, not just the muscles in my arm, but I was mentally exhausted." A lot of eggs must have cracked in the process of learning to control his robotic hand.

Ian's experience learning to control the wristband that restored motion to his paralyzed arm was the same. "Learning how to control it is extremely straining. It felt like a ten-hour exam—just mentally drained, not wanting to do anything, as well as physically drained because my muscles had atrophied quite a bit from not using them. Things have gotten much better. Now it's just normal to me," he says.

"I am just in awe at what you have accomplished," Keven says, looking into the eyes of the scientists who conducted the study on him.

Ian is even more awed. He never expected to benefit personally from having the BCI implanted in his brain, but what astonished him and his doctors even more than being able to use electronics to bypass his broken spinal cord and energize his arm muscles directly was that after undergoing so much neurofeedback training, Ian's brain learned how to move his paralyzed limb without the armband!

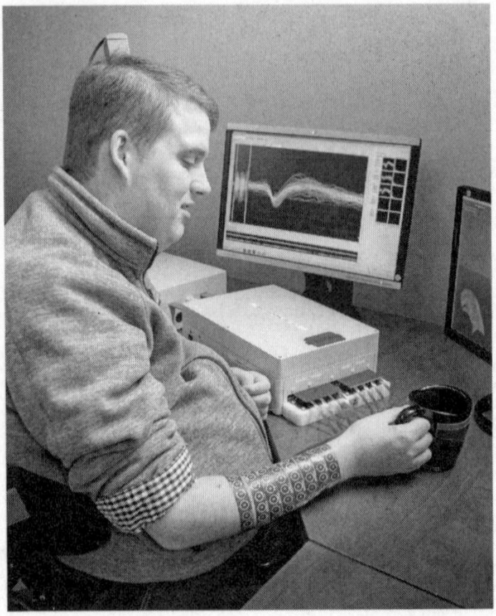

**FIGURE 30:** *Ian Burkhart using a brain-computer interface to grasp a cup with his paralyzed hand and arm after he was paralyzed from the neck down from a spinal cord injury. Electrodes implanted in his motor cortex pick up neural activity and energize electrodes in his armband to contract the appropriate muscles to make the desired hand and arm movements.*

Afterward I met with Ian as he sat in his wheelchair. I introduced myself and handed him my card. He raised his once paralyzed arm, grasped my card in his fingers, and read both sides. Although his arm had been paralyzed for years, apparently latent neural pathways that had been spared or perhaps new pathways that sprouted were engaged by electrical stimulation in the neurofeedback training, and now Ian, formerly a quadriplegic, can operate the joystick of his wheelchair, grasp and swipe his credit card with ease, and even drive a specially equipped automobile.

Working his atrophied arm muscles by electrical stimulation restored some strength to them, and likewise to his nerves. Now researchers are applying electrical stimulation to spinal cord injury patients, coupled together with physical therapy, and finding great success in what was previously considered a hopeless situation. Some patients with spinal cord injury have recovered some ability to walk after combined electrical stimulation and physical therapy.[41] Until very recently, electrical stimulation was not used for spinal cord injury because it was assumed that with connections severed in the spinal cord, it would be pointless to strengthen muscles that cannot be used. However, the purpose of the brain is to control the body and interface with the world, and if there is any way it can learn to do so, it will.

"I knew going into the study that I may not get anything out of it," Ian says. "It was really a drive to help future generations—to push the science forward. But I have been able to benefit from it. I've been able to regain some coordination in muscles that I do have control over. That has improved my life quite a bit. I have since moved out of my dad's home and I live on my own. I can be mostly independent. For me, that's what I really wanted after my injury. The constant needing to have someone do everything in your life gets old very quick. Doing things for myself really brings me a lot of joy."

The technological advances currently being developed using wireless communication will enable BCI implant patients to do even more. "I want to help bring this study to where I can use this device at home and do things independently," Nathan says. Using his BCI device, Nathan enjoys making art and playing video games.

All of the participants said how being involved in the study had payoffs beyond what they ever imagined. Beyond the physical benefits they got from taking the risks of brain surgery and enduring the hard work of training the interface, they found that being involved in the scientific study returned to them something precious that their tragic accidents had robbed from them. It gave each of them a sense of purpose. No longer isolated, they became an important part of a team. They and the scientists became close friends. Despite the ordeals, each of them said they would do it again in a minute.

"Looking back at it," Nathan says, "I would do it as many times as they would let me. If they say, 'Hey, we got something new. Do you want to pop it in?'

"Yeah, go for it. Hey, can't you just put a plexiglass cover there and a hinge?"

Ian adds, "It gives hope. Once I saw my hand move for the first time, I could see the light at the end of the tunnel. Now we just have to keep working our way there."

"This is a thing that is real," Nathan says. "I am honored to be doing this. Let's do it better."

Then something happened that does not happen at a scientific meeting. Everyone in the packed audience jumped to their feet and gave a thunderous standing ovation. It was a moving experience. To see these pioneers, undefeated by their chance accidents and grave disabilities, bravely and selflessly devoting themselves to advance science and technology for the benefit of mankind, left everyone awestruck, grateful, and humbled. And all of those in the room felt united in purpose to move ahead into the future of BCI to see the day come when:

"Hey, no problem. It's all good. And your insurance covered it."

## MACHINES WITH MINDS

Imagine a black-and-red checkerboard. Rotate it so that it forms a diamond, with the corners of the checkerboard at the top, bottom, left, and right. Now illuminate it with a flashing strobe light, disco style, and you will have the setup used for the earliest EEG-based BCI system. It was developed by a computer scientist, Jacques Vidal, at UCLA in the early 1970s.[42] Vidal used five electrodes to record the EEG of people staring at the flashing checkerboard. The brainwaves evoked by the image each time the light flashed create a visually evoked response in the brain; that is, a clear ripple in the EEG response appearing about 100 milliseconds (in this case) after the image is flashed. Vidal then asked the subjects to fixate on one of the corners of the checkerboard. Because the checkerboard pattern projects onto the retina in a different place, depending on which corner of the checkerboard is centered on the fovea of the retina, the EEG response produced in the visual cortex also differs. Focus your eye on the left corner and the rest of the checkerboard falls to one side of the center of your retina. Focus on the right corner, and now the image of the rest of the checkerboard falls on the opposite side of your retina. Thus, Vidal could tell from the computer's analysis of the person's EEG response which corner of the checkerboard the person was looking at. With this information, he was then equipped to enable the subjects to guide the path of movement through a maze using only their EEG response—look left, go left, for example. The computer determined where the subject was looking by analyzing the brainwave that was evoked and then moved a cursor through a video game–like maze accordingly. At that time, a large IBM System/360 mainframe computer was required to handle this real-time EEG detection, analysis, and control, but the BCI system worked well, with 90 percent accuracy.

Subsequent BCI devices have used other features of EEG; for example, in the P300 response, a positive ("P") voltage peak occurs 300 milliseconds ("300") after an event is presented to the subject that is novel or of significance to the person.[43] This event can be a sound or an image or even an unexpected word in a sentence. The P300 response reflects cognitive processing in the brain relevant to the task, and the P300 EEG signature can be used to signal a computer to take some action—move a virtual joystick to navigate a maze or video game, pick out a letter to spell or type using only brainwaves, or operate a robotic arm.[44]

Other types of EEG signals that are used for BCI include brainwaves that change when a movement is executed or imagined, such as the sensory motor rhythm (SMR)[45] and mu or beta rhythms.[46] As was discussed in the opening chapters of this book, this method of detecting a person's intention exploits the observation that was made the first time Richard Caton and Adolf Beck discovered brainwaves in animals a century ago and noticed that brainwave activity changed and frequently was suppressed upon moving a limb or receiving stimulation through the senses.

Other methods of noninvasively monitoring neural activity have also been used, including fMRI[47] and MEG (magnetoencephalography).[48] MEG is like EEG, except that it detects the magnetic component of fluctuating electric fields through the scalp. Because all electromagnetic fields have both an electrical and a magnetic component, brainwaves also have a magnetic vector. Near-infrared spectroscopy (NIRS), which detects changes in blood oxygenation accompanying neural activity, can also be used for BCI using an array of sensors on the head.[49]

Better detection of neural activity is possible with multielectrodes implanted in the brain[50] or placed on the surface of the brain (rather than the scalp) via electrocorticography (ECoG),[51] but such invasive procedures are only appropriate for individuals with serious conditions, such as paralysis, or those undergoing brain surgery for epilepsy, for example.

EEG-driven BCI has wide application to manipulating the world in many ways through thought alone. Gert Pfurtscheller, for example, at the Graz University of Technology in Austria, uses brain-computer interface technology to enable people to navigate through a virtual reality environment by picking up their mental activity through electrodes on their scalp. Panagiotis Artemiadis, director of the Human-Oriented Robotics and Control Lab at Arizona State University, is conducting BCI research funded by the US Air Force and DARPA. Artemiadis's results using BCI to control drones by a person's thoughts are a dramatic example of what can be done with imagination that a pilot could not possibly do with a joystick—control swarms of drones to attack an enemy. The computer detects electrical brain activity through 128 electrodes in a skullcap while the person watches multiple drones flying and thinks about making them move together. In response, the computer sends commands to the drones to control their flight. The cognitive loop between brain and drones fuses thought into action. This capability came as a surprise to Artemiadis. "We have hands and limbs . . . we don't [normally] control swarms. I was surprised the brain cares about that and that it can adapt [to control swarming activity of multiple drones]," Artemiadis says.[52] But it can. Imagination is boundless, and brainwaves flowing into computers can turn imagination into reality.

## BRAIN-TO-BRAIN TRANSMISSION

There is a long history of mind control by radio-controlled signals activating brain implants, but the ability to record and interpret brainwaves together with noninvasive methods to alter brainwave activity in a recipient's brain has brought us to the reality of achieving brain-to-brain transmission of information.

During my visit to her lab at the University of Washington, Seattle, in 2015, neuropsychologist Chantel Prat tells me, "We've found that you can use brain signals from one person to communicate with another person. We can encode information into a human brain." Prat collaborates with her husband, Andrea Stocco, a psychologist at the University of Washington Institute for Learning and Brain Sciences. Using an EEG cap to record brainwaves of a person playing a "point-and-shoot" video game, Stocco and colleagues transmitted signals from one player's brain over the internet to energize a transcranial magnetic stimulating coil positioned over the head of a person in another room across campus watching the same video game. The magnetic stimulation coil was directly positioned over the part of the brain controlling hand movements in the recipient's brain. When energized, the electromagnetic pulse alters ongoing electrical activity of neurons focused in its beam, causing the person's hand to respond. Stocco, who participated in the experiment himself, described the sensation he felt when his distant partner thought about pushing a button to shoot during the video game. "The first time I didn't even realize my hand had moved. I was just waiting for something to happen," he said. His partner's intentions, encoded in brainwaves, caused Stocco to pull the trigger "telepathically."[53] Using the same technique to activate the visual cortex, Stocco and colleagues have since used brain-to-brain transmission to allow two people to cooperate over distance in visual recognition of objects displayed on a computer screen. The transcranial stimulation induced by the "sender's"

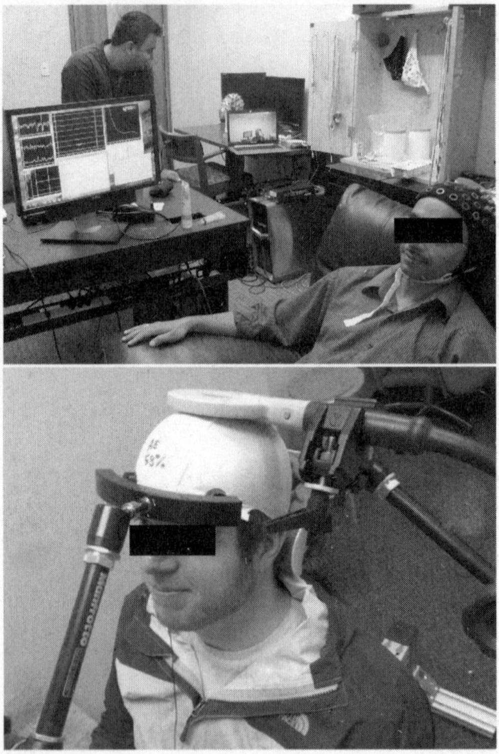

**FIGURE 31:** *Brain-to-brain communication using brainwaves. EEG signals are being recorded from the subject in the top photo, who is watching a computer game located just outside the frame of this image. The brainwave responses are used to command another person in a remote location to fire a cannon in the video game. The motor cortex of the recipient is activated by transcranial magnetic stimulation in response to the other player's brainwaves, communicating over the internet. The recipient is visible by Skype on the laptop and shown below, where the TMS coil is on his head.*

EEG response to seeing a certain object causes the recipient to perceive a flash of light. "If I stimulate your visual cortex and you are seeing [because of it], you are seeing with your brain, not with your eyes," Prat tells me.[54]

Stocco's work with brain-to-brain transmission of information was a natural outgrowth of the success of brain-to-computer interface. Brain-to-brain communication uses a computer to stimulate the recipient's brain in response to detecting the intention of another animal's brain through their neural activity. Intracortical electrical stimulation in monkeys, guided by a BCI device detecting neural activity in the motor cortex of another monkey, enabled two monkeys to operate together in a virtual reality task. Two monkeys, with electrodes implanted in their sensory cortex and motor cortex, were linked through brain-to-brain interface. In the task, the monkeys explored objects using a virtual image of an arm as they searched through a set of virtual objects and selected one with a particular texture to obtain a reward. The animals' brains were linked and cooperating in this task, demonstrating direct bidirectional communication between two primates' brains to operate an external tool.[55]

Because brains of many different kinds of animals operate using a similar neural mechanism, it should be possible to transmit information from brain to brain, not only between people or between animals of the same species, but also between the brains of different species. In 2013, researchers in Korea achieved brain-to-brain interface between humans and animals to make a rat twitch its tail when the human formed the mental intention for the rat to do so.[56] Moreover, the system used entirely noninvasive methods to detect the person's thoughts and to impose them on the rat's brain. A microelectrode could have been implanted into the rat's brain for this purpose, but if brain-to-brain transmission between humans is to become practical, people will not want stimulating electrodes implanted in their brain.

In these experiments, brainwaves from a person were recorded by EEG electrodes and analyzed by a computer to sense when the person wished to make the rat flick its tail. Then the computer sent a command to a device to stimulate neurons in the motor cortex of the rat controlling tail movement. Transcranial magnetic stimulation (TMS) and transcranial direct current stimulation (tDCS) can be used to modulate neuronal firing in a person's brain by a noninvasive method (as in Stocco's experiments), but these techniques influence relatively large areas of the brain, limiting the type of information and level of detail that can be "encoded" into the recipient's brain. Ultrasonic waves, however, can be tightly focused. Indeed, focused ultrasound is sometimes used as a precision surgical technique to ablate neurons in specific spots in the brain for neurosurgery. When less intense ultrasound energy is focused and beamed into the brain through the skull, neurons at the focal point of the ultrasonic beam fire impulses. The precise mechanism by which ultrasound causes neurons to fire is an area of active research, but the evidence suggests that minute motion of the neuronal membrane triggers stretch-activated ion channels in the membrane to generate an action potential, but local heating may also excite the neurons. Regardless, the method is believed to be harmless and highly specific, and it can be delivered through a simple ultrasonic transducer placed against the scalp.

The EEG signal that was recorded in these experiments was a brainwave response triggered by a visual stimulus—a visually evoked potential, or VEP, much like that used by Vidal in 1977. Interestingly, VEPs are generated especially well when a visual stimulus is flashed repeatedly at a rapid rate. In these experiments to unite the brains of human and rat, the image of a circle was flashed before the person's eyes on a computer screen at a rate of 15 Hz, flickering rapidly. Brainwaves are generated in response at the same 15 Hz frequency, and the amplitude

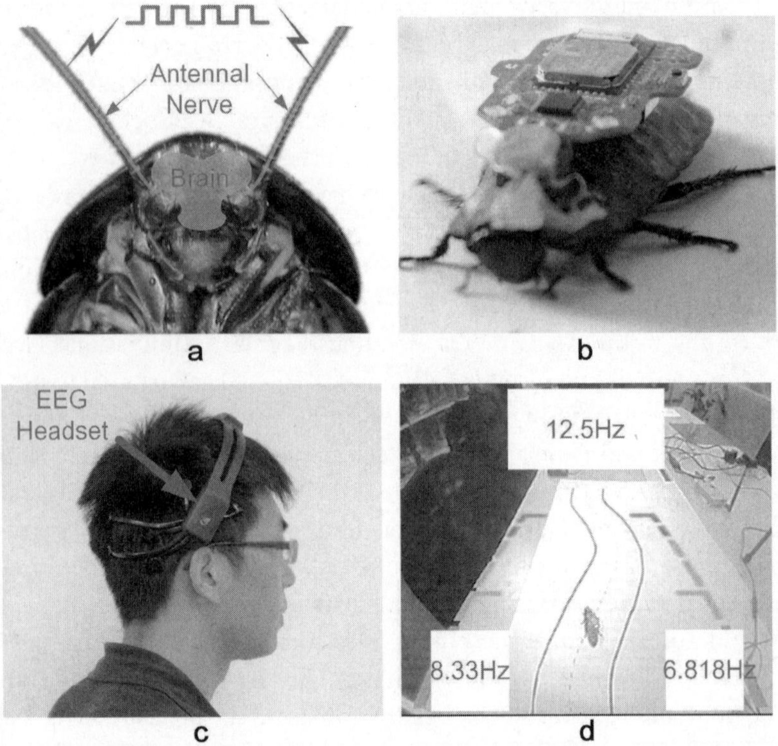


**FIGURE 32**: *Communication of human brain to cockroach via brainwaves. The intentions of a person are determined by EEG analysis and a signal is wirelessly transmitted to stimulate electrodes in the antennae of a cockroach that enable the person to steer the path of the insect.*

of these waves is very constant. This constancy makes them highly sensitive to perturbations—for example, by the person forming the intention to make the rat's tail twitch. When this occurs, the amplitude of these brainwaves peaks suddenly, and this triggers the computer to generate the ultrasonic pulse to stimulate neurons in the rat motor cortex to make the animal's tail flick. This human-to-rat brain communication works with 94 percent accuracy. A video of these experiments can be seen by accessing this paper published online.[57] This same system could be used for brain-to-brain communication between humans (or other animals, for that matter), and such signals could be transmitted over long distances over the internet or by radio transmission—rudimentary computer-assisted mental telepathy via brainwaves. Similar methods have been used to turn a cockroach into a cyborg that can be guided by EEG recordings from a human brain and wirelessly transmitting signals to stimulate the insect's antennae.[58]

## FACEBOOK'S AND TESLA'S PLANS OR PIPE DREAMS OF BRAIN-TO-BRAIN COMMUNICATION

On my visit to Princeton University to learn about Timothy Buschman's research on oscillations and how information is encoded in the brain, I ask him what he thinks about the feasibility of Mark Zuckerberg of Facebook and Elon Musk of Tesla developing brain-to-brain interface devices to communicate over the internet.

"Reading signals out of the brain, I think you can probably do. I think they are going to run into problems because the intersubject reliability is high," but still, Buschman says, he thinks detecting signals in the brain will be possible to a certain extent. But reading signals is not the same thing as full-level communication. A ripple in your ongoing brainwave activity can be used to trip a computer to activate a device, but that is a far cry from decoding thoughts. Buschman says that reading brain activity is pretty easy: "Even just reading your pupil diameter I can tell you what your cognitive load is, how many things you are holding in your mind right now. I can tell you how aroused you are. How much you are paying attention so there is a lot of potential, but if they want to put stuff back into the brain, that's where I think it gets really hard."

If BCI or brain-to-brain communication is going to be limited to EEG event detection, it will have very limited practical value, except for special situations, such as overcoming nervous system disease or damage, or responding very rapidly by detecting preconscious disturbances in EEG. A 280-character bandwidth on Twitter is enormous by comparison to such EEG event detection, which is used as a trigger. To get higher fidelity in detecting a person's thoughts, it is necessary to sample electrical activity in tens or thousands of individual neurons—simultaneously. This capability is being approached in animal experiments, but implanting multielectrodes into the healthy brains of humans is not possible for ethical and practical reasons.

Turning to the more difficult problem of delivering information, there are two overwhelming obstacles that prevent encoding information into the brain in a meaningful way. The first is that neuroscientists do not yet understand how the brain encodes and processes information anywhere near the level that would be required to download information into a person's brain from a computer interface, as, say, portrayed in the popular dystopian sci-fi television program *Black Mirror.* "I will know which brain region to target, but there is no way I will know which neuron," Buschman says. "Even if I could target the same neuron in every individual, what that neuron does will be so different in different individuals' brains."

The technical hurdles to putting information directly back into neural circuits are daunting. "How would you stimulate 300, 1,000, or 10,000 neurons at the same time?" Buschman asks. Considering this challenge mathematically, Buschman notes that if you simplify neuronal responses to two states—firing or not firing, which is an enormous oversimplification—there are two raised to the 300 states possible in a 300-neuron network. "Which is more than the number of atoms in the universe," he calculates. "It is an impossible number of states."

Nevertheless, Buschman is actively pursuing research in an effort to implant ideas into the brains of lab animals, and what he is finding suggests that it may indeed be possible, not because of our sophisticated electronics, but because of the brain's sophisticated capabilities.

"We are trying to get our electrodes to stimulate arbitrary patterns of neuronal activation [in the visual cortex of animals]. We will stick our electrodes into V1 [region of visual cortex] and we will define the target: We want this neuron to be high [firing] and this neuron to be low [silent], recording thirty-two neurons at the same time. We've been able to get the system to do that quickly."

Such stimulation produces a visual stimulus that the animal will perceive. That perception can be used to train the animal to perform a specific behavioral response. "The next step, which is what we are working on now, is that the system [of multielectrode neuronal stimulation] will be able to reproduce behaviorally relevant patterns of activity rather than just some arbitrary pattern," Buschman says. In other words, by sampling electrical activity in a network of neurons that is activated when a mouse or a monkey sees a blue square, Buschman plans to deliver that same stimulus back into the network, which should generate the image of a blue square in the animal's mind. To know that this has happened, the animal could be trained to use the blue square as a cue, perhaps to turn right in a maze.

I ask how far he thinks the field is from being able to achieve the capability of encoding meaningful information in the brain. "I would have said really far away about a year ago, but we've been making surprisingly good progress," he replies. "What we've been able to do so far is to show that our stimulation can produce patterns that are similar to what visual stimulation will do. In the end I think that that is not so amazing, because when you think about it, the system has so much dynamics built into it. Basically, what I think is happening is that we are probably putting in some kind of weird signal that the brain is then making sense of. What we are doing is just figuring out which one of these weird signals makes the most sense [to the brain] and enables it to learn from that. I think the brain really helps out a lot."

These are ambitious experiments, but Buschman and others are eagerly performing them. The results of his experiments to encode visual images into the brains of experimental animals are not yet in, but "I think this is going to work," he says with a smile.

# CHAPTER 9

# Brainwaves Reveal Your Thoughts, Strengths, and Weaknesses

Hans Berger's dream was to be able to discern specific mental illnesses, different moods, and various cognitive states through brainwave analysis. The technology and neuroscientific understanding at the time allowed for only limited success, but today, EEG recording is a standard diagnostic tool for neurological diagnosis. The diagnostic value of brainwaves goes well beyond one of the original uses of detecting the focal point of seizure in the brain to diagnose epilepsy. Many mental illnesses—depression to schizophrenia—can be assessed by abnormal brainwaves.

These extraordinary achievements will be taken up in chapter ten, but first I must share with you an even more spectacular achievement taking place right now. These new capabilities can reveal how your individual brain is wired. They can not only discern your current mental state, they can accurately predict your aptitude for learning and your future intentions, and divine your risk for specific neurological and psychiatric problems—not only diagnose a disorder you may already have. Using the latest technology, pioneering scientists are uncovering how the electric brain's neural circuits operate as a system to produce the mind. That is, how the *brain*, a bodily organ—essentially a biological machine—creates *mind*: our thoughts, perceptions, emotions, imagination, memory, personality, the ability to learn from experience, the capability of thinking, all that makes us who we are as an individual unique from every other. The interface between mind and brain has been an eternal mystery, but by monitoring functional activity in the brain by using EEG analysis of brainwaves and fMRI, scientists can know what type of brain and cognitive abilities you have. What's more, they can read your thoughts.

## BRAINWAVES FORETELL YOUR LEARNING POTENTIAL

Imagine looking out on a classroom of fresh faces and being able to spot accurately which students will learn a new language rapidly and which ones will struggle. A researcher at the University of Washington in Seattle, Dr. Chantel Prat, claims to be able to do just that by simply recording the brainwaves of a person as they sit quietly at rest. Fascinated, I contacted Dr. Prat to ask if I could visit her lab and challenged her to tap into my brainwaves and divine my ability to learn a second language. She eagerly agreed.

"I think language is the most complex feat the human brain is capable of performing," Prat tells me as we discuss her research in her office prior to her sticking electrodes on my head. By studying the brain's electrical energy intercepted by these electrodes, Prat hopes to discover how we learn to speak and how learning language changes our brain, and, most remarkably, to determine in an instant if a person has the type of innate brain wiring that will enable them to learn a second language easily or if it will be a struggle.

We all learn our spoken language as children, simply by experiencing it. "We know that *in utero*, fetuses are already starting to learn about the properties of their language," Prat says. "Because we are constantly hearing our native language and practicing it in our thoughts, internal dialogues, and in conversation every day, we fail to appreciate how complex human language really is. But when we learn a second language we usually learn in a classroom, and it becomes immediately obvious how complicated language is, and how difficult it is to learn."

Sifting through the brainwave data she collected on students learning a foreign language, Prat made a remarkable discovery: As a person learns a second language as an adult, Prat can separate students who were rapidly mastering the new language from those who are slower to learn, simply by telltale changes in their brainwaves. No need for a pop quiz—Prat found that after foreign language training, the power of beta brainwaves increased in a particular region of the right hemisphere of the cerebral cortex. "The bigger the change, the better they learned," she says.

Prat's brainwave research revealed something even more astonishing. No method that Prat applied for pretesting students was a useful indicator of a student's potential to learn a second language, but, she says, "I definitely believe that you can have someone close their eyes for five minutes, measure a pattern of brain activity, and get a much better idea of how they are going to perform."

Your brain's activity when it is idle is what provides this remarkable insight into its learning potential, because what your brain does at rest (resting-state EEG) reveals a great deal about how your brain is wired and how it operates as a system. Prat found that the higher the power of beta waves in the right temporal and parietal regions of a person's brain was, recorded as they sat still doing nothing but letting their mind wander, the faster the student would be able to learn a second language.

Prat and her graduate student Briana Yamasaki apply electrodes to my head, moistening each one with a salt solution to improve conduction of the tiny electrical signals from my brain to the electrodes. As she tests each electrode, the positions of each one on my skull are displayed on a computer monitor, changing color from red to green when the signal strength is strong. Once they are all green, she says, "Close your eyes. It'll be five minutes—remain still." As she dims the lights and slips out the door, she says, "Just relax. Clear your mind."

I try, but my mind is racing. I'm recalling my visits to other laboratories where scientists were able to pick my brain with their electrodes and expose my mind at work. I am excited and intrigued, and a bit anxious, about what Prat will find. Five minutes later Prat and Yamasaki return.

"Did you get good data?" I ask.

The two researchers click the keyboard to analyze the power of different frequencies of my brainwave activity displayed on the computer screen.

"This is a little lower than average," Prat says, looking at the feeble power of my alpha waves. (That's independent confirmation that my brainwaves are lacking in the alpha band, but whether this is an innate property of my EEG or just my high level of excitement and engagement in these experiments is up for debate.) She pulls up a recording of her own brainwaves at rest showing a sharp peak in the alpha frequency band. It looks something like a spike in a stock market chart. My brain lacks this peak and instead is charged with high-frequency gamma-wave activity.

"Am I a good second language learner?" I ask.

"No."

I'm suddenly feeling exposed and a bit infuriated. I feel as if I've been beamed up to a flying saucer and aliens are dissecting my brain.

"Your slope is about 0.5 and the average is about 0.7," Prat says decisively.

It's true! I took Spanish in high school and German in college, but neither language really stuck. This is creepier than tarot cards! Recalling all the many cognitive benefits of being bilingual that I've heard about, I am feeling horribly inadequate.

"There *must* be something good about it!" I insist.

"Sure . . . plenty of things . . ."

"Tell me one!" I demand.

"You are very entrenched in your first language."

I utter a groan of disappointment.

"You have less interference in your brain. You are going to be doing really good in America. You should not move to Europe."

Her ribbing is small consolation for having my brainpower and weak second language learning aptitude exposed.

Then she adds mercifully, "The relation of beta power to reading is the opposite. You are probably an excellent reader. Our brain is optimized, and when you get better at one thing it comes at a cost to something else."

Indeed, a week after I returned to my lab a new study by researchers at the University of Cambridge was published, reporting that monolingual people are superior to bilingual people at metacognition, which is described as "thinking about thinking," and that they excel at correcting their performance when making errors.[1]

The article described metacognition as "the need to consider a complex problem and not just go with the familiar thing." *Thinking about thinking . . . and learning from failed experiments.* That's exactly what I do in my chosen career as a neuroscientist! In my free time I enjoy mountain climbing, and reading and writing—in English. The research I just described has since been published in the scientific literature.[2]

Chantel Prat recorded my brainwaves for five minutes as I sat quietly in the dark letting my mind wander, and she instantly pronounced that I would not be able to learn a second language easily, because of the low beta-wave power in a key region of my parietal cortex. I thought it was creepy. Our minds are our most personal features—our identity—and we need to believe that our capabilities and ability to reach our goals are, in large part at least, determined by our individual commitment and effort, rather than being biologically predetermined. It is disturbing to have someone say that they can discover your cognitive potential and your future success—especially when it is based on a five-minute EEG. Was Prat's ability to predict my future success or difficulty in learning a specific type of information isolated to the rather arcane task of learning to speak a foreign tongue? Or was my experience that day just the door opening to a vast arena in which brain scientists can know your mind by tapping into your electric brain? A year after Prat had scrutinized my brain with her EEG machine, I would find myself north of the Seattle area to discover the answers to my questions.

## LEARNING TO READ

On entering the massive auditorium at the Convention Center in Vancouver, British Columbia, the din of thousands of people fervently conversing in cocktail party–level volume hit me as if I had been rocked by a roaring gust of wind. The crowds were abuzz, swarming around banks of bulletin boards that spanned the entire length of the cavernous space in row upon row, the scene harshly illuminated from glaring mercury lamps beaming down from the steel girders above. The rows of bulletin boards were covered with different printed posters and throngs huddled around them, as if ogling a spectacle. Like rapt visitors in an art museum in front of a famous painting, where no one except those in the front row can see, the spectators strained to hear authorities as they were gesticulating and discoursing on their poster board displays. It was difficult to thread through the thick flood of people; most were wearing daypacks or shoulder bags, and all of them were frantically taking notes. The thicket between the rows of poster boards was sporadically punctuated by empty spots where no one was gathered. In these clearings a solitary person stood with their back to their poster, eyeing each passerby expectantly like a restaurant hostess on the sidewalk, hoping to beckon in a potential customer for dinner. I walked up to read the title of the poster as the woman standing there politely eyed me, watching to see if I found the "menu" she offered interesting enough to engage in a conversation.

The woman, who wishes to remain anonymous until her research is published in a peer-reviewed journal, is an assistant professor at an East Coast university. She and the thousands of people gathered in this auditorium are scientists who had come from around the world to share their work and learn about the latest research on brain imaging from other scientists.

Much of the research discussed at the Organization for Human Brain Mapping meeting was still in progress and would not be published in any scientific journals for at least a year or two. At a meeting of 10,000 to 30,000 neuroscientists, it is impossible to give each participant the podium to deliver a lecture on their latest research, so most of the exchange at scientific meetings takes place at "poster sessions." Here scientists display the results of their new research on a four- by eight-foot poster, and stand by it attentively to present the findings to anyone who is interested. Poster sessions are superior to lectures in many ways, because there is only scant time after a lecture for a few members of the audience to ask a question. Also, from the perspective of the speaker, poster sessions can be far more rewarding than disgorging a preplanned lecture and then walking away from the podium at the end of your spiel to let the next speaker have the floor.

At a poster session, you can engage one on one in a detailed conversation for as long as you like.

After quick introductions, she began to go through her poster. Her interest, she explained, is in the neural circuits that are required to learn to read. Reading is essential to modern life, but it is not a natural phenomenon. It must be learned, and learning to read requires years of practice. It is easy to forget how strange and difficult the process is for decoding shapes of ink on paper into meaning, emotion, and imagination. Some children have considerable difficulty learning to read, and this can be a severe impediment to their general education. What are the circuits in the brain that are engaged as we drag our eyes robotically from left to right (in English) across the page and conjure meaning and evoke virtual experiences by automatically, unconsciously, and effortlessly attaching meaning to strings of letters, words, and sentences? Are these brain circuits strengthened during the process of learning to read, and do intrinsic differences in these circuits in different individuals predispose different people to learn to read efficiently or to struggle with it?

The technique the scientist used to identify and monitor brain circuits that are engaged in reading is a type of functional brain imaging called functional near-infrared spectroscopy (fNIRS). Like functional MRI, this method measures changes in blood oxygen that are caused by neural activity in localized regions of the brain, but fNIRS uses the principle of spectroscopy, not magnetic resonance imaging. Spectroscopy is a technique that breaks down light into its composite spectrum of different colors, and because chemical reactions and other physical phenomena are often associated with changes in color, the technique can be very informative. In astronomy, for example, spectroscopic analysis of distant starlight can enable scientists to discern what elements are present on a remote celestial object, light-years away from Earth and beyond man's ability to explore directly. Similarly, fNIRS uses light to determine a subtle change in the color of blood in tissue at specific spots inside the brain. Blue light, the highest frequency of visible light, does not penetrate tissue well, but longer-wavelength deep red light does. If you have ever held your palm over the beam of a flashlight in darkness and seen your hand turn translucent red to reveal the bones and knuckles within, you have seen the penetrating power of red light through your

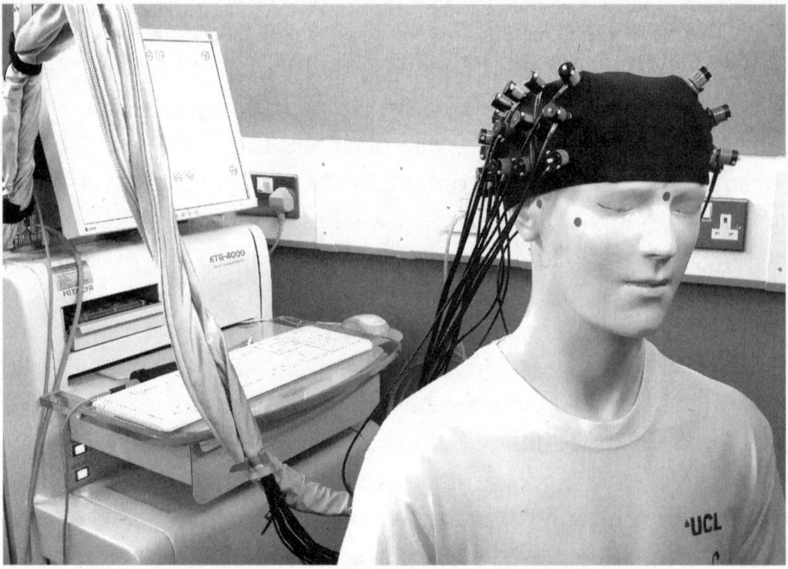


**FIGURE 33:** *Functional near-infrared spectroscopy (fNIRS) can be used to monitor brain activity by shining infrared laser light into the brain through the scalp. Actively firing neurons deplete oxygen from the bloodstream in the vicinity, changing the color of the blood from bright to dull red. The infrared light reflected back out of the brain is detected by sensors in the fNIRS cap to monitor neural activity in a manner complementing electrical measurements by EEG and functional brain imaging (which is much slower) by fMRI.*

own body. The deepest red, infrared, cannot be seen by the human eye, but it penetrates tissue even more deeply than visible red light. Infrared wavelengths, like those used in television remote controls, can only be detected by electronic sensors.

The beauty of this technique is that instead of sliding a person into the imposing maw of the massive doughnut-shaped magnet in a room-size MRI machine, fNIRS uses only a simple headband to record brain activity, or, in the case of this experiment, a device is used that is similar to a swimming cap with multiple sensors to cover a broader area of the brain. The headband is studded with diode lasers that shine invisible infrared beams through the scalp, penetrating the skull and entering brain tissue. The light is reflected out of the brain at a specific color and is detected by sensors on the headband. Oxygenated blood is bright red, but deoxygenated blood is dark red, almost purple, as we can see in the shallow veins of our legs and arms where blood depleted of oxygen is returning to the lungs to be refreshed. The sensor measures the shift in blood color intensity caused by depletion of oxygen in tissue. In the brain, actively firing neurons draw oxygen out of the blood in the local vicinity to fuel their high metabolism. By using infrared light to probe the brain, centers of neural activity that are engaged in any given cognitive task are revealed as spots in brain tissue where oxygen is depleted from the blood's hemoglobin. These changes can be tracked over time, much like an EEG, to reveal fluctuations and responses in brain activity at discrete points inside the brain. Also, by correlating patterns of activity in pairs of brain regions, active circuits can be tracked in the brain. In contrast to an EEG or an fMRI, both of which require that the subject remain very still to eliminate movement artifacts, fNIRS is much easier to use because it is less susceptible to interference by muscle contraction and bodily movements. Although brain development is one of the most interesting areas of neuroscience to explore, comparatively little functional brain imaging is done on children because it is difficult to get a three-year-old to cooperate in sticking his or her head into a scary MRI machine and for them to stay still long enough to image activity in their brain during an experiment, which typically lasts about thirty minutes.

The researcher explained that fNIRS is great for kids and she uses it to monitor neural activity inside the brains of young children between the ages of three-and-a-half to four-and-a half years old as they listen to words—both real and nonsense—being read aloud to them. The brightness of the fNIRS signal in each region of the brain indicates how active neurons are in each spot. Comparing the brain responses detected in both conditions, she subtracts out networks of neural circuits that were activated when she read the children nonsense words. The result reveals the networks that are active in the brain when the child is cognitively processing and understanding real words. She reasoned that making sense of words is fundamental to both speech and reading, so this approach might show the specific neural connections in a young child's brain that are necessary for learning the meaning of words and being able to read. After all, a child cannot learn to read and comprehend the word "cat" without already knowing the word we use for that animal in English. Her analysis showed that the functional connection between the left inferior frontal gyrus to the right superior temporal gyrus was increased in the children's brains in response to hearing real words.

To explain this anatomy briefly, the human cerebral cortex is deeply wrinkled like a dried prune. The ridge of each wrinkle is called a gyrus, and the valley between wrinkles is called a sulcus. Of course, anatomists named these mounds and valleys in the same way geographers name topographic features on Earth to identify location and communicate those locales with others. The inferior frontal gyrus is simply Latin for "below and toward the front" and refers to a prominent convolution in the frontal lobes of the brain known to be involved in language and other functions. The famous region critical for language processing and speech, Broca's area, is situated along this ridge, but it is on the left side of your brain if you are right-handed and the right side if you are left-handed. The French physician Pierre Paul Broca (1824–1888) observed that his patients with damage to this spot suffered speech disorders. The superior temporal gyrus is another ridgeline of the cerebral cortex near the top of the temporal lobe of the brain. The temporal lobes, like fleshy earmuffs on the sides of the brain, contain the auditory cortex, which is responsible for processing sounds. This region is also important for processing speech, especially at one spot called Wernicke's area. German physician Carl Wernicke (1848–1905) is credited with discovering that in addition to damage to Broca's area, damage to this brain region also resulted in difficulties in language comprehension. So it is not surprising that there is increased neural activity detected by fNIRS in these children's brains at these spots as they listened to real words and not nonsense words. But what the research discovered next is truly astonishing.

The researcher then followed these same children on their long path of learning to read. She found after the children entered school and began to learn to read, their reading test scores matched up with how well connectivity in the brain circuits she had monitored in the children's brains *a year earlier* had responded to hearing meaningful words.

Consider what this means. Only ten minutes of sampling brain activity in these young children was all that was needed to predict approximately how well that child would be able to accomplish the vital skill of reading in the future.

Attracted like seagulls to bait, a flock of eager scientists had gathered, listening intently to her research findings, as she shared her experiments that are still in progress.

I asked her what she thought about applying her method in schools to predict which students will excel at reading and which will struggle, and if that could be used to place schoolchildren into different educational tracks.

"That's a slippery slope," she said, betraying her alarm. Apparently, she had not conceived that her work to understand how children's brains learn to read might be used in this way. "We don't know the brain well enough yet," she warned defensively.

I pushed her by drawing an analogy to biomarkers, genetic analysis, and other medical diagnostic tests, which are routinely used in physical medicine. For example, we accept that high cholesterol in one's blood puts one at increased risk of heart attack and stroke, and we can then act on that biological information to change behavior: improving one's diet and exercise, or taking drugs to correct the abnormality.

She was still uncomfortable with the idea. The prospect of using brain analysis to predict a child's learning potential pits biological predetermination against self-reliance. Acting on this biological information could restrict individual freedom and replace it with authoritarian planning of a child's fate. But she agreed that her brain analysis to predict reading ability would be similar to a cholesterol test to predict heart attack, only applied to the brain instead of another bodily organ. Still, she balked.

Then I asked, "If you had a child and he or she was low on this scale of brain connectivity at age three, what would you do?"

Framing my question in a personal way was unfair, perhaps, and it clearly unsettled her for a moment. The scenario was troubling, and it caused her to pause and deliberate. She appeared conflicted, but, after contemplating for a minute, she said earnestly, "I would give my child every neuroimaging method available. Then if there were indications [of potential reading problems], I would increase my reading with her to help build this pathway."

Like it or not, the genie is out of the bottle, and we had better get to know her.

## READING YOUR MIND

Imagine peeking through the skull and seeing what makes one brain smarter than another, or identifying the brains of schizophrenics, dyslexics, autistic children, or people suffering from Alzheimer's disease—even sorting good readers from poor, as we have just seen, and natural musicians from the tone deaf. What if we could watch the brain changing as a new skill is mastered, or—the ultimate fantasy—pick a person's brain and read their thoughts?

At a closed meeting of neuroscientists held inside one of the United States defense department offices, I first heard psychologist Marcel Just report his astonishing success in reading a person's mind. Just and his colleagues at Carnegie Mellon University reported that they could reliably predict, for example, whether a person is thinking of a chair or a door, what number from one to seven they have chosen, even what emotion they are feeling, all by simply looking at the person's brain's activity via fMRI.

Mind reading had not been the goal of the scientists' research; rather, they were simply looking to test the conclusions of their investigations into brain function. Just and his colleagues had amassed a large body of data on how specific types of information flow through the brain's networks when people in an fMRI scanner are presented with various images, sounds, and words, or when they are moved to feel certain emotions or to think specific thoughts. Using this approach, Just and many other researchers around the world have traced functional "activity maps" of the human brain. The most rigorous way to validate their findings about the specific neural circuitry involved in specific cognitive operations is to attempt to reverse engineer the process. Having observed that a specific thought gives rise to a specific pattern of brain activity, it should be possible to analyze the patterns of brain activity flowing through neural networks of a subject inside an fMRI machine, and then conclude what that person's brain was "thinking."

In practice, a subject in an fMRI scanner would first be shown a blank screen and then, say, a picture of a house. The researchers would repeat this sequence over and over, and then use the combined data from all the fMRI images from the person to form two composite images of the brain's average activity in each of the two states. As in visual puzzles where two scenes are shown and the task is to find what has been subtly changed in one photo, the computer compares the two brain scan images. "Subtracting" the two images reveals what has changed in the brain while performing the task—in this case, seeing a house. The task of viewing a house would naturally boost neural activity in the visual cortex (in addition to other brain regions), but in addition, by analyzing the synchrony of activity in different brain regions with fMRI, researchers can detect populations of neurons that are functionally interconnected during a given task—viewing an image of a house—just as EEG researchers analyze the synchrony of EEG signals from different scalp electrodes to map out networks of interconnections in the brain.

In studies attempting "mind reading," after showing the person a house, researchers would then show the subject a different object—for example, a hammer—and the activity maps in the brain provoked by perceiving the two different objects would be compared. While seeing either object excites the visual cortex, the cognitive representations of a house and a hammer are quite different. Our home is a refuge, where we enjoy the company of friends and family, eat and sleep, and so on. A hammer is a tool. It is heavy and must be gripped in our clenched hand, and it is used to pound nails and other objects. Thus, perceiving these two objects should activate a somewhat different constellation of activity across the brain. And it does.

In a series of papers, Marcel Just and his colleagues have shown that they can reliably identify what object a person is seeing or thinking about, what number they are thinking of, and what emotion they may be feeling through exactly the process described above, because different clusters of spots throughout the brain increase activity with each of these concepts or emotions. Remarkably, these spots in the brain are consistent from person to person.

Their latest research goes further still, showing that the phenomenon also applies to learning new abstract concepts: for example, scientific concepts. Working with a classroom of students learning about physics, the researchers were able to identify which of thirty physics concepts the student was learning and thinking about simply by analyzing fMRI images of their brain activity throughout the course.[3] In a particularly fascinating development, the data shows that different abstract scientific concepts map onto brain regions that control what might be considered related or analogous functions. Learning and thinking about the scientific concept of the periodic motion of wave propagation, for example, engages the same regions of the brain that are activated in dancing. It seems that our abstract scientific concepts are anchored in discrete bodily functions controlled by specific circuits in the brain. Perhaps this is the neuroscience of metaphor.

Marcel Just has suggested that this insight into brain function could be exploited to transform teaching methods. Albert Einstein, after all, had his epiphany that time, space, matter, and energy were interchangeable by imagining himself riding on a beam of light. This imagination probably activated his sensory motor cortex and striatum, regions of the brain that control bodily motion. In Just's research, all physical concepts of radiation—light, heat, sound—activated similar brain regions. These subjects are not typically taught together in physics courses, but perhaps they should be. Why not present subject matter in terms of how the brain organizes concepts naturally, rather than according to an artificial outline that collects material in a way we have deemed "logical"? College physics, typically considered a difficult course of study, might not be so difficult if the material were taught according to what our brain was built to do. Clearly, the human brain evolved to control the body and its interactions with the environment, not to solve differential equations. Humans have taken our brain's structural and functional framework and adapted it to create, compute, and analyze abstract concepts, but none of those things are in our brain's organization or function intrinsically.

Fascinated by the insights generated by his research, I asked Just if I could visit his laboratory to learn more about his latest work, and he kindly agreed.

## LANGUAGE OF THE MIND

As I drive west to Pittsburgh, Pennsylvania, through a cold morning rain, frothy fog whips up from green hollows and rises from the glistening wet pavement threading its asphalt track through the Allegheny Mountains. Patches of brilliant autumn color tear through the shroud cloaking the hillside along Interstate Highway 76, exposing large swaths of saffron and mustard, cinnamon and rust, speckled with splotches of moss green where a few trees are reluctant to lose their summer color. Towering oaks stipple the summit ridgeline, standing bare and soaked, and the air smells of wet autumn leaves. It is a quiet Sunday morning as I drive to Carnegie Mellon University, where psychologist Marcel Just is engaged in a quest to use the latest brain imaging technology to understand how the couple of pounds of flesh inside our skull can think.

The rural countryside is a picturesque landscape of red milk barns with gleaming steel roofs anchored to towering silos and framed by white fences like those you see in children's farmhouse toys. Preachers and country music stations dominate the FM radio dial until, nearing Pittsburgh, the highway widens and the flickering sounds of honkytonk piano and Conway Twitty are crowded out by Fleetwood Mac and Queen. From time to time, strange broadcasts blast through the familiar—loud, frenzied polkas and a Slovakian announcer who jabbers enthusiastically before setting a reedy accordion frenetically pumping. Winding now through the city streets, I find Carnegie Mellon situated in the hilly Shadyside district, a neighborhood of stately avenues lined with Victorian mansions dating from Pittsburgh's days as an industrial giant. In 1910 Pittsburgh was the eighth largest city in the United States, producing nearly half the nation's steel, but with the steel mills long closed, Pittsburgh rusted into a comparative ghost town, now a distant sixty-third on the list of the country's largest cities. But the financial legacy of Andrew Carnegie and other wealthy philanthropists of the industrial era lives on in Pittsburgh's outstanding academic institutions. Today, Carnegie Mellon and the nearby University of Pittsburgh are universally recognized for their science and engineering research, including flourishing programs in robotics and neuroscience.

The next morning, I meet Marcel Just and his colleagues Tim Keller, Robert Varsa, Vladimir Cherkassky, and Rob Mason at their laboratory in Baker Hall, a spacious three-story yellow brick building with large arched windows built in 1914 and now designated as a historical landmark. Our lively discussion begins in the early morning, extends through lunch, and continues into the afternoon as they share their research in progress and look into the future to contemplate where the field may be heading. Just's team has been attempting to use fMRI to decipher what a person is reading. I remember at a scientific meeting I'd attended two months earlier, Just concluded, after describing his most recent work, that "we are getting at the beginnings of a universal model of the language of the mind." In research he published in the journal *Science* in 2008, his group monitored the brain activity of volunteers as they read a list of sixty different words. After training their computers to recognize the patterns of fMRI activity evoked by each word, the scientists were then able to identify which word a person had read when they were tested by fMRI later. Not only is this an impressive accomplishment on its own, this line of research is beginning to uncover how the brain organizes semantic information. Semantics refers to how meaning gets conveyed to our mind through language. It is what you are doing now—painting pictures in your mind and comprehending information from looking at abstract symbols, be they squiggled codes (written words in the proper sequence) or, in spoken language, stereotyped mouth sounds (in the proper sequence).

Interestingly, in analyzing how the brain responds to reading specific words, they found that the activity patterns evoked by different words were systematically related according to the similarity of the concepts; for example, "houses" and "closets" are related conceptually as being enclosures, and reading these two words evokes rather similar patterns of activity in the brain. Words for different foods are also similarly represented in the brain's activity patterns. This is not surprising, perhaps, but the word "spoon," too, was found grouped with the "food" words. It activates similar spots in the brain because all these words are related to the concept of eating.

This line of research has advanced greatly, and Just has found that, in fact, these groupings are a consistent form of organization. "The neural representations of these concepts are common across different people," he tells me.

Others are doing similar research. The group headed by Jack Gallant at UC Berkeley, as described in a 2016 paper published in the journal *Nature*, imaged the brains of people listening to the public radio program *The Moth Radio Hour.* Using fMRI, they were able to make a map of the human cerebral cortex in which the meaning of specific words in the English language were found. The locations were consistent from one person to another—the brain's system of organizing the semantics of human speech.[4]

So as I read the word "hotdog" now, typing it, my brain activity pattern is identical to the activity pattern that was evoked in your brain as you read it. Hotdog! There, I did it again! With this mastery of single-word recognition, would it be possible for scientists studying your brain while you read this book inside an fMRI machine to decode the sentences you are reading?

The problem that arises, of course, is that the act of reading and understanding a complete sentence is far more complicated than simply identifying a specific word. Another daunting challenge is that fMRI is very slow. How could this method possibly keep pace with the average rate of 200 words a minute at which we read? If the brain read text by simply decoding a vast collection of words, the way a computer might, the mind reading of what a person is reading would indeed be impossible. But research suggests that the brain does *not* work by translating word by word. The human brain deciphers the words and stores the information it collects as integrated concepts.

"If you study the eye movements in reading, as I once did, you see a pause at the end of the sentence of about 200 milliseconds, so I think if you image what occurs at the end of a sentence, all of the concepts are sitting there," Just tells me. "At the end of a sentence you gather together all the appropriate concepts [in that sentence] into a structure." That complete thought, lingering after a sentence is read and comprehended, should be sustained by the complex pattern of brain activity that sentence conveys, and therefore be available for detection by fMRI.

To test this theory, Just and his colleagues took a list of 240 complete sentences—for example, "The old man threw the stone into the lake"—and had people read them while inside an fMRI machine. Then they tried to determine whether they could identify which of the sentences a person was reading by monitoring their pattern of brain activity. They found that they could—with a high degree of accuracy. From this remarkable achievement they moved on to see if they might be able to decode *new* sentences on the basis of what they had learned from their fMRI data on how the brain had decoded the 240 sentences on their list. To accomplish this, they averaged all the brain imaging data from reading sentences that contained a particular word—"man," for example. Doing so would cancel out all the activity from the other words in the sentences, with the result being the imprint left in the maps of brain activity by the concept "man." Recognizing that the word "man" is more than a noun—it has a discrete thematic meaning in any given sentence, linked to a complete thought—the researchers also mapped brain activity according to the specific role played by these various words, for instance, whether "man" was the subject or the object of the action.

Using this information, Just and his colleagues were able to use fMRI analysis of brain activity maps to decode completely new sentences being read by their subjects with an accuracy of 86 percent. Even relatively subtle nuances of meaning were able to be discerned. Reading "The engineer walked through the peaceful park," for example, was easily distinguished from the task of reading the similar sentence "The old doctor walked through the hospital." The team found that sentences containing exactly the same words, presented in different order, could be parsed by fMRI. A sentence is not just a bag of words, Marcel Just explains, and the fMRI data shows that the brain doesn't decode it that way. His method handily distinguished "The dog chased the cat" from "The cat chased the dog." "If you were doing it by identifying a bag of words, you wouldn't know which is which," he says.

Perhaps especially illuminating is the study of cases in which this fMRI analysis fails and is confused in decoding what a person is reading. "The flood damaged the hospital," for example, could not be reliably distinguished from "The storm destroyed the theater." This reveals something fundamental about how the brain works to perceive the world.

"It is not done literally," Just says. "A hospital doesn't look like a theater, but it is a similar concept—a building, destruction, and a force of nature. That's the level of coding that occurs that we can capture." As further proof of this important insight—the primacy of concepts over words—Marcel Just and colleagues found that they could use the same brain activity maps that they had developed from people reading in English to decode sentences in Portuguese or Mandarin, for example, when read by those for whom these languages are their native tongues.

"The destruction of a hospital is not an English concept," explains Just. "It is a universal human concept." By directly interrogating the brain in this way, the particular language used to communicate becomes superfluous. This process could be automated, Just says. Imagine the potential applications of this capability. In theory, it could enable a universal translator freely operating among all languages, dispensing with the scores of human translators and interpreters frantically at work at the United Nations, in hospitals, and elsewhere. Of course, while conceptually valid, this would obviously be technically problematic; however, there are other, more immediate applications of the science beginning to uncover the universal language of the mind, as Marcel Just is well aware. "You can see how for a brain-computer interface, you wouldn't want an English-only interface. You would want a universal one."

For now, of the ability to read a person's mind with fMRI, the psychologist says, "It is fun—it is a nice parlor trick." What Just ultimately wants to accomplish is the opposite: to predict how the brain's activity map would change with any given sentence read, or thought or emotion evoked. Again, the motivation behind this goal is the scientific verification it would provide that their brain/mind decoding via fMRI is correct.

"The accuracy is very similar in the two directions," says Just. With the data and formulae he's developed, he can give the computer a sentence and have it predict the pattern of neural activity that would be evoked in the brain of a subject reading that sentence in an fMRI machine, then confirm that prediction with a high degree of accuracy by having a person read the sentence and matching the evoked neural activity map with the predicted one. This reverse engineering, incidentally, is exactly what would be required to plant a thought or sensation into a person's mind, or to provoke a response. If the computer could stimulate the appropriate pattern of brain activity in someone by electrical or magnetic stimulation, information could be directly fed from a computer into the brain. Or a thought or sensation from one person's mind could be read by a computer and then transmitted "telepathically" via computer, implanted into another person's mind, as we have already seen in the research by Andrea Stocco and Chantel Prat earlier (see page 253). For now, let's turn to another question—whether there might be a way to use brain activity detection methods to go beyond reading a word in someone's mind to seeing their future actions reflected there.

## KNOWING YOUR HIDDEN INTENTIONS AND PREDICTING YOUR FUTURE

"White male, with long blond hair, white t-shirt, and blue jeans," the caller reported, describing a suspicious person to the Georgia Tech campus police. "Possibly intoxicated, holding a knife, and possibly armed with a gun on his hip."

Campus police responded quickly, and confronted a suspect matching the description, later identified as twenty-one-year-old Georgia Tech student Scout Schultz.

"Come on, man, let's drop the knife," shouts Officer Tyler Beck, his gun drawn, in the graphic video captured at the scene.[5]

"Shoot me!" the suspect yells.

The officer backs off. "Nobody wants to hurt you, man."

Schultz is now surrounded by officers. "Drop it!" one of them shouts.

Schultz pauses, and then takes three slow steps toward another officer, not in a threatening manner, but deliberating, moving hesitantly like a tightrope walker, almost zombielike, with both arms held stiffly to his sides. He takes one more step, and gunshots erupt.

Schultz was pronounced dead shortly afterward at Grady Memorial Hospital.

Police found that Schultz was not carrying a gun; the knife the caller reported him brandishing proved to be a multitool. Its blade was not extended.

"Why did you have to shoot?" pleaded Scout's grieving father at a news conference. "Why did you kill my son?"

The police had been set up. Further investigation determined that the phone call reporting the dangerous man had been made by Scout Schultz himself; three suicide notes were found in his dormitory room.

Schultz's anguished parents reported that their child had a long history of anxiety and depression and had undergone therapy after attempting suicide by hanging two years earlier. But Schultz seemed to have overcome his mental health issues. His friends all asserted that Schultz had seemed fine in recent weeks. In a statement to reporters, Scout's mother, Lynne Schultz, said, "We had no clue."[6]

This tragic case is complicated, but one thing it illustrates dramatically is how difficult suicide can be to predict and prevent. Close friends, parents, and even psychiatrists are often completely unaware of a person's suicidal intentions. We like to think we would know if a close loved one was truly struggling. But the many high-profile suicides—celebrity chef Anthony Bourdain, actor and comedian Robin Williams, fashion designer Kate Spade, and others at the top of their fields who were also supported by friends and family—make it clear that unless the person having suicidal thoughts tells someone, it is impossible to know that they are intent on killing themselves.

Or maybe not impossible, Marcel Just realized. If he could read a person's mind, then he might be able to determine if an individual was having suicidal thoughts by looking at the patterns of activity generated by their brain around such thoughts.

In a study involving seventeen people who had clinically documented suicidal ideation and seventeen controls, Just and colleagues presented the test subjects with a series of words that might be associated with self-destructive feelings as well as those that might be associated with the opposite condition, of personal satisfaction. These words included "death," "funeral," "suicide," "cruelty," "trouble," "carefree," "good," and "praise," among others. Just tells me that the brain representation for "death" in those who had suicidal ideation and in the controls is obviously different. From their functional brain scans alone, he was able to correctly identify every individual in the study as being in either the suicidal ideation or the control group, with only two exceptions. Moreover, nine of the seventeen in the suicidal ideation group had made a previous attempt at suicide, and Just could spot those people who had actually acted on their suicidal thoughts with 94 percent accuracy. All this because certain words activated a different cluster of neural responses across the brain from that found in non-suicidal people.

Even more useful than simply identifying those with suicidal ideation by their neural responses is what Just found when he sought to look at what emotions might drive these people to contemplate suicide in the first place. In previous research, his team had been able to identify what emotion a subject was feeling by the pattern of neural activity seen when people undergoing fMRI experience different emotions.[7] He applied the same method to analyzing the emotional responses of these two groups to this list of words as well. From their brains' signatures of emotions alone, represented in neural activity patterns in regions of the brain responsible for emotional responses, Just was able to identify the individuals who were currently contemplating suicide with 85 percent accuracy. Comparing the control group and group of people with suicidal thoughts, he notes, "They are different with respect to the emotions that these various concepts [the words presented] evoke. In general . . . concepts such as 'death' and 'trouble' evoke more sadness and shame but less anger in the group with suicidal ideation. So you see specifically how the thoughts of the suicidal ideators have been altered. You see that suicidal ideation can be directly assessed." This capability, according to Just, could be used to direct treatments, and to monitor how effective those treatments—whether medication, counseling, or neurofeedback—have been. This astonishing accomplishment might give psychiatrists and other behavioral health professionals something that at the moment they can only imagine as a fantasy—the ability to truly know what their patients are thinking.

## WHEN YOUR PRIVATE THOUGHTS AREN'T PRIVATE

It is not difficult to see the ethical questions that arise from these technologies. What are the implications of being able to get inside a person's head, to know not only what they are thinking but also how their thinking may compare alongside what is considered "normal"? If a person insists he has no plans to harm himself but his fMRI tells a different story, do we have him committed? What would this capability have revealed about the hidden murderous thoughts of the two teens who committed the mass murder in Columbine High School, or the depraved sniper in Las Vegas who murdered fifty-eight innocent people and injured 869 others attending an outdoor music concert before he committed suicide as the police descended upon him?

The new ethical horizons that come with widening our window into human thought will be examined further (see chapter twelve), but it is reassuring to remember that we have faced troubling ethical challenges from new technology and scientific advances in the past. For example, the advances in knowledge and technology that allowed us to monitor bodily functions with medical tests and to ascertain genetic risks have already forced us to confront similar dilemmas about our health: Do you want to know if you have high cholesterol levels, increasing your risk of stroke, or that you carry a gene that increases the risk of Alzheimer's? What's more, do you want the government or your insurance company to know? Mind reading, of course, is thornier territory still. "Our thoughts are our most private and last refuge," Just says. "There is nothing more private than a thought."

As mentioned, while the technical capabilities to rapidly and easily control thought are likely to exist in the future, as a practical matter, fMRI is currently a cumbersome, slow, and complicated method. Just notes, "One thing that lets me sleep better at night is that we need extreme cooperation from our participants. If they don't want us to know what they are thinking, it is extremely easy for them to disable our method." The participant can do this because they must concentrate on the thought, object, or concept while the fMRI is recording their brain activity. They cannot let their mind wander or think of something else. "All you would have to do is space out in our studies and we are helpless," his colleague Vladimir Cherkassky elaborates. "We need the person to think about an apple six times. So all they have to do is think about a red apple the first time, a green apple the next time, maybe a Macintosh computer is an apple the fourth time, and we are done."

The ability to monitor and decode the language of thought raises the possibility of altering thoughts by specifically altering neural circuit activity. Currently this is only possible to a very feeble extent in people, but some success has been achieved in animal studies. Activity in neural circuits can be modulated by transcranial stimulation, in which electricity is passed into the brain from electrodes on the scalp, as well as by inducing electrical activity inside the brain with transcranial magnetic stimulation (TMS) from electromagnets positioned outside the head. But to achieve the fine control of neural circuits necessary to implant or manipulate a thought, arrays of electrodes would need to be tucked inside the skull and directly attached to the surface of the brain, or impaled deep into brain tissue. Both methods are currently used in humans for controlling prosthetic devices and for providing sensory input to the brain. In animal studies, however, scientists have succeeded in using multielectrodes to deliver appropriate stimulation to generate network activity in scores or hundreds of neurons that encode a perception or cognitive process. Even with current methods, such as TMS, human thought could be manipulated, Cherkassky suggests, if not yet outright controlled. "I think you can at least get to a level of modification," he tells me. "Say, for example, thoughts of an apple. You could facilitate thoughts of deliciousness and make the person think more about the deliciousness of an apple" using TMS—although the method of stimulation lacks precision, it could activate the general region of the brain that gives us the sensation of something being delicious or foul.

In working to crack the eternal mystery of how the human brain forms a thought, it is important to remind ourselves yet again that this organ was not engineered. It does not work like a computer or indeed like any other human-made contraption for calculating or performing tasks of any kind. This is an organ that evolved to orchestrate the myriad bodily functions necessary for life, and to meet the survival need of animals to sense and interact with their environment.

"The language of mind emerged from the structure and nature of the brain," Just explains when discussion in the lab turns to the more existential questions raised by his research. Subsystems of the brain that evolved to perform very specific information processing functions—sensory, motor, and emotional responses, for example—become coordinated. "Coordination is a key thing . . . So if I think 'apple,' it is all the various things that compose the concept of an apple and the neural circuits acting together that create the thought."

Now that we can see thoughts emerging inside people's brains and begin to comprehend how the brain generates the mind, the implications are, well, thought provoking. "It suggests that the human brain constrains the way we think," Just observes. We like to think of imagination as being boundless, but abstract concepts and thoughts beyond the neural functions our brains evolved to perform may well be inconceivable to the human mind. In other words, there may be vast areas of reality that the human mind is incapable of comprehending. Just hopes that his research is telling us something new about human thought. "Getting to the essence through this incredibly complex brain system that *generates* thoughts, *understands* thoughts, and lets us be the human beings that we are."

## THE NETWORKED BRAIN AND ASSESSING YOUR BRAIN'S INFRASTRUCTURE

Similar to resting-state EEG (recording brainwaves while a person does nothing but let their mind wander), resting-state fMRI detects the spontaneous low-frequency fluctuations in functional activity in the brain while a person is not engaged in any task. This resting-state brain activity reveals the intrinsic functional organization of a person's brain. Differences in resting-state fMRI have been reported that can predict a person's age[8] and IQ,[9] for example, but many other normal and abnormal neurological and psychological features are reflected in the resting-state fMRI. This capability has come about only in the last decade from a transformation in the way neuroscientists think about how the brain operates.

### YOUR BRAIN'S ELECTRICAL GRID REVEALS WHO YOU ARE

At the time Hans Berger was undertaking his studies of EEG, brain science was engaged in what would become a century-long search to determine which parts of the brain controlled specific cognitive and biological functions. Working at the turn of the twentieth century, the anatomist Korbinian Brodmann, for example, was able to divide the human cerebral cortex into fifty-two distinct areas, based on subtle differences in cellular structure of tissue forming a patchwork that covered the entire cerebral cortex. These anatomical regions continue to be used by scientists, and decades of research has tied specific cognitive functions to distinct cortical regions. Speech, as mentioned previously, is dependent on Broca's area, spanning Brodmann's areas 44 and 45 of the left cerebral cortex. But this reductionist approach, or focusing on the parts of a whole that was the thrust of scientific research on the brain in Berger's time, is a primary reason why his studies of electrical activity emanating from the entire head attracted no interest.

Reductionism was the path taken by nearly every scientist until recently, in a search to identify the particular spots in the brain or specific cells responsible for a particular cognitive function. This approach also stemmed from the techniques in vogue. Histology (that is, determining the cellular structure of brain tissue) was the primary approach

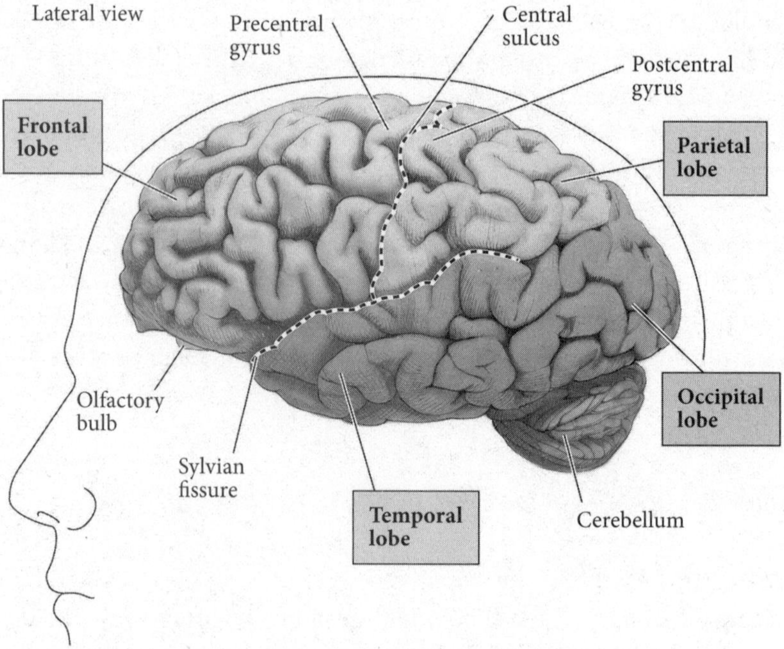


**FIGURE 34**: *The human brain showing the major regions of the cerebral cortex, the cerebellum, and olfactory bulb.*

to brain research in the late 1800s and early twentieth century because of advances in microscopy and new techniques to stain brain tissue. Notably, the Spanish neuroanatomist Santiago Ramón y Cajal, who perceived that neurons were individual cells separate from all other cells and communicating across an invisible gap of separation called the synapse, laid the bedrock on which all of cellular neuroscience is built.[10] Likewise, the Italian neuroanatomist Camillo Golgi, who developed methods to use silver chloride to stain neurons, emphasized the reductionist approach down to the level of brain cells. Interestingly, in terms of how our observations and thinking are constrained by our preconceived assumptions, Golgi rejected Cajal's "neuron doctrine" in favor of the view, shared by many at the time, that nerve cells he saw in brain tissue formed a complicated meshwork of interconnected fibers. This structure would enable neural signals to be sent broadly through a rich complex of pathways, much like the postal service delivers discrete messages over a complicated and interconnected network of roads. Throughout the nineteenth and twentieth centuries, while anatomy was undertaking reductionism, so too was physiology. The physiological pursuit to record electrical signals generated from individual neurons in specific spots in the brain was driven by the acceptance of Cajal's neuron doctrine, but it was also a consequence of new techniques afforded by electronic amplifiers and microelectrodes to sample electrical responses of individual neurons. (Note that Golgi and Cajal shared the Nobel Prize in 1906 for work on the structure of the nervous system.)

This view of brain functions being localized to specific spots in the brain has been expanded by a modern concept of neural networks, in which cognitive functions are regarded as being dependent on integrating information across large expanses of neural networks, which can span the entire brain.

Is this a more enlightened view or simply more fashionable? There is truth in both perspectives, but the transformation in thinking about brain function in terms of neural networks stems directly from new techniques, just as the reductionist view of the past stemmed from using histological and electrophysiological methods. In other words, the modern network view of the brain, popularized by President Obama's Brain Initiative in 2013 to map the human brain, is an outgrowth of modern methods, such as EEG and fMRI that enable recording functional activity in large populations of neurons. Our modern view of brain function is now more aligned with the thinking of Hans Berger, because we are now applying the same approach he used: monitoring functional activity of the brain as a whole.

This new network view of brain function has been extremely enlightening. One of the most surprising findings from neural network studies arose from recording brain-wide activity while the person did no task at all—they simply sat quietly and let their mind wander as the intrinsic activity in their brain flowed and was monitored by EEG or fMRI. By monitoring how signals are coordinated in time from one spot in the brain to other regions of activity, it is possible to map the functional connectivity among all regions in the brain.

When the brain is at rest, the EEG continues to fluctuate, as do signals detected by fMRI. Brain-wide networks traced during these resting-state recordings are associated with distinct cognitive functions. By simply monitoring activity in a person's brain at rest, researchers can determine a great deal about how that person's brain is wired. This very personal information about your brain gives researchers scientific and

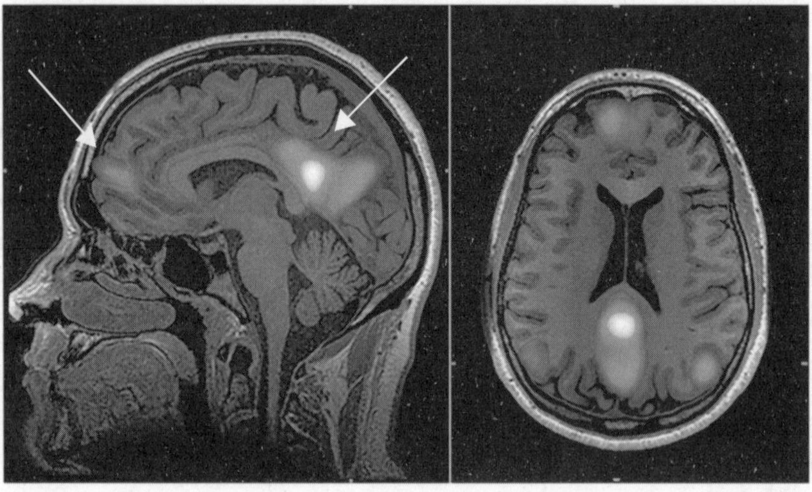

**FIGURE 35**: *The default-mode network shown by fMRI is one of the networks that are active in the resting state when a person is not engaged in a task; the mind is instead left to wander. This and other networks, notably the salience and executive control networks, provide insight into an individual's cognitive abilities and personality.*

reliable insight into your innate cognitive abilities, personality traits, learning capabilities, and psychological or neurological dysfunctions you may be experiencing, as well as to predict the future trajectory of your brain function and of your mind.

Some of the primary networks of functional activity in the brain that have emerged from resting-state fMRI and EEG analysis are described on the following pages. It will be helpful to refer to the figures on page 290 to see how these functional networks map out in the brain.

### *DEFAULT-MODE NETWORK (DMN)*

Active during introspection, daydreaming, and mind wandering, the default-mode network is disrupted when a person engages in a goal-oriented task. DMN is the neurobiological basis of self. Ignoring sensory input, the default-mode network recalls autobiographical information, such as memories of events and facts about oneself and thoughts about the future and envisioning events that might happen. The DMN carries self-referencing thoughts about one's personal traits and descriptions of one's self and reflections on one's own emotional state as well as value judgments and thoughts about others. This judgment of others is the "theory of mind," the highly developed capability of the human brain to understand the emotions and intentions of other people, and to empathize with other people's feelings. The DMN also carries thoughts about morality, right and wrong, and social evaluations about judgments of other people and social concepts.

If you return to the description of the thoughts swirling through my mind while I was having my EEG measured for three minutes by Jessica Eure and Robin Bernhard during neurofeedback training (see pages 139–149), you will see that the default-mode network was operating in my brain during this period. Recall my autobiographical thoughts about my recent travels to skin dive in the Caribbean, experiences with my family, and views about others, including Jessica, Robin, and even Edgar Allan Poe. I also had thoughts anticipating events in the future and making personal judgments, such as how the guitar I was building for my daughter will turn out and how it will be perceived.

The hubs of the DMN account for why these types of thoughts fly

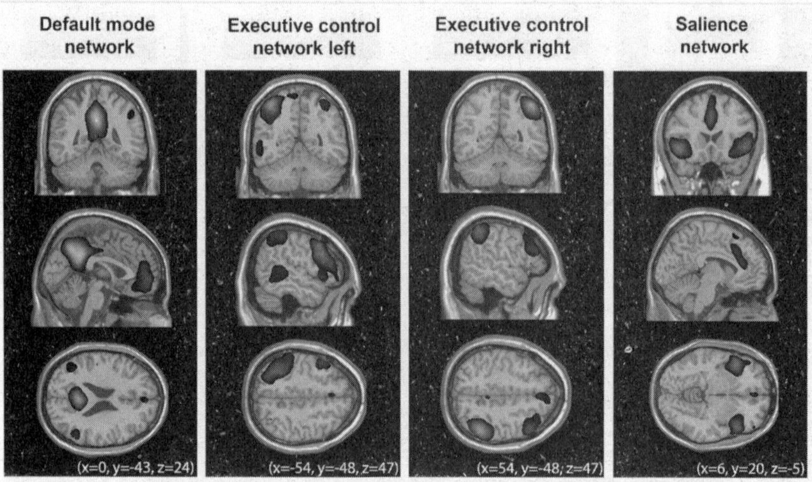


**FIGURE 36**: *Four of the major long-range networks of activity. Normally shown in pseudo color, the active regions in the brain are seen here as areas highlighted with a gray cloud. The* ***default-mode network****, active when a person's mind is wandering, involves interconnections between the prefrontal cortex, temporal lobes, and hippocampus. The* ***executive control network*** *is active when you are planning, making decisions, holding thoughts in working memory, reasoning, problem solving, and thinking abstractly. It regulates impulsive behavior, aggression, and antisocial behavior. The* ***salience network*** *monitors situational information from external and internal input.*

through our brain at rest. (Hubs or "nodes" are the major intersections of a neural network, much as Wall Street in Manhattan is a financial hub, interacting with others around the world.) Thoughts about one's self engage the posterior cingulate cortex and precuneus. These cortical regions are where information from memory and incoming perceptions are combined to focus attention. The ventral posterior cingulate cortex is activated in all tasks that relate to thoughts about the self and others, remembering the past, thinking about the future, processing concepts, and spatial imagery. The temporal lobes, hippocampus, and parahippocampus are engaged in remembering the past, imagining the future, and recognizing places and scenes. The dorsal posterior cingulate cortex is activated by involuntary arousal, such as when a fire alarm goes off and you drop all other thoughts from your mind and focus on the potential threat in your environment, or instead turn inward to your own thoughts. The precuneus is involved in visual and sensorimotor integration as well as attention. The medial prefrontal cortex is engaged in making decisions about one's self, such as personal information, autobiographical memories, future goals and events, and decisions regarding people who are personally very close to you, such as family. The ventral part of the prefrontal cortex is involved in positive emotional information and internally valued reward.

### SALIENCE NETWORK

The salience network activates the anterior insula and dorsal anterior cingulate cortex, regions engaged in detecting and filtering out salient stimuli; that is, stimuli that are important and need to be noticed. This network also connects with substantia nigra, ventral tegmental area, ventral striatum, amygdala, dorsomedial thalamus, and hypothalamus—brain regions that detect and integrate emotional and sensory stimuli and modulate the abrupt switch between internally directed cognition in the default-mode network to the externally directed cognition about one's situation and location, and interactions with the executive control network (described next). The abrupt thoughts that intruded into my daydreaming, bringing situational information to my conscious mind (*Melanie is waiting outside . . .*) was my resting brain switching transiently from the DMN to the salience network. This switch can be provoked by something in the environment, but the brain at rest periodically takes a peek at what's happening, and then shifts back to other network functions. An animal's survival depends on a high-functioning salience network, and large swaths of the brain are engaged to be always aware of where we are and what we are doing, on guard for dangers that may be lurking, all carried out involuntarily and automatically.

### EXECUTIVE CONTROL NETWORK

The executive control network, centered on the left and right prefrontal cortex, is engaged in processing and integrating complex information. It is active when you are planning, making decisions, holding thoughts in working memory, reasoning, problem solving, and thinking abstractly. It regulates impulsive behavior, aggression, and antisocial behavior. Three subregions of prefrontal cortex are specialized in each of these cognitive functions, although there is overlap: The dorsolateral prefrontal cortex is engaged in processing information streams and reasoning, among other functions. The anterior cingulate cortex is involved in emotional drives, sensory integration, and making decisions about behavior. The orbitofrontal cortex is concerned with controlling impulses and aggression. My executive control network switched on to critique the offerings of my default network (*It's a wind chime, not a marimba!*) and to command my DMN to stop daydreaming and make the bells ring.

Several other networks are responsible for sensory and motor control, such as the visual and auditory networks, but the DMN, the salience network, and the executive control network are the most insightful networks in terms of higher-level cognitive ability, personality, and behavior (both normal and abnormal). These are the significant networks, uniquely tailored in each of us by genetics and our personal experience. They can reveal what kind of brain and mind you have from only a few minutes of eavesdropping on your electric brain while you sit quietly and let your mind wander where it will.

Let me give you some examples:

**PERSONALITY:** Our individual personality is a core feature of our identity, and it stems from neural circuits in our brain that predispose each of us to interact somewhat differently with our environment. Some of us are bold and brave, others are mild and meek, some crave novelty and exploration, and others cling to routine. Psychologists divide general personality traits in responding to the environment into three categories: harm avoidance, novelty seeking, and reward dependency (roughly equal to "I'm scared," "What's that!" and "I want it"). Where you fall on this three-axis scale of motives indicates your particular psychological temperament. Researchers Rongtao Jiang and colleagues at the Chinese Academy of Sciences report that they can determine a person's personality from the pattern of functional connections in their brain. This was made possible by the Brainnetome Atlas,[11] a large Chinese database of anatomical and functional connections in the human brain that includes annotations about the functional and cognitive characteristics of the individuals in the database.[12]

In their fMRI study of 360 healthy undergraduates, the researchers were able to accurately predict temperament scores based on the functional connectivity determined by resting-state fMRI. Connections between the amygdala, thalamus, parahippocampus, and orbitofrontal gyrus are known to be important in the neurobiological basis of personality traits,[13] and dysfunctions in these connections underlie certain psychiatric disorders.[14] Their analysis showed that circuit activity among the amygdala, thalamus, cingulate gyrus, inferior temporal gyrus, and parahippocampus predicted the tendency for harm avoidance. All of these circuits are core areas of the limbic system.

Novelty seeking was correlated with functional connectivity in the orbitofrontal gyrus, parahippocampus, and thalamus. The personality attribute toward reward behavior corresponded to connectivity between mesolimbic-dopamine centered networks, including the amygdala, hippocampus, basal ganglia, prefrontal areas, and inferior temporal gyrus, which are circuits in the frontal-limbic system.

Knowing the anatomical terminology is not essential to understanding the important message from these studies: A person's perception and regulation of fear, anxiety, reward processing, and attraction to novelty and exploration, which are fundamental personality traits largely (but not entirely) genetically determined, can be ascertained by fMRI performed on a person doing nothing but letting his or her mind wander while the resting-state activity in their brain is scanned.[15]

Other studies of EEG power in response to a stimulus have revealed that asymmetrical responses in alpha power in the left versus right prefrontal cortex relate to the approach-withdrawal behavior of the individual to that stimulus. Decreased alpha activity indicates higher alpha desynchronization and increased activation of cortical circuits. Suppression of alpha rhythm across the left prefrontal cortex (relative to the right) is associated with a propensity to approach a stimulus, while the relative power suppression of the alpha rhythm across the right prefrontal cortex is associated with a propensity to withdraw from the stimulus.[16] This insight has been used to investigate the effectiveness of different TV commercials to better target the advertisements to the appropriate age and gender of the viewer.[17]

**BRAIN AGING:** In a study of 693 healthy older adults (ages fifty-five to seventy-five) who were scanned at a single research site and 2,018 healthy adults (ages eighteen to eighty-one) scanned at multiple sites, researchers in Germany, in a team led by Deepthi Varikuti, were able to use machine learning to accurately predict the age of a person by their fMRI.[18] (Machine learning is a type of data analysis performed by a computer using mathematical methods and artificial intelligence to train the computer to recognize patterns and make decisions. This is the same analytical approach used by Marcel Just and his colleagues.)

**GENDER:** The brains of men and women, just like the rest of their bodies, are somewhat different. But new research can determine the gender of a person simply by monitoring their brain at rest using fMRI. A study by Chao Zhang and Andrew Michael of the Geisinger Health System in Pennsylvania, composed of 366 males and 454 females between the ages of twenty-two and twenty-seven, was able to predict the gender of the individuals with 80 percent accuracy from resting-state fMRI.[19] The sex-specific functional connectivity differences were found predominantly in the default-mode network.

**INTELLIGENCE:** Intelligence is a thorny subject, fraught with cultural (and historically racial) bias, disparities in definition, and difficulties in measurement. IQ is, however, a well-established metric of cognitive ability that is associated with academic success and a broad range of attributes and abilities. There are many different types of brain-imaging studies that have yielded significant differences tightly linked to IQ. For example, resting-state fMRI from the Human Connectome Project,[20] which includes extensive behavioral data, including IQ on each test subject, can predict fluid intelligence from the patterns of functional connectivity.[21] (Fluid intelligence is the facility of thinking abstractly and rapidly, accurately identifying patterns, and quickly reasoning to solve problems. This is in contrast to crystallized intelligence, which is the ability to use acquired knowledge and reasoning to recognize patterns and find correct solutions to problems. This distinction is roughly analogous to "quick thinking" versus "wisdom.")

In 2000 Norbert Jausovec and Ksenija Jausovec reported that differences in resting-state EEG taken with eyes closed could predict intelligence and creativity.[22] Weak correlations were found with the power and frequency of EEG activity, but coherence measures were strongly related to both creativity and intelligence. With the eyes open, the differences were observed mainly over the right hemisphere.

In another study by Péter P. Ujma at Semmelweis University in Budapest, Hungary, differences in EEG spectrum during sleep were found to correlate with intelligence, but in a gender-specific manner.[23] In these studies, the Raven's intelligence test, which measures nonverbal reasoning and logical ability, was used to evaluate fluid intelligence. A clear association between intelligence and beta power (reflecting increased spindle activity) was found in the frontal region during REM sleep for females but was not evident in males. The authors concluded that EEG activity during sleep and wakefulness is associated with intelligence, but intelligence may have different neural substrates in males and females.

Slow-wave sleep, and thus the accompanying high EEG delta power, is associated with neural plasticity. Age-related cognitive decline and impairment of slow-wave sleep may go hand in hand. Individuals with higher general intelligence showed less age-related cognitive decline, and a 2017 study found less age-related decrease in slow-wave sleep EEG delta power in the group of people with higher intelligence (IQs equal to or above 120).[24] Another study found that autistic children with normal IQ had fewer sleep spindles in stage two NREM sleep than children without autism. The duration of sleep spindles was correlated with verbal IQ in children without autism, but this correlation was not found in children with autism. The researchers concluded that differences in organization of neurons in the cerebral cortex may exist in these two groups.[25]

In another study of school-age children, spindle power was correlated with full-scale and fluid IQ, concluding that "sleep EEG can be a marker of intellectual ability in school age children."[26] These findings are in alignment with earlier research, published in 2006, showing that children with learning disabilities had higher-power delta, theta, and alpha power than control subjects without any learning issues. Viewing these differences against the changes in EEG power that are seen during average childhood development, the researchers conclude that these features of EEG that correlate with lower IQ and learning difficulties may reflect less developed brain maturation required for complex information processing in the children with learning disabilities.[27]

Another possibility is that intelligence correlates with faster thinking. Faster oscillations indicate faster processing of information, so the speed of information processing may be gleaned from EEG measurements. The dominant frequency in the adult human brain in a relaxed state is about 10 Hz (alpha band), but this differs among individuals. Several studies have found an association with higher peak frequencies and higher intelligence, but other studies find no such association, and conclude that "smarter brains do not seem to run faster."[28] Finally, it is interesting that brilliance in children may come with some costs. In a longitudinal study of a cohort of children tracked into adulthood, higher childhood IQ at age eight (compared to others within the normal range) was associated with manic tendencies when they reached the age of twenty-two to twenty-three, suggesting that high IQ in childhood is a risk factor for developing bipolar disorder later in life.[29]

**CREATIVITY:** The ability to think creatively is not only valuable for those drawn to professions such as art, science, writing, or music but it is important for everyone so they can effectively engage in problem solving. New studies are tracing the brain circuitry that produces creative thinking. Functional MRI studies, published in 2018, find distinct features of effective connectivity between three brain networks in highly creative people: the default-mode, salience, and executive brain networks, which were described earlier in this chapter.[30]

In these studies, participants were engaged in a divergent thinking task while their brain activity was monitored by fMRI. In this task, the subjects were presented with the image of a common object, perhaps a hammer, and then asked to think of an unusual use for it. In a contrasting task, the same test subjects were asked to think of common uses for the object presented to them. Their verbal answers in these tests were then evaluated and scored for the level of creativity, and the functional brain responses during these two tasks—one creative and the other not creative—were compared. The results, which will be described in more detail below, enabled the researchers to not only tell when a person was having a creative or unimaginative thought simply by the pattern of functional activity in their brain, but also to accurately predict creative thinkers from people with low creativity simply by observing the patterns of functional connectivity seen in their brain using resting-state fMRI. By monitoring your electric brain idling at rest for only eight minutes, while you are engaged in no specific task at all, the researchers can determine if your brain is wired in ways that make you highly creative or lacking in imagination.

The study found that highly creative individuals show brain activation patterns that can simultaneously engage all three of these brain networks (default, salience, and executive), presumably facilitating creative thinking. This is an interesting finding because typically these networks operate in opposition. For example, the default-mode network is active when one is left to their internal musings and mind wandering, but activity in this network stops when a person begins to execute a deliberate act or when the salience network rapidly surveys the external environment. By measuring the strength of functional connections among different brain regions, the researchers found especially dense functional connections between the nodes of these three networks in highly creative individuals. The centers of these creativity networks are the posterior cingulate cortex (default-mode network), right dorsolateral prefrontal cortex (executive network), and left anterior insula (salience network). The researchers suggest that the default network generates ideas by calling on memory; the salience network selects candidate ideas that fit the situation or context; and the executive network evaluates and elaborates on that concept. This dynamic interaction need not be serial, the researchers state, but in highly creative minds the interactions can reverberate simultaneously among all three networks.

From this new information we can see scientifically and objectively that creativity engages nearly the whole brain. Moreover, the parts of the brain engaged in creative thinking are where the highest-level cognitive functions are performed. The connection between creativity and cognitive function is a concept that should come into play when evaluating education, particularly in light of schools curtailing art and music programs to save money in favor of classes that are perceived as being more academically rigorous. The crayon doodling of a kindergartener is easily dismissed as a charming distraction or practice to develop motor control—stay between the lines—but actually, inside that child's brain the sprouting networks of memory retrieval, mental stimulation, judgment, and cognitive control are growing and strengthening along with the brilliant streaks of color and the soothing smell of crayon.

In addition to fMRI, there is a large body of EEG research connecting brainwaves and creativity.[31] A 2006 study by Austrian researchers found that during a verbal creativity task, individuals with the most original responses have EEG patterns indicative of lower levels of cortical arousal; that is, sharp increases in alpha power are seen from just before the creative act, and the most original responses were accompanied by stronger synchronization of alpha oscillations in the posterior cortex (centroparietal region).[32]

Once again, correlation is not causation, and the surge in alpha waves during a creative flash could be the result, not the cause, of creative thinking. A study published in 2015 tested causality by using transcranial alternating current stimulation at the alpha frequency (10 Hz) applied to the frontal cortex.[33] The results showed that boosting alpha activity in this way also boosted creativity, as measured by the Torrance Test of Creative Thinking, the most frequently used assay of creative strengths. There was no effect when 40 Hz stimulation was applied, supporting the conclusion that alpha waves promote creativity, rather than some other aspect of brain stimulation, such as novelty or arousal, affecting the results. The researchers believe that the surge in alpha activity energizes neural networks that activate inhibitory top-down control (from the prefrontal cortex) and working memory to critically assess the idea under consideration. (As discussed previously, top-down control refers to behavior that is initiated willfully and with deliberation, and bottom-up control is a reflexive behavioral response to an external stimulus.)

**MUSIC:** A study by researchers at Wesleyan University, published in 2017, compared EEG responses of nonmusicians, improvisational jazz musicians, and non-improvising musicians to hearing chord progressions, and the researchers found that they could spot the more creative music makers by their brainwaves.[34] Baseline testing showed that jazz musicians favored more unusual musical progressions, rather than the expected sequence of chords. The EEG response (event-related potential), which was the brainwave response provoked by hearing an unexpected chord in a progression, was much larger in the improvisational jazz musicians than in the others in the study.

Another study by researchers at the University of Western Ontario in Canada, published in 2017, yields new insight into the creative process by analyzing EEG activity while musicians listened to, played back, and improvised jazz melodies.[35] The results showed increased frontal alpha-wave activity during the more creative task of improvisation compared to the less creative activity of playing back the music the musicians heard. Moreover, the effect was greatest for musicians with formal jazz improv training, suggesting a link between creativity and frontal alpha power. Indeed, when the improvised jazz licks were later ranked for creativity, the power of alpha-wave activity at the time the piece was being improvised correlated with the degree of creativity of their performance. This suggests that creativity is a distinct mental state and that it can be measured by the power of frontal alpha activity generated during the creative act.

A similar study also found power differences in the alpha and beta bands in experienced musicians playing learned melodies in contrast to improvisations of those melodies.[36] The researchers pinpointed the effects to neurons in five regions: the left superior frontal gyrus, supplementary motor area, left inferior parietal lobe, right dorsolateral prefrontal cortex, and right superior temporal gyrus. These networks indicate that musical improvisation requires heightened real-time creativity involving neural networks engaged in moment-to-moment decision-making, monitoring one's performance, to spontaneously create new variations on the melody. Musical training also boosts EEG responses to hearing a note played slightly out of rhythm, an ability that musicians excel in. The brainwave evoked over the frontal and frontal-central areas of the brain right after a delayed note is played is significantly larger in musicians compared to nonmusicians, indicating the heightened ability of their brain circuitry to discriminate rhythmic variation in music.[37]

**VIDEO GAMES:** Researchers reported in 2017 that they were able to identify differences in brain connectivity in males who are addicted to internet gaming.[38] After sampling only ten minutes of resting-state EEG, the team of researchers at Seoul National University in Korea found that brainwave activity in the 30 to 40 Hz band (gamma waves) was more highly synchronous between the left and right brain of gamers than in people who are not addicted to internet gaming.

Considering internet gaming disorder as an obsessive, compulsive, or addictive behavior, the researchers hypothesized that people addicted to alcohol might share similar intercerebral hyperconnectivity in the gamma band in association with craving alcohol. This is reasonable, considering that brain circuits involved in addiction, reward, and impulsive behaviors are involved in both alcohol dependence and in a compulsion to engage in excessive internet gaming. However, the experiments found something different. The brains of alcoholics did not exhibit higher intercerebral interconnectivity in the gamma band, but there was a tendency for somewhat higher intercerebral interconnectivity in the lower frequency theta band of brainwaves (4 to 8 Hz) in alcoholics. That difference between brainwaves in alcoholics compared to healthy controls was not as strong as the finding of higher intercerebral connectivity in gamma brainwaves in people with internet gaming disorder. This finding makes sense when one considers that people dependent on alcohol are not necessarily also compulsive internet gamers, but this in itself shows how insightful EEG analysis can be in identifying the neurophysiological basis of psychiatric and other mental attributes. Interventions to correct alcohol dependence and internet gaming disorder will change brain connectivity, and this recovery should be reflected in changes in brainwave oscillations and synchrony.

**READING DIFFICULTY:** The hereditary component in dyslexia suggests that underlying neural network properties at a preliterate age might be detected by resting-state EEG, and several studies support this conclusion. For example, consistent with the study that opened this chapter showing that measuring brain activity by using near infrared spectroscopy in young children predicted their proficiency in reading once they reach school age, a 2014 study used resting-state EEG to make a similar prediction. Researchers in the Netherlands and Italy measured resting-state EEG in children three years of age with their eyes open as they sat on their parent's lap.[39] After the children reached nine years of age and were learning to read in the third grade, the researchers found that they could identify the children who had become fluent readers from those who had difficulty reading on the basis of the resting-state EEG activity that was measured in the children's brains when they were preliterate toddlers. The resting-state EEG recordings of the nonfluent third graders, measured when they were three years old, showed significantly lower levels of delta activity and significantly higher alpha power compared to the toddlers who became effortless readers. Lower alpha and beta activity have been reported in a number of studies in children with dyslexia during reading.[40] Researchers in Australia reported similar findings in 2016, in that a dominance of theta activity in eyes-closed resting-state EEG prior to learning to read suggests that children with dyslexia have delayed maturation of linguistic networks in the brain.[41]

Brainwave patterns change during childhood. Delta-wave activity dominates the human EEG during early development (between birth and age two) and decreases in power over the course of development. The researchers studying dyslexia speculate that delta power relative to the power of other brainwave frequencies can be considered a marker of brain maturation, and, consistent with this interpretation, correlations between delta-wave activity and gray matter volume have been shown.[42] Theta-wave activity increases between ages two and six and alpha-wave activity increases between ages five and eight. Beta waves increase between ages eight and twelve,[43] whereas gamma-wave activity peaks between ages four and five.[44] Much of the speeding up of brainwaves as a child matures is associated with myelination, the electrical insulation on axons that enables high-speed transmission of impulses. Different pathways in the brain are myelinated at different times in fetal and postnatal development, and in many parts of the brain myelination continues through adolescence and into early adulthood.

A study comparing connectivity networks in the brain determined from resting-state EEG that those with dyslexia had reduced network integration in the theta band compared to controls. This analysis supports other compelling evidence from fMRI data showing differences in brain networks in people with dyslexia that make it more difficult for these individuals to learn to read.[45] Dyslexics have also been reported to have reduced interhemispheric coherence of EEG,[46] altered evoked potentials to speech sounds (a transient EEG response that is evoked by a sound),[47] slow EEG amplitude oscillations during NREM sleep,[48] and increased spindle activity and power during NREM sleep.[49]

Dyslexia is commonly referred to as a dysfunction. While it is certainly a serious disability in the modern world, I would not consider it a dysfunction, because reading is not a normal brain function. Reading is something that human beings never did until very recently in our history. Almost no one could read a few hundred years ago. Difficulty in learning to read is no more a brain dysfunction, in my view, than is difficulty in learning to play the piano a brain disorder. Some can excel at the keyboard, but others will struggle or fail. From this perspective, dyslexia is an educational dysfunction that fails to recognize and accommodate the biological fact that individual brains are different, and they develop different abilities at different rates and to different extents. Indeed, an interesting article by Matthew Schneps, published in *Scientific American* in 2014, finds evidence that people with dyslexia have a talent for science. Possibly when sentences are not automatically decoded, and instead each one represents a puzzle, the young brain becomes extremely proficient at analysis and extracting meaning from fragmentary information.[50] In this respect, the features of resting-state EEG described earlier may be considered predictors of a three-year-old's increased potential to become a scientist.

Like all things in biology, your brain's electrical grid, which is the backbone of your unique cognitive abilities, personality, and psychology, is the result of genes, experience, and chance events. A good example of how experience affects the brain's electrical network and EEG is provided by a long-term study on Romanian orphans. These children suffered severe neglect in orphanages in Romania under the brutal regime of Communist dictator Nicolae Ceaușescu in the 1980s and '90s.[51] Research by Charles Nelson, Boston Children's Hospital; Nathan Fox, University of Maryland College Park; and Charles Zeanah of Tulane University found that the orphans' brainwaves were significantly different from normal.[52] The neglected orphans had low alpha and beta brainwave power and higher theta brainwave power than normally reared children of the same age. This altered EEG signature has been associated with attention-deficit/hyperactivity disorder (ADHD), learning disorders, disruptive behavior disorders, and psychosocial risk factors. However, orphans who were subsequently placed in foster homes developed normal EEGs when assessed twelve years later, while those who remained institutionalized still had these abnormal features in their EEG.

While this tragic example illustrates the detrimental effects of adverse experience on brain development, which is revealed by resting-state brainwave activity, it also demonstrates how positive experience can change the brain's neural networks and brainwaves. This is a subject that will be taken up in chapter eleven, but next let's move from the ability of EEG and fMRI analysis to discern differences in normal cognitive function and personality to identifying abnormal function.

# CHAPTER 10

# Detecting Brain and Mental Disorders

Given the insights that monitoring electrical activity in the brain by EEG and fMRI can yield about personality and cognition, it should be no surprise that these techniques can also reveal neurological dysfunctions of all kinds. Research has not reached the point where EEG and fMRI can be used diagnostically, although in some cases, such as epilepsy, it can. Instead, the connection between brain activity and neurological and psychological dysfunction are based on population studies. This is much like how high blood pressure clearly predicts increased risk of suffering a stroke, but it does not mean that a person with high blood pressure will have a stroke, although the biological basis for the correlation is well understood. In time, however, the number of diagnoses of neurological disease or psychological disorder made by recording electrical activity in a person's brain will increase.

## NEUROLOGICAL DISORDERS

Neurological disorders are caused by disease or injury in the brain, spinal cord, sense organs, muscles, and nerves throughout the body. When neuronal function is altered, so too can the pattern and transmission of electricity in the brain change. As a consequence, many neurological abnormalities can leave traces of dysfunction in the EEG, and this direct measure of the brain's electrical activity can be a powerful means of diagnosis and treatment.

## EPILEPSY

The importance of EEG in the study and treatment of epilepsy is widely appreciated; indeed, the electrical brain storm during a seizure that sets the EEG recordings wildly erupting is most likely the image that comes to mind when one mentions EEG recording, but this was not always so. One of the pioneers in the field of EEG research as applied to epilepsy was Herbert Jasper, and his career intersected with those of some of the most well-known giants in the field of neuroscience. His example illustrates what I hope is evident from the quest undertaken in this book to understand the electric brain today, that science is a highly collaborative process that spans the globe to share ideas and experimental results

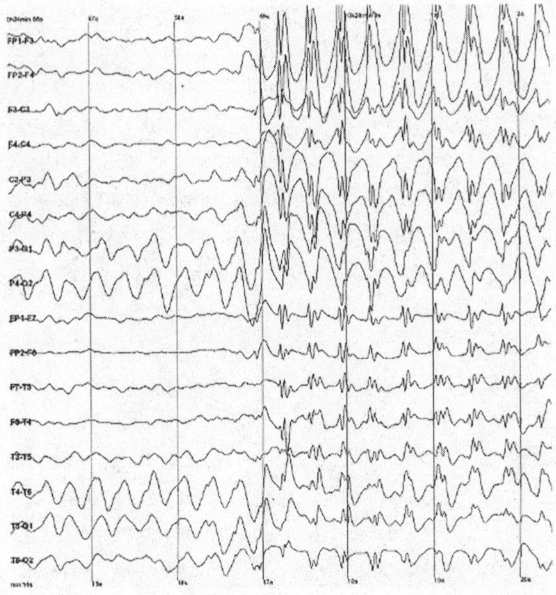

**FIGURE 37:** *Highly synchronized hyperactivity of brainwaves accompany seizure.*

among a relatively small community of people who are passionate about discovering nature's mysteries.

Herbert H. Jasper, born in Oregon in 1906, was the first person in America to publish an EEG recording from the human brain. Jasper's intrigue with the technique was originally motivated by its possible application to diagnosing epilepsy. Hans Berger's discovery of the human EEG was widely viewed with skepticism, and it was largely ignored until 1934, when Nobel Prize winner Edgar Douglas Adrian repeated Berger's experiments and published the findings in a paper coauthored with Bryan Matthews in the scientific journal *Brain*. The following year, 1935, Jasper traveled to Germany to meet Hans Berger, and he also visited Adrian and Matthews in Cambridge, England. Jasper published his EEG recordings of the human brain in the prestigious journal *Science* in 1935, recognizing Berger for this important discovery. In his paper, titled "Electrical Potentials from the Intact Human Brain," Jasper wrote:

> Dr. Hans Berger, of Jena, has published a series of papers in which he reports that changes in electrical potential which are correlated with human brain activity may be magnified and recorded by the use of a suitable vacuum-tube amplifying system and an oscillograph. These potential changes are obtained from needle or surface electrodes placed on different points of the head . . . He holds that the records secured by the use of this general technique show, among other phenomena, two characteristic forms of rhythmic electrical oscillations. The waves of greatest magnitude he calls alpha waves. Smaller oscillations which are sometimes observed alone and sometimes as superimposed upon the alpha waves he calls beta waves . . . Dr. Berger has shown in these experiments that the alpha waves diminish in magnitude under certain types of anesthesia, during an epileptic seizure and, it may seem at first sight paradoxically, when the subject is given sensory stimulation or does a "mental" problem.[1]

Wilder Penfield, the famed neurosurgeon known for his work on epilepsy and for using electrical stimulation during brain surgery to guide his surgical treatment as well as to explore human cognition, gave a seminar in the Psychology Department at Brown University at the invitation of Jasper in 1935. According to a historical review written by Massimo Avoli, Penfield described the meeting with Jasper as follows: "He (Jasper) could, he said, localize the focus of an epileptic seizure by the disturbance of brain rhythms outside the skull. I doubted that, but hoped it might be true."

Penfield agreed to operate on two patients that Jasper had examined by EEG to pinpoint the focus of their seizures, and upon opening the skull, Penfield found that Jasper's EEG analysis had accurately located the epicenter of the seizures. This sparked a collaboration beginning in 1938, in which Jasper would load a portable EEG machine into his car and drive to the Montreal Neurological Institute where Penfield worked. Jasper soon relocated to Montreal, where he pursued a fruitful collaboration with Penfield that lasted thirty years. Today, EEG analysis is the most powerful tool guiding physicians and neurosurgeons in treating epilepsy.

## ALZHEIMER'S DISEASE

Xiaojing Long and colleagues at the Chinese Academy of Sciences in Shenzhen, China, have used structural MRI data to detect anatomical differences in the brain that can be used to predict Alzheimer's disease and to distinguish with a high degree of certainty people who will get Alzheimer's disease (AD) from others who have the medical condition of mild cognitive impairment (MCI).[2] (Mild cognitive impairment is a more serious degradation of cognitive ability from the normal decline expected with aging. MCI can progress to dementia and may be an indicator of developing Alzheimer's disease, but not all people with MCI develop Alzheimer's disease and some people with MCI can improve.) Currently it is difficult for doctors to distinguish mild cognitive impairment from early-stage Alzheimer's disease; the ability to do so is highly desirable so that appropriate treatments can be applied early in the course of disease.

In these studies, 227 people in the Alzheimer's Disease Neuroimaging Initiative (ADNI) database were analyzed.[3] The ADNI is a global database of brain imaging data on AD patients that includes genetics, cognitive tests, and biomarkers for the disease for every person in the database. The researchers compared subjects who had mild cognitive impairment and had not converted to AD within sixty months to those who suffered progressive MCI and converted to AD within thirty-six months of their baseline visit. Based on anatomical differences in the MRIs, the researchers were able to discriminate people who will only have MCI from those who will suffer progressive MCI leading to AD. They were able to accomplish this prediction with an accuracy of 89 percent using an algorithm detecting changes in hippocampus and amygdala.

The structural differences seen by brain imaging that are predictive of developing AD should be accompanied by functional differences that might be detected with EEG, and there are many such studies.

Resting-state EEG analyses show reduced power in the alpha band in early stages of AD, as well as a slowing of the alpha rhythm and increased presence of lower-frequency bands.[4] The typical EEG features in AD patients are increased delta or theta rhythm and decreased alpha or beta rhythm activities. Enhanced gamma-rhythm power in addition to the increased delta and decreased alpha power has also been reported in AD patients.[5] The pathological increase of ongoing gamma-band power may result from disruption of the GABAergic interneuron network in AD patients, the researchers conclude. (Recall that GABAergic interneurons are the neurons in the cerebral cortex that inhibit other excitatory neurons from firing, and they are particularly important in generating gamma oscillations.) Furthermore, the cross-frequency enhanced synchronization in the gamma band seen in AD patients could indicate reduced complexity of the neuronal network, and to compensate, AD patients may have to use more neural resources to maintain resting brain state function, they propose. EEG oscillations are not independent but instead interact and join across frequencies to regulate multiple brain networks.[6] For example, theta-gamma phase and amplitude coupling in the hippocampus varies with working memory load,[7] and impairments in theta-gamma-phase coupling is associated with cognitive deficits in a mouse model of AD.[8]

A study published in 2018 has linked biomarkers (molecules in cerebral spinal fluid) indicative of neurodegeneration in AD with global EEG power and synchronization.[9] Decreased amyloid beta42 levels in cerebral spinal fluid (CSF) was found to correlate with increased theta and delta global EEG power and synchronization, whereas increased p- and t-tau protein correlated with decreased alpha and beta power and synchronization. These changes paralleled cognitive decline in the subjects.

Resting-state MRI studies show disrupted default-mode network (DMN) function in AD, while, as mentioned above, resting-state EEG studies consistently report reduced power in the alpha band. A recent study simultaneously recording resting-state MRI and EEG in AD patients finds a strong correlation between resting-state EEG power and fMRI signals in control subjects in the DMN, but the correlation between EEG power and fMRI is diminished in the DMN and numerous other brain regions in AD.[10]

In addition to resting-state studies of brain activity, changes in brainwaves while engaged in a cognitive task can diagnose AD and provide more detailed insight into the specific neural networks that are impaired. For example, a prominent feature of AD is a progressive decline in language ability. A recent study of patients with amnestic MCI compared with elderly controls in word comprehension tests, while recording EEG activity, found telltale indicators of dementia and the ability to predict who would progress from MCI to AD.[11] The researchers presented the test subjects with a word test while recording their EEG responses. For example, the test subjects would be asked "type of wood," and they would have to respond by identifying the appropriate picture associated with that query—for example, an oak tree. Normally when the test word is presented to healthy controls there is a rapid increase in theta activity over the posterior brain regions, which is related to processing words. Immediately thereafter, alpha-wave activity is suppressed, which is associated with processing the meaning and context of the word after first perceiving it.

The alpha suppression was observed only for healthy age-matched controls, and not seen in either MCI patients or those who later developed AD. The EEG results showed that individuals with MCI who later went on to develop AD had diminished posterior-parietal theta activity induced by presenting the target phrase. Moreover, these individuals showed different oscillatory signatures when processing the test words compared to healthy controls, in terms of joining different brainwave frequencies. The researchers did not detect any significant cross-frequency connectivity between theta and alpha/beta power in either MCI patients or those who would go on to develop AD, a connectivity that is normally seen in healthy controls during this word test. The occurrence of weaker alpha suppression in this test provides a useful indicator for people with MCI and those with MCI who will go on to develop AD. The differences between the two latter groups may reflect impairments in networks important for both implicit and explicit memory processes.

Together these studies of signals emitted by the electric brain can predict the presence of early dementia and pinpoint the specific neural networks that are becoming dysfunctional. MCI patients who progress to Alzheimer's disease, as a group, have subtle deficits in language comprehension that are revealed by oscillatory changes in their EEG during word processing. These brainwave differences are thought to indicate degradation of the brain network providing language comprehension, and by monitoring EEG it may be possible to foretell who will develop AD and memory failures years before the disease is evident, permitting early intervention.

## PARKINSON'S DISEASE

Any neurodegenerative disorder impairs normal brain network function, and therefore the electric fields in the brain would be altered in strength, coherence, and location. Such effects will, in principle, alter brainwaves and, if these changes are detectable by scalp electrodes, EEG recordings can provide useful diagnostic information through a completely harmless and noninvasive method.

Many neurodegenerative disorders begin without symptoms and may progress for years or decades before becoming fully evident. Parkinson's disease (PD) is a good example. Early intervention is desirable in any degenerative disorder, and if EEG recordings can detect early signs of disease, the benefits could be life-changing. In addition to impaired movement, PD is often associated with cognitive deficits. Increased delta and theta band power, with a decrease in alpha and beta wave power, has been observed in PD, which is reminiscent of AD.[12] These two disorders are very different, and indeed a recent study using high-density EEG recordings (that is, using many more EEG electrodes—256 channels as opposed to the 20 electrodes used in the study described earlier) could accurately distinguish these two groups. Twelve minutes of resting-state EEG recording with eyes closed was all that was required. Similar differences in EEG are evident between patients with AD and PD; however, more pronounced slowing of the EEG rhythm was clearly seen in PD patients as compared to AD patients.[13] The alpha/theta ratio in the central left brain region and the theta power in the temporal left region were the two most important features in distinguishing PD from AD or MCI by EEG.[14]

A study published in 2017 on PD patients who had donated their brains to a PD brain bank, and who had also undergone EEG recordings while undergoing treatment, found a strong association between delta band power and pathological protein deposits in PD brain tissue, called Lewy type synucleinopathy, the same protein deposits found in Lewy body dementia.[15] Thus, resting-state EEG may substitute for biopsy and histological analysis of brain tissue, which is usually only done postmortem. In addition to resting-state EEG, task-specific EEG responses are also indicative of PD. For example, the time to peak responses to evoked brainwave potentials are longer in PD than in controls.[16]

## PSYCHOLOGICAL DISORDERS

That neurological disorders alter electrical activity in neural circuits is self-evident—if the hardware of the brain is malfunctioning so too will its activity exhibit abnormalities that should be detectable with EEG or other methods. At one time the notion that psychological disorders had a biological basis was a revolutionary concept, but today this is well understood. Just as monitoring the electric brain by EEG reveals aspects of personality that are rooted in our brain's biology, so too should EEG and fMRI methods illuminate psychological disorders. Current research is actively exploring, and proving, this connection between electrical activity in the brain and mental health. It is not possible to cover this enormous body of research in detail here; but to illustrate the exciting research in the field of EEG analysis related to psychological disorders, which promises to greatly advance how psychological disorders are diagnosed, understood at a biological level, and treated, let's consider a few examples.

## ATTENTION-DEFICIT/HYPERACTIVITY DISORDER (ADHD)

In 2013 the FDA approved an EEG marker, the theta/beta power ratio, as a diagnostic for attention-deficit/hyperactivity disorder (ADHD). This condition, in which individuals demonstrate a persistent pattern of inattention, hyperactivity, and impulsivity, was the first EEG method for diagnosing any psychological condition approved by the FDA. The FDA approval was based on a body of research associating ADHD with an increase in theta power (4 to 7 Hz) relative to a decrease in beta power (13 to 30 Hz). This distinctive EEG signature is observed in the frontal brain regions that are well sampled by an electrode at the Cz position.[17] A review of studies in the scientific literature, published in 2006, found that the method was 94 percent accurate in diagnosing ADHD,[18] but recent studies suggest that while a large fraction of ADHD patients show this characteristic change in theta/beta ratio, others do not, making the measure an insufficient diagnostic method on its own at this time.[19]

While the scientific underpinnings are sound, practical considerations make the application of EEG analysis for diagnosing ADHD (and other mental disorders) problematic. So many factors can influence the power of different brainwaves, such as fatigue, attention, stimulants, and depressants, and the technical reliability of the EEG recordings vary greatly with the vast differences in instrumentation and expertise used in EEG analysis. EEG can be diagnostic in scientific research studies using the best equipment, expertise, and procedures available, but it may not yet be reliable enough for routine use by mental health practitioners in diagnosing their patients, which is the conclusion of the American Academy of Neurology published in 2016. Considering the risk of harming people by misdiagnosing them with ADHD, the Academy recommends that this EEG analysis should not replace standard clinical evaluation, but that the theta/beta analysis can be useful in combination with other diagnostic methods for ADHD.[20] That is the current state of affairs, but as the technology and science of EEG research advances rapidly, this situation is certain to change in the near future.

Another important consideration in ADHD is that this disorder is quite heterogeneous, and here EEG analysis can be helpful. For example, there is a subset of adults with ADHD who have high IQ but experience difficulty in focusing on tasks and organizing, tend to procrastinate, and exhibit emotional instability. A study published in 2017 of forty adults with ADHD who have high IQs (above 120) found that these individuals had increased beta and theta power in eyes-closed resting-state EEG. The researchers conclude that increased beta-wave power at frontal, central, and parietal brain regions is associated with decreased impulse control, and that this resting-state EEG pattern may be a risk factor for delinquent behavior in adults with ADHD.[21]

## AGGRESSION AND PSYCHOPATHOLOGY

In the 1970s a theory for antisocial behavior was put forward by Emeritus Professor of Psychology Robert Hare of the University of British Columbia, proposing that low arousal levels in individuals predisposed them to crime and violence. Early EEG research in the 1950s found correlations with theta and delta slow-wave brainwaves in adult psychopaths in prison[22] and violent criminals.[23] Supporting this concept, slow-wave brain activity is associated with low states of arousal. Since that time, higher prefrontal EEG activity at these slow-wave frequencies has been found in several studies on males with aggressive or violent behavior and in violent psychiatric inpatients. In theory, a poorly aroused brain may seek higher arousal through risky, impulsive, and possibly violent activities. Other studies have found that high alpha power in children is predictive of criminal activity. Most of these studies are on males, largely because aggression and violence are far more common in males than in females. Both antisocial behavior and alpha power have a strong genetic component, suggesting that innate, inherited aspects of brain connectivity, and thus electrical activity in the brain, could underlie aggression and antisocial behavior.

A study published in 2015 by psychologist Sharon Niv and colleagues at the University of Southern California examined this correlation more closely. Previous studies were derived from cross-comparisons (between controls and psychopaths, for example), but a longitudinal study design, which performs repeated tests on the same individual over time, can provide results that are more reliable and insightful. In this study published in 2015, 900 twins were given EEGs in late childhood (ages nine and ten) and again in mid-adolescence (ages fourteen and fifteen), and the level of antisocial behavior in each person as a child and adolescent was assessed by psychological tests.[24]

The results showed a strong ability to predict adolescent aggressiveness in males from their childhood EEG recordings, which showed higher levels of alpha power in the frontal region of their brains. This association was not evident in girls. The researchers conclude that "brain patterns are a potentially useful marker for identifying children who are susceptible to difficulties [antisocial behavior] later in their lives, and may serve as useful tools for early detection."[25]

Frontal lobe involvement in antisocial behavior makes sense, because this is the part of the brain that inhibits impulsive behavior and evaluates risk versus reward. An interesting outcome of this study was that children with frontal alpha power did not currently exhibit aggressive behavior, but it was well correlated with aggressive behavior that would emerge when the children became adolescents. The high frontal alpha power shown at a younger age somehow predisposes these children to engage in aggressive behavior years later. Nonaggressive antisocial behavior, such as rule breaking, was not at all correlated with high alpha power, indicating distinct neural circuitry involved in these two types of antisocial behavior and criminality. The reason the relationship between high frontal EEG power and antisocial behavior holds for males but not females helps explain why antisocial behavior takes different forms in the two sexes. In contrast to men, women, having lighter and physically weaker bodies than men on average, tend to engage in indirect aggression (sabotage, ganging up, gossiping, poisoning, and so on) rather than get into outright physical fights, which men (and male vertebrates) generally will do. This finding from EEG analysis fits with other measures of arousal, such as skin conductance (to measure perspiration) that correlates with nervousness. In psychopathic individuals, this response is greatly muted, consistent with them experiencing lower arousal.

## AUTISM SPECTRUM DISORDER (ASD)

A study by Akihiro Yasuhara of 1,014 autistic children in Osaka, Japan, found micro-seizure activity evident in the EEG recordings in 37 percent of the study participants.[26] Yasuhara concludes that children with ASD have partial epilepsy and the seizures (evident as discharges in the EEG recordings) impair function in the brain's mirror neuron system. These neurons in the frontal lobe react when an animal moves and when the animal observes the same action performed by another animal. In humans, the mirror neuron system is thought to have an important function in higher cognitive processes such as language and in the ability to interpret intentions from other people's actions, as well as in the ability to empathize with the feelings of others.

## SCHIZOPHRENIA

Diagnosing schizophrenia is typically based on self-reports and abnormal behavior, but patients with schizophrenia can be distinguished from controls by whole-brain functional network analysis. Researchers Yu Wang and colleagues at Fudan University, Shanghai, China, and Fengzhu Sun, at the University of Southern California in Los Angeles, developed computational methods to identify schizophrenia patients by using fMRI data. In their studies involving 398 schizophrenia patients and 375 healthy controls, they were able to accurately identify those people with schizophrenia from diagnostic features of their fMRI.[27]

Another study used functional brain imaging to identify people with schizophrenia based on the "oddball task."[28] The oddball task is a psychological test of responses to a novel stimulus presented within a string of repeated stimuli. For example, a series of snare drum strikes might be heard, but in the midst of that series a piano key is struck. That novel sound of the piano key is the oddball event. An event-related potential is observed in the EEG in the parieto-central area of the head that rises about 300 milliseconds after the stimulus is presented, but when an unexpected stimulus is perceived, this "P300" wave surges to much larger amplitude than the waves evoked by the repeated stimuli. ("P" refers to positive polarity of the evoked brainwave response, and 300 refers to the time in milliseconds after the stimulus when the EEG response peaks.) An fMRI indicates that the dorsolateral prefrontal cortex, engaged in attention-dependent cognitive states, is involved. This finding makes sense, because individuals with schizophrenia have attentional difficulties preventing them from focusing and staying on task, and they have a poor ability to inhibit responses to distractions. The EEG oddball response to auditory stimuli is greatly impaired in people with schizophrenia (meaning they do not react as strongly), but interestingly, their oddball response to visual stimulation, while impaired, does not progress with the disease in the same way as deficits in the auditory oddball task.[29] Because visual and auditory information are processed through different neural circuits, this difference in responding to novelty in sounds versus images over the course of the disease should be instructive as to the aberrations in neural networks in this serious mental illness.

Other research shows that genetic risk for schizophrenia is correlated with differences in the P300 brainwave response;[30] that fast sleep spindle activity is reduced in schizophrenia;[31] and that theta-gamma coupling during working memory is impaired in schizophrenia[32]—all lines of research on the electric brain that are bringing us closer to a better understanding and treatment for this disorder.

## ADDICTION

Recent studies using EEG and fMRI to analyze patterns of brain activity are being conducted on addiction and substance abuse disorders involving tobacco, alcohol, and heroin. A 2018 study found increased beta-band activity in the anterior cingulate cortex during resting state associated with alcohol craving in people dependent on alcohol, and fMRI showed hypersynchronization and connectivity between this brain region and others at the theta frequency band. "The results of our study show a consistent pattern of alcohol craving elicited by external cues and internal desires," the researchers conclude.[33]

Another study published in 2017 has used resting-state EEG to successfully screen patients for alcohol use disorder, which is the technical term for alcoholism.[34] EEG was recorded for five minutes with the eyes open and five minutes with the eyes closed, and researchers could identify people with alcohol dependence from control subjects with 80.8 percent accuracy by the differences in interhemispheric coherences and EEG power. "Based on the results, it is concluded that EEG data (integration of the theta, beta, and gamma power and interhemispheric coherence) could be utilized as objective markers to screen the alcohol use dependent patients and healthy controls," the researchers concluded.[35] Similar studies using resting-state EEG have been used successfully in heroin addiction.[36]

Grace Wang and colleagues at the University of Auckland in New Zealand reported in 2015 that opioid users and addicts being treated with methadone had higher gamma power in resting-state EEG compared to controls.[37] In reviewing the literature on opioid addiction and resting-state EEG and fMRI published up to 2017, Hada Fong-ha Ieong and Zhen Yuan, at the University of Macau, concluded that strengthened brain connectivity in the reward system, accompanied by enhanced beta and alpha power, is evident in people addicted to opiates, and that oscillations in the temporal areas become more normal after methadone treatment.[38]

A particularly remarkable capability of analyzing resting-state brain function is in predicting how people will respond to treatments. There seems to be no way of knowing who can and who cannot break addiction to cigarettes, but smoking, like all behaviors, and especially addictive ones, is driven by neural networks. Therefore, you would expect to see differences in functional neural networks in the brains of people who become hopelessly addicted to nicotine and people who are able to kick the habit. Xiaoyu Ding and colleagues at the National Institutes of Health (NIH) in Baltimore and Duke University report that they can predict who will be able to quit smoking by examining their brain using fMRI.[39] The researchers analyzed the functional connectivity maps of twelve large-scale brain networks in every person's brain in the study, while the subjects did nothing but allow their minds to wander. (As previously mentioned in explaining the networked brain, several neural networks have been identified by fMRI in addition to the default-mode, salience, and executive networks. It is not necessary to explain all of these networks, but the analysis in this study on smokers included the anterior and posterior default-mode networks, left and right executive control network, salience network, striatum network, attention network, auditory networks, sensorimotor network, and three visual networks.) The smoking cessation program these people were enrolled in was effective for half of the participants, but by using only resting-state functional connectivity measurements, the researchers were able to look inside the smoker's brain and predict who would be successful in stopping smoking with an accuracy of 67.5 percent. What they found was that the functional neural networks of smokers who relapsed tended to stay within the same state; that is, there were no changes developing in their brain that were evident in others after undergoing the smoking cessation program. Functional networks in the brains of those who were able to quit smoking were inclined to transition between two particular functional connectivity states the researchers identified. Another study using EEG analysis, published in 2017, found that increased coherence in delta-wave and theta-wave activity between the right frontal region and the left posterior region, observed in resting-state EEG, predicts the reduction in cigarette craving in smokers undergoing a smoking-cessation program. This finding suggests that this increased connectivity in this brain circuit has an important role in reducing cigarette craving, and the researchers suggest that properly targeted repetitive transcranial magnetic stimulation to boost this circuit function might help reduce nicotine addiction.[40]

## DEPRESSION

EEG biomarkers for major depressive disorder have been reported. For example, Sebastian Olbrich and Martijn Arns surveyed the scientific literature in 2013 and summarized EEG markers for major depressive disorder in EEG power and event-related potentials. This body of research indicates that in addition to helping diagnose and understand the underlying pathological mechanisms of depression, EEG analysis of the P300 response and frontal theta activity are predictive of a patient's response to treatments with SSRIs. However, better optimization and standardization of EEG recording methods are necessary before psychologists can rely on EEG analysis for diagnosing depression.[41]

This brief survey illustrates how EEG analysis is rapidly advancing from its roots in epilepsy diagnosis to becoming a tool for diagnosing and treating psychological illnesses. This cutting-edge research illuminates not only how EEG analysis can help in diagnosing psychological disorders, but also how monitoring brainwaves can uncover the neurological basis for psychological impairments and open new avenues of treatments: from using precision electrical stimulation to neurofeedback training. Learning how to manipulate brainwaves to alleviate neurological and psychological disorders—and to enhance normal cognitive function—is the subject of the next chapter.

## CHAPTER 11

# Mastering Your Own Brainwaves

Because aspects of personality, mental ability, and neurological and psychological features of a person can be detected by differences in their brainwaves, then changing brainwaves may provide a way to alter these aspects of brain function and dysfunction. A large body of evidence supports this conclusion, providing a relatively safe way to change our brains for the better, and at the same time providing additional support for a functional connection between brainwaves and brain function. Meditation, biofeedback, and several methods that can directly alter the electrical fields in brain tissue are being used to enable a person to control their own brainwaves and thus their mental state and cognitive abilities, as well as to treat neurological and psychological disorders. Improving mood, increasing mental focus, promoting relaxation and meditation, alleviating chronic pain, and boosting mental and physical performance are all possible if you learn how to control your brainwaves.

Some of these techniques are being used by elite athletes. For example, wrestlers on the United States Olympic team listen to sounds of appropriate frequencies to control their brainwaves to sharpen mental focus and speed reaction time, and to relax and promote sleep in preparation for the next match. These techniques use biofeedback, rhythmic sound, and rhythmic lights to entrain brainwaves to the desired frequency and synchrony. A number of companies are beginning to market these techniques.

## NEUROFEEDBACK FROM THE FRINGE

It's been a long strange trip for neurofeedback from its roots in the lava lamp and flower power days of the 1960s counterculture quest to expand consciousness, its walk through the new-age mysticism of the 1970s, and now its jump onto the bandwagon of "brain training" with the zeal of "neuroplasticity," the touted miracle that the brain can heal itself. The fringe science and fleeting fads of EEG-guided meditation as a cure-all for sundry psychological problems as well as an electromagnetic elixir amplifying creativity and cognition evolved in parallel with scientific progress in neural networks and brain oscillations. But the practitioners of both realms operated in their separate domains, having little common ground to connect them and operating with considerable fear of interacting and becoming tainted by the idiosyncratic, antagonistic beliefs and practices of the other. Neurofeedback was, and still is, regarded by many as science versus quackery.

Today, many reject EEG neurofeedback entirely, citing little proof in the scientific literature and expressing concerns motivated by the obvious potential for commercial exploitation of the feckless and vulnerable. Indeed, the published studies that are available on neurofeedback are often of poor quality, with inadequate experimental controls, lacking objective outcome measures, and with results that are open to alternative interpretations.[1]

The methods and standards of practice for neurofeedback treatments are not well established. How should the feedback session be run? How many treatments are required? What changes in brainwaves should be attempted and for what disorder or desired benefit? What equipment is required and how are the EEG data to be analyzed? These questions are generally left up to the individual practitioner to sort out.

A review of the scientific literature between 1994 and 2010 on neurofeedback for treating pediatric ADHD (most but not all of the studies used theta/beta neurofeedback with a unipolar-electrode placed at Cz) found the body of literature lacking in requisite scientific rigor but, nevertheless, concluded that neurofeedback for pediatric ADHD was "probably efficacious."[2]

Another review article published in 2017 surveying peer-reviewed publications on neurofeedback draws similar conclusions and states that "ubiquitous placebo influences" likely drive these treatment outcomes.[3] A study was published in the journal *Lancet Psychiatry* in 2017, which was conducted by researchers in Germany who carried out a blinded randomized controlled trial of neurofeedback for treating adults with ADHD over fifteen weeks. Of the 118 subjects in the study, 38 received theta/beta neurofeedback training, 39 received sham neurofeedback, and 41 received meta-cognitive therapy (talk therapy). Self-reported ADHD symptoms decreased for all three groups, providing no evidence that neurofeedback provides added benefit to treating ADHD in adults.[4] This study takes the desired experimental approach that is badly needed, including control subjects, but, with so few studies, it is unknown if the optimal neurofeedback procedure was used. Even a drug study with a sample size of only thirty-nine subjects receiving treatment might well lack the power to show a statistically significant benefit, especially given the variation in this psychological disorder and its wide-ranging prognosis in different individuals.

Neurofeedback began in the late 1950s and early 1960s. Dr. Joe Kamiya at the University of Chicago and Dr. Barry Sterman at UCLA are recognized as pioneers in the field. Today neurofeedback is offered for an extreme range of psychological and cognitive purposes, which in itself raises skepticism of neurofeedback being the latest incarnation of snake oil—a cure for anything that ails you. Neurofeedback has been used for all of the following: relaxation; promoting sleep; sharpening awareness and brightening cognition; improving performance in sports; improving impulse control; enhancing memory; treating ADHD; increasing IQ; boosting creativity; improving mental focus and concentration; improving arithmetic skills; lowering blood pressure; reducing stress; reducing anxiety and treating PTSD; treating obsessive-compulsive disorder; altering mood and personality; enhancing musical performance; improving spelling; treating dyslexia; controlling seizures; treating Asperger's syndrome and autism; managing pain; treating headache and migraine; treating drug and alcohol addiction; relieving depression; reducing tinnitus; treating bipolar disorder; promoting recovery from traumatic brain injury and stroke; improving balance; treating incontinence, fibromyalgia, chronic fatigue, and Lyme disease; and improving performance in golf.

The most compelling case for neurofeedback as a potentially effective approach to treating neurological and psychiatric disorders is emerging from basic scientific studies investigating brain oscillations and brain function/dysfunction, in which neurofeedback is applied as a technique to test specific scientific hypotheses about brain function. As previously mentioned, such studies show that neurofeedback can regulate neural network function in association with changes in cognition, mood, and behavior, but they do not, in general, aim to devise a therapeutic clinical protocol. Research into neurofeedback as a therapy does not compete well with other areas of neuroscience for funding, tainted as it is for being a faddish fringe science, and there is an extreme mixture of variable practices and applications of the technique, which is available to anyone. You can purchase neurofeedback equipment on the internet.

Certification in neurofeedback treatment is earned from the Biofeedback Certification International Alliance.[5] Several organizations, such as the Association for Applied Psychophysiology and Biofeedback (AAPB) and the International Society for Neurofeedback and Research (ISNR), host national meetings and bring scientists and practitioners together to advance the field. Neurofeedback treatment is not yet covered by most insurance companies.

What is happening now, however, is that the thrust of research in neuroscience to understand brain oscillations, neural networks, and the neural circuit dysfunctions underlying brain illnesses is converging with the aims and territory of neurofeedback practitioners. This confluence will replace pseudoscience with science, something that always occurs at the cutting edge of knowledge. Practitioners care only that it works to help their patients; they don't care how. Scientists have the opposite view. Given time, research will uncover the true potential and the mechanisms to use neurofeedback effectively, if possible, to relieve human suffering.

## ENHANCING LEARNING AND COGNITIVE PERFORMANCE

A study published in 2017 used fMRI-guided neurofeedback to counter cognitive decline in aging.[6] The study included ten patients with early-stage Alzheimer's disease, sixteen healthy elderly subjects, and four healthy subjects who received sham neurofeedback as a control. The subjects were trained to recall a footpath while fMRI neurofeedback was applied to enhance activation of the left parahippocampal gyrus, and then the subjects were given several standardized neuropsychological tests of cognitive ability and memory. The results showed improvements in visuospatial memory performance after neurofeedback training in the healthy elderly subjects and in patients with early-stage AD, but no significant changes were detected in brain activation over the course of training. Structural MRI following the neurofeedback training showed increases in gray matter volumes in the precuneus and frontal cortex. None of these changes were found in the control group given sham neurofeedback. The researchers concluded that cognitive decline in healthy aging populations and in early-stage AD can be counteracted by using fMRI-based neurofeedback. Two other studies using fMRI-guided neurofeedback report improvement in working memory that trained the dorsolateral prefrontal cortex.[7] Needless to say, you cannot purchase an MRI machine over the internet, so while this body of research supports the efficacy of neurofeedback training with objective data, the practical application of neurofeedback will require a much simpler approach.

A study using EEG-based neurofeedback training to improve attention also used diffusion tensor imaging (DTI) to determine if the neurofeedback training altered the structure of white matter tracts in the frontal and parietal cortical areas. (DTI is an MRI method that can reveal how tightly structured microscopic features in major connections in the brain—white matter tracts—are. This method is sensitive to differences in nerve fiber branching, myelination, and other cellular properties.) The test subjects were university students who were randomly assigned to the experimental group undergoing neurofeedback, or to a control group, which received mock neurofeedback. The EEG-guided neurofeedback training sought to increase the amplitude of beta waves at electrodes F4 and P4 (in the frontal and parietal regions). Higher scores on visual and auditory attention were found in the experimental group after neurofeedback training, and the white matter pathways analyzed by DTI showed increased organization of the fibers. The volume of gray matter in the cortical regions connected to these white matter pathways, known to be involved in attention, was increased, indicating that the neurofeedback had changed the structure of the subjects' brains.[8]

These results are consistent with another study by researchers at the Université de Montréal published in 2013 using EEG-guided neurofeedback to increase attention and working memory. The test subjects were randomly assigned into either a neurofeedback training group, attempting to increase theta activity in the frontal-midline brain regions, or a group in which neurofeedback was based on EEG activity in another region unrelated to the task, as a control.[9] Significant improvement in performance on attention and working memory tests were found relative to the control group, and the benefits were evident in older individuals, leading the researchers to conclude that neurofeedback training to increase the power of frontal-midline theta brainwaves can be an effective intervention for cognitive aging.

As discussed in chapter ten, EEG gamma-wave activity has been associated with fluid intelligence and memory, but, with aging, fast gamma-wave activity decreases while the power of the beta wave frequency band increases. That is, brainwave oscillations, like so many other bodily functions, tend to slow down in aging. These EEG changes may underlie the cognitive decline in normal aging. Indeed, loss of gamma-band synchronization has been found in dementia and mild cognitive impairment. A study by Sabine Staufenbiel and colleagues published in 2013 used a double blind experimental design to test whether EEG-guided neurofeedback could alter these changes in EEG power associated with aging and improve cognition.[10] In one test group, a tone sounded whenever gamma power detected by one electrode (F2) increased, and in another group the reward tone sounded when the beta power of this electrode exceeded a specified limit. Eight thirty-minute neurofeedback sessions were given within a twenty-one-day period. The twenty elderly test subjects were also given standard tests of IQ and memory before and after neurofeedback. The results showed that the neurofeedback training increased the gamma-wave and beta-wave power in these elderly participants. The changes were evident during the training sessions, but they did not persist as changes in the resting-state EEG.

The results show that neurofeedback training can change the power of beta and gamma waves even in the elderly brain. However, no measurable improvement in intelligence or recollection was found. The researchers note that neurofeedback training in young healthy persons has been shown to improve cognitive abilities, including memory performance, fluid intelligence, relaxation, and creativity, but they caution that the relatively small number of subjects in their study make it more difficult to correlate changes in cognition with the changes in brainwave power that were produced by neurofeedback.

Neurofeedback training has been shown to increase creativity. For example, several studies by J. H. Gruzelier and colleagues at the University of London have shown that alpha/theta and sensory-motor rhythm guided neurofeedback enhances musical performance by novices and professionals, as determined by two experienced judges of musical performance.[11] The training raises theta-wave power over alpha power in the eyes-closed state, which resembles the hypnagogic state in twilight sleep (see page 192).

## NEUROFEEDBACK FOR NEUROLOGICAL DISORDERS

Managing neurological disorders by neurofeedback based on EEG or functional brain imaging is cutting-edge research, taking place in experimental laboratories, with results just now coming out at scientific meetings and scientific journals, but the potential for neurofeedback to become a valuable and much needed new tool for physicians to help their patients is clear. Published experimental studies on human subjects have demonstrated that with practice, reinforcement, and neurofeedback, people can learn to volitionally control neural activity in their own brain circuits on the microscale of single neurons,[12] on the mesoscale of local field potentials (LFPs) surrounding small groups of neurons or detected by electrocorticography (ECoG),[13] and on the macroscale across large cortical networks, as measured with EEG or fMRI.[14] That type of control of brain function is something that physicians strive to achieve with surgery and drugs, and as technology and understanding improve, neurofeedback seems likely to become a low-risk and effective method of treating a wide range of neurological conditions.

## PARKINSON'S DISEASE

This neurodegenerative disorder is caused by the death of neurons controlling voluntary movement. Located in a part of the brain called the substantia nigra, these neurons use the neurotransmitter dopamine to transmit signals across their synapses. Any number of factors can cause these neurons to die, including genetic risks, exposures to toxins, and other forms of disease. Approximately one million new cases per year are reported in the United States.[15] Symptoms typically begin with tremor in the hands or limbs, but as the disease progresses, the ability to control movement, speak, walk, write, and swallow can become severely impaired. There is no cure, but drugs that increase levels of the neurotransmitter dopamine, brain surgery, and deep brain stimulation[16] (discussed later) are used to treat the symptoms. All of these treatments can have significant side effects and risks. EEG-based neurofeedback is being used experimentally as a therapeutic intervention that carries minimal risk, with the goal of changing the functional connectivity in motor control circuits to relieve symptoms.

Brainwaves in the motor cortex (M1 cortex) show characteristic changes associated with voluntary movement, and these EEG signals can be used to guide neurofeedback or to control BCI devices. In experimental studies, electrodes implanted into the motor cortex of humans show strong beta-wave desynchronization during movement tasks, but abnormal beta oscillations in the motor cortex (13 to 30 Hz) are associated with Parkinson's disease (PD). Both deep brain stimulation and drugs that increase dopamine levels to treat PD reduce beta-band power in these regions, suggesting a causal link between abnormal beta waves and this movement disorder. Increased coherence in 10 to 35 Hz oscillations between cortico-cortical regions is also observed in PD, in addition to the increased power. This abnormal oscillatory coupling between populations of cortical neurons is thought to impair computation by neurons in the primary motor cortex. Because this cortical region is the final output area of movement generation in the brain, one strategy to overcome the excessive synchronization of beta frequencies in motor areas may be to reduce the power of beta waves in the M1 region by neurofeedback. A recent study showed that PD patients can learn to modulate this abnormal neural rhythmic activity in their cerebral cortex by using a chronically implanted neurofeedback device.

In a study published in 2017, three PD patients had a therapeutic deep brain neural stimulator implanted in their brain, but the device could also record cortical activity (ECoG) and wirelessly stream neural data to a computer.[17] (The approach of stimulating the brain with electricity to relieve neurological ailments will be discussed later in this chapter.) The high-resolution recording of oscillations in the cortex, which ECoG provides, was used to guide a neurofeedback game that enabled the PD patients to modulate the beta-band power in their sensorimotor cortical areas. This is the same idea that I used to reduce the power of delta waves in the targeted region of my brain in my neurofeedback session described in part II of this book (see page 139). In my case, electrodes on my scalp sensed my EEG and a computer delivered a reward tone when it detected that the desired changes in EEG power of a specific region of my brain was occurring. After playing the neurofeedback game for one to two hours, each of the three patients in this study significantly increased their performance in controlling their cortical beta oscillations. This neurofeedback-guided improvement was seen regardless of what stimulation was applied to their brain through the therapeutic deep brain stimulation electrodes, which is a standard method for treating severe Parkinson's disease. Stop a moment and consider what was just stated: These people changed their brainwaves by neurofeedback, a nonsurgical, drug-free intervention. The researchers state that this is the first study to use a completely implanted neural recording system to guide neurofeedback, although I would note that Delgado's "stimoceiver" was a quite similar device, but Delgado did not study the effects of neurofeedback (see page 227). The research demonstrates the feasibility of therapeutic neurofeedback training from a fully implanted, self-contained neurofeedback device, which could be used at home rather than the cumbersome equipment presently used in the laboratory setting. Research is underway to determine if neurofeedback guided by this implant reduces motor impairment in PD patients. There is good reason to think that it will, based on results from animal studies.

A placebo-controlled study in nonhuman primates (marmoset monkeys) with Parkinson's disease induced by the neurotoxin MPTP[18] found that sensorimotor rhythm (SMR) neurofeedback training[19] reduced parkinsonian symptoms. The SMR is an oscillatory thalamocortical rhythm of synchronized brain activity in the 12 to 17 Hz frequency band measured above the sensorimotor cortex. The SMR becomes depressed when moving limbs on the contralateral side of the body, or even when imagining moving the limb.[20]

The study of the marmoset monkeys, published in 2017, found a significant reduction in PD symptoms after SMR neurofeedback, both when the animals were on L-DOPA, the standard drug treatment for PD, and when they were off the drug. Similar neurofeedback training of the SMR has been previously shown to reduce susceptibility to epilepsy in cats.[21]

The monkeys in this study were implanted with EEG electrodes to monitor their brainwaves by telemetry (that is, by wireless communication between computer and brain electrodes). Neurofeedback training was accomplished by giving the monkeys food rewards when their EEG showed increased SMR activity. Control monkeys received the same number of food rewards, but the treats were delivered without regard for EEG activity. After neurofeedback training, the SMR activity was greatly increased in the trained monkeys compared with control animals. The researchers confirmed that in both groups the neurotoxin had killed over half of the neurons in the substantia nigra, thereby causing PD, but the PD symptoms were significantly reduced in the group receiving SMR neurofeedback. If monkeys can do it, why not humans? Other studies in human PD patients using SMR neurofeedback combined with respiration-based biofeedback (feedback guided by breathing rhythm rather than EEG recording) have reported a reduction in PD symptoms, and that the dosage of L-DOPA drug could be decreased.[22]

Two recent studies have used neurofeedback guided by real-time fMRI, rather than EEG, in Parkinson's disease. The first study used fMRI-guided neurofeedback in five PD patients to increase activity in the supplementary motor areas (SMA) and improve motor speed (finger tapping). Clinical ratings of symptoms were found in comparison to five PD patients who received no neurofeedback as a control.[23] The second study randomly assigned PD patients into two groups, fifteen of whom underwent fMRI-guided neurofeedback to increase activity in the supplementary motor area (SMA) in addition to receiving motor training, and the second group received motor training alone.[24] The patients in the neurofeedback group were able to increase activity in the SMA, but the improvements in motor control and clinical scores were not different between the two groups.

## BRAIN STROKE

A common cause of serious and sudden disability, brain stroke affects 5.2 million Americans a year. Neurofeedback may be helpful in guiding neuroplasticity during the long healing process following a stroke. A 2017 study on three stroke patients who were fitted with EEG electrodes for wireless transmission and recording reported that neurofeedback training at home improved upper limb motor function in these stroke patients, and improvements were associated with structural changes in brain circuits. After neurofeedback, MRI revealed increased white matter integrity in the cortical regions controlling bodily movements.[25]

In an EEG-based neurofeedback approach, the stroke patients were presented with visual feedback of their EEG, much the same way I experienced when watching the colored bar graphs instead of the auditory tones ("make the bars green") (see page 135). The real-time EEG signal informed the stroke patients when the abnormal balance in EEG patterns between their left and right cortex was improved. Training was given every other day for four weeks, and the results indicate that neural plasticity in the motor cortex was improved by neurofeedback-guided alterations in EEG activity toward a more normal pattern of brainwave activity after the stroke. Without any intervention, the brain will repair its damaged circuits to the extent possible after a stroke or other trauma—depending on the extent of the damage—but neurofeedback can help guide that process of neural plasticity to speed and improve the repair of neural circuits.

## CHRONIC PAIN

Chronic pain is profoundly disabling for millions of people who have suffered an injury or have a nervous system disease. Unfortunately, the most powerful drugs to manage chronic pain (narcotics) can lose their effectiveness over time as tolerance to them increases, reaching a point at which the high doses cause significant side effects and the drugs lose their potency to suppress the pain. A recent study, conducted by researchers in Germany and Switzerland, set out to determine whether neurofeedback can reduce pain, and to explore why individuals have various levels of pain tolerance. Twenty-eight healthy subjects were given a harmless painful stimulus while undergoing fMRI in real time, so that brain imaging could guide neurofeedback training.[26] A thermoelectric probe was placed on each subject's right forearm and the temperature was slowly increased. As the probe became progressively hotter, the test subjects reported the level of pain they were experiencing on a scale of one to ten. Researchers, mercifully, did not heat the metal probe beyond the point where the subject's pain response reached an intensity of seven. The temperature of the probe was recorded, providing a quantitative measure of the reported intensity of the painful stimulus. When the temperature of the probe was compared with the level of pain that different subjects experienced, the data documented the well-known fact that different people have quite different tolerances for pain.

When the pain was induced, fMRI showed increased activity in areas of the brain known to be involved in pain perception (circuits in the primary and secondary sensory cortex and the posterior insular cortex). The consequences of pain (for example, arousal and emotion) activate regions of the prefrontal cortex where higher-level cognition takes place, including the anterior cingulate cortex, the anterior insula, and prefrontal cortical areas. Subcortical areas including the basal ganglia and thalamus are also activated by pain in association with their function in shuttling signals to the cerebral cortex to grab attention and evaluate whether the stimulus is positive or aversive. Pain can be shut off from our conscious awareness, as every nurse knows who distracts a child in order to rip off a bandage painlessly, or when a soldier in the midst of a life-or-death gun battle can be unaware that he has been shot. Deeper in the brain still, in the brain stem's periaqueductal gray and ventral tegmental areas, transmission of pain signals from the spinal cord to higher-level brain regions is regulated. During baseline studies, perception of pain was found to be associated with deactivation of the striatum, the anterior cingulate cortex, and anterior insular cortex. The data showed that subjects with higher pain sensitivity had increased activity of neurons in the pain-related areas than people with lower pain sensitivity tolerance. This finding tells us that their heightened sensitivity to pain was not an exaggeration; their perception accurately reflected the level of activity in pain circuits in their brain, and these circuits are more sensitive to pain than in people with a higher pain tolerance.

The subjects were then asked to reduce the increase in activity of these brain regions by using neurofeedback, guided in real time by the fMRI analysis of their brain's activity in these regions. During neurofeedback, increased ability to cope with pain was clearly observed, and it was associated with predicted changes in brain activity in the anterior cingulate cortex, prefrontal cortex, hippocampus, and visual cortex.

Neurofeedback guided by real-time fMRI is a relatively new approach, and one that has tremendous potential because it provides the ability to clearly pinpoint spots of neural activity in the brain activated in cognition, and because, in contrast to EEG, which records activity from the cerebral cortex, fMRI can detect neural activity throughout the brain. Application of fMRI for neurofeedback, however, is currently limited to experimental studies because of the enormous expense of these instruments and the sophisticated expertise required to operate them. Neurofeedback guided by fMRI has been used for studies to reduce tinnitus[27] and in previous studies of chronic pain,[28] as well as in treating psychological disorders such as schizophrenia[29] and depression.[30] Although MRI-guided neurofeedback is not practical as a medical therapy, its use in scientific research will increase our understanding of how to use EEG-guided neurofeedback treatments most effectively.

## HUNTINGTON'S DISEASE

Another recent example of the potential for fMRI-guided neurofeedback to treat severe neurological disorders is research by Marina Papoutsi and colleagues from University College London who reported their findings at the 2017 meeting of the Organization for Human Brain Mapping in Vancouver, where I presented research from my lab on a new mechanism of nervous system plasticity. Huntington's disease (HD), which took the life of folk guitarist Woody Guthrie, is an inherited genetic disease that impairs motor and cognitive function, slowly progressing and eventually causing death ten to thirty years after symptoms first appear. This team of researchers found that real-time fMRI neurofeedback training in HD patients could enable them to learn to control activity of neurons in the regions of their brain affected by the disease, and that the amount of plasticity induced in these brain circuits by fMRI-guided neurofeedback strongly correlated with the level of improvement in motor performance.[31]

## NEUROFEEDBACK FOR PSYCHOLOGICAL DISORDERS

Neurofeedback to increase alpha/theta power have shown promise in relieving several anxiety disorders. In reviewing the literature, Corydon Hammond at the University of Utah concludes that depression, anxiety disorders, and OCD are improved by alpha/theta enhancement by neurofeedback.[32] For example, PTSD is notoriously difficult to treat effectively, because the incidence of relapsing is so high. A randomized controlled study by Eugene Peniston and Paul Kulkosky reported in 1991 that all fourteen Vietnam veterans who received standard treatments for PTSD relapsed after thirty months, but only three of fifteen who received fifteen hours of alpha/theta neurofeedback training relapsed.[33] The veterans in the group that received neurofeedback were able to reduce their medication, and improvements in their anxiety scores were significantly higher than the control group receiving standard treatment for PTSD.

A recent assessment of the efficacy of EEG-guided neurofeedback in children by Elizabeth Hurt and colleagues of Wright State University concludes that EEG-based neurofeedback for ADHD is recommended as a treatment for families who have tried conventional treatments, based on the outcome of twelve randomized controlled trials in the scientific literature.[34] They found that neurofeedback training in these studies resulted in measurable improvements in sustained attention, sensory/cognitive awareness, communication, and sociability in children with autism.

However, a similar analysis of the scientific literature by Samuele Cortese and colleagues at the University of Southampton, United Kingdom, published in 2016, comes to the opposite conclusion, because in their analysis, many of the studies are small and not sufficiently rigorous.[35] The quandary here is that the neurofeedback methods and experimental protocols vary widely. Researchers on all sides of the debate agree that more studies and optimization of methodology are required to reach definitive conclusions on the efficacy of neurofeedback training for ADHD. A poorly controlled small study can lead us astray, but at the same time, a little bit of bad data can ruin a lot of good data, so more research is needed.

## BRAIN STIMULATION

Several methods of directly stimulating the brain are being used in experiments to enhance cognitive performance, to treat neurological and psychological disorders, and as an experimental tool to investigate how neural circuit function controls behavior.

### DEEP BRAIN STIMULATION (DBS)

Neurosurgeon Itzhak Fried of UCLA cares for patients who suffer life-altering neurological disorders. At a scientific meeting in Israel in 2018, he showed a video clip of one of his patients, a teenage girl in a wheelchair, who was slumped in disarray as if she were a marionette tossed into the chair, with her arms and legs tangled and floppy, her head drooped inertly to one side, and her long black hair trailing to the floor. She is suffering from a condition called dystonia, which impairs the brain's ability to control muscles. In the next scene the young girl walks with a gleeful smile down the hallway, just like any other teenager, because of electrodes in her brain placed there by Dr. Fried and his colleagues that were stimulating the globus palladus region of her brain. This part of the brain is known to be involved in motor control, but not much is known yet about the mechanics of deep brain stimulation (DBS)—where to put

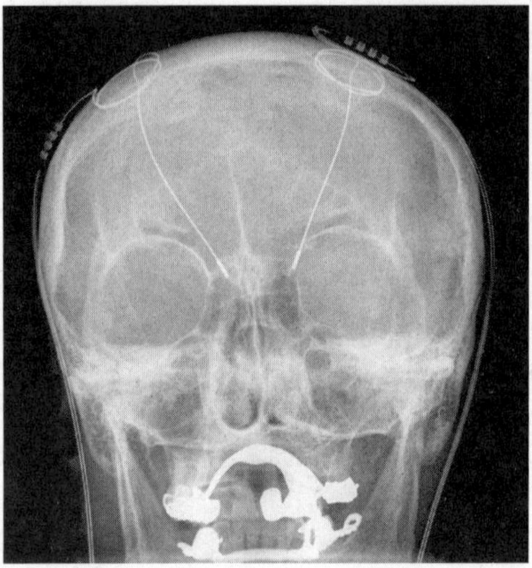

**FIGURE 38:** *Electrodes implanted in a patient with Parkinson's disease for deep brain stimulation (DBS) to improve control of bodily movement.*

the electrodes and what stimulus patterns to use—or even how it helps. But as Fried told me, "We don't fully understand how it works, but we can't wait for the *Drosophila* and mouse research; we have to jump to humans."

Introduced for treatment of Parkinson's disease in the 1990s, DBS is now a well-accepted therapy for patients who fail to respond to medications. Even though how it works is not clear, DBS can be extremely effective. Susan Mollohan, who has Parkinson's disease, described her experience with DBS in her blog on the Michael J. Fox Foundation website on September 2016: "I've not had one tremor, and my toes no longer curl. In the past I was taking five Sinemet (levodopa-carbidopa) a day, and I'm now taking only two. I'm back on my bike preparing for the New England Parkinson's Ride . . . I can truly and sincerely say that DBS surgery was worth it."[36]

Unfortunately, doctors cannot predict how effective DBS will be for any given patient, or even if it will work for them at all, because, the truth is, they are not sure how it works in the first place, although there are several good theories. Additionally, the procedure itself is hit or miss, which is why the patient must be awake when the electrodes are implanted into their brain. Physicians quickly found that if the stimulating electrode that is targeted on a cluster of neurons that regulates movement (the subthalamic nucleus) penetrated slightly deeper, the

Parkinson's patient was instantly racked with severe dysphoria, began crying, and felt profoundly suicidal, but the psychological meltdown stopped when the stimulation was terminated.

Imagine having surgery on your brain, especially while awake: "I remained partially awake in a 'twilight sleep' to ensure the surgeon positioned the electrodes in the right place," Mollohan wrote. "I was extremely anxious . . . The doctors placed the electrodes in various places in my brain to see how my body would react. At one point they hit a spot, and my tremors started going ninety miles an hour. Other spots made me lose my speech and made my voice sound unfamiliar and slow. The doctors placed the electrodes in an area of the brain that controls movement, and I became emotional when I saw that my right hand, where I experience tremor, was able to open and close almost as fast and freely as my left. It's been eight years since I've been able to do that."

After DBS was approved by the FDA for treating Parkinson's disease, the technique began to be more widely used medically for treating many other neurological and psychiatric disorders.[37] Some of the extremely varied conditions that have been and continue to be treated by DBS include dystonia, Huntington's disease, multiple system atrophy, epilepsy, pain, cluster headache, depression, obsessive-compulsive disorder (OCD), and Tourette's syndrome, as well as impulsive and violent behavior. In early research into using DBS for treating the psychiatric illness of OCD, electrical stimulation was applied to the same brain regions where surgery is used to produce therapeutic effects[38] in the frontal-basal brain networks (orbitofrontal cortex, anterior cingulate, the caudate, and thalamus). DBS of the caudate nucleus was attempted to treat depression and anorexia in the 1950s,[39] and stimulation of the septal region was first used in 1963 to treat schizophrenia.[40] Today, DBS is used for OCD, major depression, schizophrenia, dystonia, anorexia, and many other disorders.[41]

It is remarkable that the approach of directly stimulating brain circuits with electricity can benefit such a range of neurological and psychiatric conditions, but in most cases, there is not a good understanding of how the stimulation works to reduce symptoms of the disorder, precisely where the electrodes should be placed, and what type of electrical stimulus (the pattern and strength) would be most beneficial. There is no basis for assuming that the stimulus parameters used to treat Parkinson's disease (approved by the FDA) would be effective in treating obsessive-compulsive disorder, for example, which involves completely different circuits in the brain.

Moreover, there is little animal research to guide surgeons on how to use DBS in an attempt to help their desperately ill patients. Psychiatric disorders in particular are not well modeled by the mouse brain, although certain aspects of psychiatric conditions can be modeled in the animals. (Is there such a thing as a mouse with schizophrenia?) Physicians applying DBS to treat disorders other than Parkinson's are, in effect, doing human experimentation without a clear understanding of the mechanism of the treatment or what neural circuits are being affected and how. But they are doing it because it is helping, sometimes transforming their patients' lives for the better.

As it flows through tissue, electricity will excite some neurons and inhibit others, so it is not always clear how the stimulation is altering function in the neural circuits. Another problem is that neurons that are far removed from the tip of electrodes are also stimulated in addition to the targeted neurons if their axons pass by the vicinity of the electrode. In addition, electrical currents from implanted stimulating electrodes excite neural impulses that travel down the axon in the normal manner to stimulate other neurons connected by synapses, but impulses triggered at a spot along an axon also travel backward (antidromically) into the neuronal cell body and dendrite, and this has very different effects, some of which were discussed in the chapter on brainwaves during sleep (see page 209).

Finally, not all benefits of DBS may be due to the electrical stimulation itself, but rather due to a healing response to the electrode implant. Chronic stimulation and electrode placement cause cellular reactions in non-neuronal cells (astrocytes and microglia) that produce neuroactive substances, like neurotransmitters and growth factors, in response to injury. The biological and electrophysiological responses of the neurons to their prolonged artificial firing may alter their structure, gene expression, and function, which could reduce the disease symptoms by indirect means, rather than by DBS directly driving normal patterns of activity in the circuits to compensate for any dysfunctional firing.

Finally, while there is a good understanding of Parkinson's disease at a cellular level, in many other neurological and especially psychological disorders a clear understanding of the cause for the disease at the neural circuit level is far from known.

Despite the benefits, DBS is not without risks. In addition to the risks one accepts with any brain surgery, 10 percent of patients who have undergone DBS for treatment of Parkinson's disease experience problems, including cognitive, language, and memory impairment; alteration in mood and behavior; and sleep disturbances—in part depending on where the stimulation electrode is placed.[42]

Nevertheless, the approach of manipulating brain function by directly stimulating electrical circuits in the brain is effective in a wide range of conditions that cannot be treated in other ways, and, while the current knowledge is weak and current methods are crude, there is enormous potential in directly and precisely modulating circuits in the electric brain to improve brain function and relieve human suffering. Surgery to remove masses of brain tissue and drugs used to treat psychological and neurological conditions are horribly blunt tools. Surgery doesn't have the microcircuit-level precision that is required to treat these issues. Drugs not only affect the whole brain but impact the body wherever cells that are sensitive to the compound are located. SSRIs, for example, affect the gut as well as the brain.

With improved understanding of the brain and better technology to manipulate electrical activity in specific brain circuits, electrical stimulation will become a much more precise and effective way to correct neurological disorders, mental illnesses, and even learning disorders.

### THE ROOTS OF DBS

The history of DBS as previously outlined (the party line) is, in fact, revisionist history. Marwan Hariz, a neurosurgeon and deep brain stimulation expert at University College London, criticizes this oft-repeated historical scenario that the safe and effective use of DBS in treating Parkinson's disease opened its application to psychiatric illness. That account may be true with a limited historical vision that looks back only to the 1990s, but not if one looks back further to the roots of DBS. The truth is that the original motivation for using DBS was a desire to influence psychiatric disorders. That history is clearly preserved in the scientific literature, but such experiments on humans came to be regarded by many as unethical, and so the true origins of DBS get pushed under the rug.[43] José Delgado's life work with deep brain stimulation (see page 217) clearly shows that his motivation was to understand and manipulate human behavior, particularly deviant behavior.

Prefrontal lobotomy and other forms of psychosurgery that were undertaken in the 1930s and '40s were, according to Marwan Hariz and others, the genesis of DBS. Although psychosurgery was a major advance in that it regarded mental illness as a brain dysfunction that could be treated, the surgical approach came with serious side effects. More troubling was that prefrontal lobotomy was increasingly abused to subdue unruly patients, notoriously by American neurosurgeon Walter Freeman, who devised an ice pick procedure to simplify the lobotomy to where it could be performed by almost anyone. Freeman performed the surgery on 2,900 patients, including 18 children! By using his simplified method of hammering an ice pick through the eye sockets into the forebrain and then slashing it side to side to amputate the frontal lobes, he did up to twenty prefrontal lobotomies per day.[44] The procedure left many in a vegetative state, and it killed an estimated 490 others.[45] President John F. Kennedy's sister, Rosemary, who was born with mild learning difficulties, was given a lobotomy by Walter Freeman in 1941. She was left incapacitated and institutionalized for the remainder of her life.

In an attempt to find a better intervention to cure mental illness by the direct delivery of electricity into the electric brain, electroconvulsive therapy was developed. In view of the harmful effects of psychosurgery, electroshock therapy (ECT) became a much more attractive alternative to treat mental illness. These two approaches (targeted surgery and electrical stimulation) then converged to treat mental illness. Using stimulation by DBS at specific spots in the brain would provide the necessary precision to correct psychiatric illness with fewer side effects.

In addition to Delgado, who worked extensively with DBS in the early 1950s, psychiatrist Robert Heath conducted similar experiments for three decades at Tulane University in New Orleans, beginning in the early 1950s. Heath used DBS to stimulate patients with schizophrenia and to search for the brain's pleasure center. The latter experiments extended to efforts to promote heterosexual behavior in a homosexual male.[46] In 2000, psychologist Alan Baumeister published an influential paper reviewing the history of the electric brain stimulation program at Tulane as an historical case study in medical ethics.[47] In it he writes, "The Tulane electrical brain stimulation experiments had neither a scientific nor a clinical justification . . . these experiments were dubious and precarious by yesterday's standards." Heath's 1972 article on orgasms and his earlier work with electrical self-stimulation of the brain to identify the brain's pleasure center and to treat homosexuality are cited as ignoring all standards of ethics.

In their 1972 paper, Charles Moan and Robert Heath described the homosexual test subject and reported his father's view that "he considers his son to be a deeply disappointing young man who was a failure and coward during childhood and adolescence."[48] The doctors hoped to cure the twenty-four-year-old man of homosexuality by stimulating electrodes in the pleasure centers of his brain in association with heterosexual experiences and stimulating the aversive regions of his brain in association with homosexual experiences.

"[Electrodes] were implanted into the following brain regions: right mid septal, right hippocampus, left and right amygdalae, right anterior hypothalamus, right posterior ventral thalamus, left caudate nucleus, and at two subcortical sites within the left lobe of the cerebellum,"[49] Moan and Heath wrote, describing how they stuck electrodes in regions all over the man's brain.

EEG recordings were performed before, during, and after the experiments. "Conventional scalp EEGs showed an abnormality characterized by bitemporal slow-wave activity, maximal on the left. Paroxysmal delta activity also appeared over the right temporal region with Chlorolose activation."[50] (Chlorolose is a general anesthetic.)

The researchers stimulated the electrodes implanted in various parts of the man's brain, increasing the stimulus current from 0.5 to 6 milliamps, until they zeroed in on one spot, the septal region that evoked an intense sensation of sexual pleasure in the man when it was stimulated. With this preparation the researchers were ready to begin their experiment:

> At this time, [the patient] viewed a 15-min 8 mm "stag" film featuring sexual intercourse and related activities between a male and female. There was continuous EEG recording of the patient, as well as observation through a two-way mirror. He showed no obvious verbal or gestural response during the presentation, but was highly resentful, angry and unwilling to respond at its conclusion. Base rate (pre-film) EEG indicated the presence of tension and hyperalertness prior to viewing, reflected by low voltage, low amplitude activity. No significant changes in EEG were apparent during the time the patient watched the film; but high amplitude alpha activity was noted at the end of the session, indicating some relaxation because the movie was over.[51]

In the following days, they handed over controls of the three-button stimulation device to the man so he could deliver electrical stimulation to his own septal region whenever he wanted.

"[The patient] stimulated himself to a point that, both behaviorally and introspectively, he was experiencing an almost overwhelming euphoria and elation and had to be disconnected, despite his vigorous protests. His post stimulation EEGs were unremarkable," they noted.

Then they showed the man the same stag film, and this time he became aroused and was provoked to onanism.

Next, the researchers proceeded to perform their ultimate experiment:

> Arrangements were made for a 21-year-old prostitute to spend 2 hours with him in a laboratory specially prepared to afford complete privacy . . . On the afternoon of their meeting, the patient's electrodes were attached to the encephalograph via an extension cord for increased mobility and recordings were obtained for 45 min with an interruption for delivery of passive stimulation of the septal region for 20 sec. [The patient]

> was then introduced to the prostitute, and EEGs were obtained throughout his relationship with her.[52]

The description of events during this encounter are graphic, but to summarize, two "successful" intercourses were documented: "Delta waves appeared at several of these sites as sexual arousal increased and that immediately prior to orgasm striking changes in recordings from septal leads occurred resembling epileptiform discharge . . . However, shortly after its onset the recordings were overwhelmed by the effects of movement; and, although it was impossible to separate the meaningful from the artifact, septal seizural activity seemed to endure throughout the orgasmic response."

Little is known of the man's eventual fate.

In his 2000 paper, Baumeister wrote that "the ethical responsibility of the editors who accept reports of this kind for publication should also be discussed."[53] Editors, peer reviewers, and funding agencies took notice, and the funding and publication of this line of research was curtailed.

DBS was primarily used for behavioral and thought disorders well into the 1970s, but ethical criticism brought this line of research to a close. Treating neurological illness with DBS faced far less daunting ethical issues than those presented in treating mental illness where even informed consent can be dubious, so DBS research shifted to treating PD and other neurological disorders as well as for use in prosthetics, but not for psychological conditions. Now, mental illness is once again being treated with DBS, and hopefully we are enlightened by experiences of the past.

## RECALLING MEMORIES WITH ELECTRICAL STIMULATION

It is quite ghoulish to see a young girl's face as she chats in the rapid fire, incessant adolescent monologue of teenage girls, except that the top of her head is open and her glistening bloodshot brain is exposed and pulsating as doctors insert metal probes into it and take out globs of tissue, leaving holes like pits dug in wet sand. She would not be undergoing brain surgery if her life had not become seriously dysfunctional from a neurological illness. Her conscious cooperation is required to aid the surgeon in placing electrodes and removing diseased tissue while sparing intermixed healthy tissue and carrying out vital functions. Unlike a heart surgeon, who is guided by accurate anatomical maps that are consistent from person to person, everyone's brain is a little bit different, and it is those differences that make us who we are. Any surgeon risks losing his patient, but a neurosurgeon risks saving his patient's life but losing the person. The revelation that brain surgery can be done on someone who is awake because the brain has no pain sensation has yielded, and continues to provide, the most insightful look into how the human mind works in terms of neuronal electric circuits.

Famed neurosurgeon Wilder Penfield, mentioned earlier for his work with Herbert Jasper on epilepsy (see page 307), was the first surgeon to realize that brain surgery could be performed using local anesthetic while his patient was fully conscious and able to engage with the medical team, helping to guide the surgical instruments and the placement of electrodes in the patient's own brain.

"The brain has no sensation of its own. Consequently, the patient has no possible means of knowing when the electrode is applied, unless he is told or unless he is aware of a positive physiological effect," Penfield described.[54] Stimulation of most brain regions, however, does not evoke a conscious thought, because most of the brain carries out unconscious brain functions. However, electrical stimulation is more likely to short-circuit normal neural network function. Describing the results of his experiments that involved applying mild electric stimulation to different regions of his patients' cerebral cortex during surgery, Penfield wrote:

> There are many things that stimulation can never produce. No constructive thinking is produced, no willed or purposeful behavior. In general, application of this crude electrical current seems to interfere with the normal functional employment of the cortex itself in the area to which the electrode is applied. For example, if it is applied to one of the speech areas of the dominant hemisphere, such as Broca's convolution, the patient

> is silent. No words can be produced that way. If the patient is asked to answer while the electrode is still applied, he discovers, to his surprise, that he is aphasic. That is, he can no longer find words to express his thought. But they come with a rush when the electrode is lifted and he says, then, the things he was trying to say while the electrode was interfering with his employment of the speech area of cortex.[55]

Disruption of brain function by electrical stimulation was also the most common response observed by Delgado pursuing his research in the 1950s and '60s, but Penfield was startled by what happened when he stimulated temporal lobe regions of his patients' brains: Very vivid memories were evoked. The memories played out as if in real life, with the proper sequence and in cinematographic detail of the day that the long-forgotten event occurred. He described his bewilderment when in 1931 he stimulated the temporal lobe of a woman's brain and suddenly the woman experienced a vivid memory of giving birth to her baby girl. "That had happened years before," Penfield wrote, "and meanwhile the girl had grown up. The mother was now lying on the operating table in my operating room, hoping that I could cure her attacks of focal epilepsy."[56]

This mysterious response was bewildering to Penfield, and, at first, he attributed the woman's reaction to an even greater mystery: "This, I thought, was a strange moment for her to talk of that previous experience, but then, I reflected, women are unpredictable, and it was never intended that men should understand them completely."[57]

That explanation was sufficient to satisfy Penfield for the next half decade, until "It was more than five years later when a somewhat similar psychical state made its appearance during electrical stimulation. This time, however, it seemed certain that the stimulus had somehow summoned a past experience."[58] This time he was operating on a fourteen-year-old girl with severe epilepsy.

He recounted: "Under local anesthesia, I mapped out the somatic sensory and motor areas for purposes of orientation, and I applied the stimulator to the temporal cortex. 'Wait a minute,' she said, 'and I will tell you.' I removed the electrode from the cortex. After a pause, she said:

'I saw someone coming toward me, as though he was going to hit me.' It was obvious also that she was suddenly frightened."[59]

This type of response was repeated in several other patients.

"I hear music again," another patient said when Penfield stimulated her temporal cortex. "It is like radio." About this patient, Penfield wrote:

> Again and again, then, the electrode tip was applied to this point. Each time, she heard an orchestra playing the same piece of music. It apparently began at the same point and went on from verse to chorus. Seeing the electrical stimulator box, from where she lay under the surgical coverings, she thought it was a gramophone that someone was turning on from time to time.[60]

Because the energized electrode evoked a piece of music, Penfield realized that the woman's memory of this event was playing out in real time from her temporal lobe brain circuits: "She was asked to describe the music. When the electrode was applied again, she began to hum a tune, and all in the operating room listened in astonished silence. She was obviously humming along with the orchestra at about the tempo that would be expected."[61]

A year later, the woman described the odd experience she had on Penfield's operating table: "It was as though it were being played by an orchestra. Definitely it *was not* as though I were imagining the tune myself. I actually heard it. It was not one of my favorite songs, so I don't know why I heard that song."

Penfield concluded that our brain has a permanent record of streams of consciousness from our experiences locked in the temporal lobes of our brain.[62] Recalling past events may be easy or impossible, but they are there and released by electrical stimulation, he concluded. "Many a patient has told me that the experience brought back by the electrode is much more real than remembering. And yet he is still aware of the present situation. There is a doubling of consciousness, and yet he knows which is the present."[63] Penfield concluded:

> There is a permanent record of the stream of consciousness within the brain. It is preserved in amazing detail. No man

> can, by voluntary effort, call this detail back to memory. But, hidden in the interpretive areas of the temporal lobes, there is a key to a mechanism that unlocks the past and seems to scan it for the purpose of automatic interpretation of the present. It seems probable also, that this mechanism serves us as we make conscious comparison of present experience with similar past experiences.[64]

In recent years the electrodes (and computers) have become much more sophisticated, enabling researchers to tap into multiple brain circuits simultaneously. The location of the electrodes can be pinpointed precisely with MRI, so the approach of experimenting on humans, pioneered by Penfield, is now yielding startling new insights into how the mind and brain work.

Using these modern methods, neurosurgeon Itzhak Fried and colleagues had his patients on the operating table explore a computer-generated virtual town as he recorded the firing of individual neurons in their brains. All of our understanding of place cells in the hippocampus and spatial navigation comes from experiments on rodents and other experimental animals, but does this understanding also apply to the human brain? We know that there are major differences in place cell behavior in bats and rats, for example; only by recording the human brain will we ever know if the same mechanisms uncovered in animal experiments similarly work in people.

After recording electrical activity in 317 neurons in the temporal and frontal lobes of his patients, Fried and his colleagues found that there were specific neurons that only responded to the patient seeing specific landmarks in scenes presented on a video display. These neurons, like place cells in rodents, were located primarily in the hippocampus and parahippocampal brain regions.[65] Likewise, the large body of research on rodents implicating theta rhythms in locomotion and navigation was confirmed in humans in 2017 by Fried's and colleagues' studies on their patients who had electrodes planted in their brain for controlling epilepsy, while they walked around freely as the firing patterns of their neurons were recorded and analyzed by computer. As in rodents, the electrodes in the human brain showed neurons firing at the theta rhythm, and the power of the theta rhythm increased during bodily movement just like it does in rats.[66]

Current thinking is that place cells are not unique, but rather they are doing what other kinds of neurons do in the higher-level cerebral cortex—responding to very specific features in the environment. For example, Fried and others have found neurons that respond to very specific faces, such as those of Bill Clinton, Halle Berry, or Jennifer Aniston.[67] But the remarkable thing is that these neurons are not performing facial recognition—they respond to the specific face no matter how it appears, in profile, full face, new photo or dated image, and even to the spoken or written name of the person. These neurons encode a fully integrated concept of very specific people. This is in line with what Marcel Just and colleagues are finding in their "mind reading" by fMRI experiments. This coding of information in the brain as integrated concepts occurs because the neuron that is firing the action potentials that the electrode is sampling is connected to a vast network of other neurons that encode the full concept of Jennifer Aniston, for example.

At a scientific meeting in Tel Aviv, Fried showed the results of experiments in which his patients watched brief video clips of different TV programs. One of the neurons that his electrodes tapped into fired when his patients viewed a scene from the TV show featuring the comedian Jerry Seinfeld. The neuron started firing as soon as the *Seinfeld* clip began to play, and it continued to fire throughout the entire twelve-second video clip until the researchers cut to a different TV program. When they cut back to *Seinfeld*, the neuron started firing again.[68] Other neurons responded specifically to video clips of the cartoon show *The Simpsons*.

It would appear that this neuron encodes the concept of Seinfeld, in all of the rich features that we attribute to him, but how can we know for sure? In an animal experiment, it is very difficult to answer this question, but in human experiments all you have to do is ask. After the experiment, the researchers asked the patient what she remembered seeing during the tests. She accurately recalled all the snippets of different TV shows that she had been shown, and when she said "Seinfeld," that same neuron that researchers had seen firing when they played the *Seinfeld* clip started firing as the woman recalled having seen it.

In the experiments in which patients negotiated a computer-generated environment, Fried and colleagues found that if they stimulated neurons in the hippocampus with a short burst of impulses at the theta frequency, just as researchers use to increase synaptic strength in brain slices of rats, called LTP (see page 205), they could enhance memory. When they stimulated neurons at the theta frequency in association with specific locations in the virtual environment, or if they presented a series of faces and delivered a theta burst stimulus while the patients viewed some faces but not others, the objects that were seen when the hippocampal neurons were simultaneously stimulated by their electrodes were remembered much better in memory tests afterward.[69]

The key, Fried told me, is to stimulate the white matter tracts connecting with the hippocampus, rather than to stimulate the gray matter itself. (Gray matter is the dense network of interconnected neurons, but white matter, lying beneath gray matter, is formed from bundles of axons that connect neurons in gray matter into circuits.) Firing of neural circuits can be better controlled by specifically boosting activity in connections to specific neurons that encode the concept being stored in memory and recalled, rather than stimulating gray matter tissue where there is a complex jumble of highly interconnected neural circuits. Stimulating white matter works for the same reason that plugging in a television works, but zapping the TV set directly with a bolt of electricity would not.

This body of research shows that electrical stimulation of the hippocampus can improve recall, and this method of brain stimulation may one day help Alzheimer's patients or others with cognitive and memory problems. Electrical stimulation of the proper neural circuits involved in memory could do for the weakened brain what a pacemaker does for the weakened heart. Such stimulation, depending on when and how it is applied, could improve encoding memory, consolidating short-term memory into long-term memory, or retrieving memories. A 2019 study using TMS to synchronize theta and gamma waves in the frontotemporal cortex reports significantly improved working memory in older adults.[70]

Penfield must have glimpsed this potential of brain stimulation when he randomly applied his electrodes to different parts of his patient's temporal lobes and the electrical stimulus evoked random, but vivid, long-forgotten events. Today the technology is much improved, and there is much better technology on the way. The checkered past of brain stimulation experiments from the 1950s to 1970s is an enormous impediment to accepting brain stimulation as a therapeutic approach to cognitive dysfunctions, but seeing the benefits to his patients, Fried believes that electrical stimulation of the electric brain is a potentially powerful therapeutic approach that needs to be developed.

## TRANSCRANIAL BRAIN STIMULATION

The work of António Egas Moniz, who launched psychosurgery in 1935 with his Nobel Prize–winning idea of severing connections to the prefrontal cortex (lobotomy) of psychiatric patients, and Delgado's deep brain stimulation that began in the 1950s were motivated to find an effective intervention to control deviant and especially violent behavior in psychiatric patients. Although harshly criticized and eventually stifled in applying their methods, both scientists were boldly breaking new ground, blazing a trail in the right direction, but using primitive technology and a primitive understanding of brain function that were simply inadequate to achieve their goal. New methods to modulate electrical activity in brain circuits, by passing a weak DC current into the brain through the skull, together with functional brain imaging (fMRI), are bringing about new understanding of the biological basis of anger and aggression, and, in experimental studies at least, providing a new approach to controlling anger and aggression by directly manipulating electrical activity in these circuits.

Devonte Washington, fifteen, heading to a barber shop to get a haircut for Easter Sunday church service, was waiting on the platform of a Washington, DC, Metro subway station with his mother and two sisters. The young man glanced up at a stranger, seventeen-year-old Maurice Bellamy, who instantly took offense, pulled out a .38 caliber pistol, and shot him to death. Three weeks earlier, Bellamy had murdered an off-duty Secret Service agent in a robbery. The teenage gunman had a long history of extreme impulsive anger and violence, which his family tried to control through appropriate medication and counseling, but to no avail. In January 2019, Bellamy was sentenced to sixty-five years in prison for the two murders.[71] The judge reduced the sentence from the eighty years the prosecution requested because Bellamy was a minor at the time of his crimes.

This tragic example, like many others that fill the daily news, highlights the obvious fact that most violent crimes are committed by individuals, primarily males, who are prone to bursts of pathological anger and violence. Seemingly the only option in response to their behavior is to remove such people from society permanently. Science may have an answer for better options.

It is known that the limbic system is a central hub of threat detection and anger, but the prefrontal cortex can suppress anger and aggression through descending connections that modulate the limbic system. Temper tantrums are common in children because these connections from the prefrontal cortex are not fully developed until after adolescence. This is the biological underpinning of why the Supreme Court holds that juveniles are not held criminally responsible to the same extent as adults—they do not have the circuitry an adult has to control impulsive anger. These circuits can be disrupted through genetics, life experience, drugs, or disease, and fail to perform their normal function to squelch violent impulses, leading to bursts of anger, assault, and violent crime.[72]

By comparison, while the prefrontal cortex suppresses impulsive aggression, severing it entirely by prefrontal lobotomy pacifies patients by destroying motivation. Severing all connections from the brain's higher-level cognitive control center, which spans throughout the brain, incapacitates patients.

New research is identifying the neural circuits of anger and aggression, and the findings offer hope of understanding how violent impulses are triggered as well as suppressed in normal individuals, and how abnormalities in these circuits can result in pathology.[73] A study recently published in the journal *Cortex* by Gadi Gilam and colleagues at Tel Aviv University reports that anger and aggression can be suppressed by targeting a specific region of the brain by using weak direct current stimulation through the skull.[74] Using the technique of transcranial direct current stimulation (tDCS), an electrode is placed on the forehead (in this case) and a weak DC current (1.5 mA) is delivered for twenty-two minutes. Unlike electroshock therapy, which drives powerful electric currents through the brain to induce a massive seizure, there are no obvious effects of applying this weak voltage to the forehead, except that neurons in the brain beneath the positive pole of an electrode will become more excitable. Conversely, electrical activity in neural circuits beneath a negative electrode becomes suppressed. In this way, activity in neural circuits of the human brain can be stimulated or inhibited by this noninvasive technique. This new study used tDCS to modulate a neural circuit that has been implicated in impulse control while test subjects were in an fMRI machine so that changes in neural activity in these brain circuits could be monitored while they were provoked to anger.

A particular region of the prefrontal cortex, the ventromedial prefrontal cortex (vmPFC) becomes engaged during anger and aggression, as can be seen by using fMRI. In this study by Gilam and colleagues, the researchers were able to show by using functional brain imaging that tDCS effectively increased neural activity in the vmPFC, and when this circuitry was activated by the applied voltage, anger was suppressed.

The subjects in the study participated in a game in which a monetary payout was distributed fairly or unfairly. Naturally, being "cheated" provoked anger in the test subjects, which was rated by the participants on a scale of one to ten. When participants in the game felt cheated, their level of anger was reduced when tDCS was applied to activate the circuits from the vmPFC that inhibit the amygdala in the limbic system. Fake stimulation designed to produce similar sensations on the skin but that did not manipulate activity in neural circuits served as a control.

The purpose of the study is to identify and understand the neurobiological circuitry of anger and aggression, but the authors state that the findings have potential clinical use. Anger and aggression provoked by interpersonal challenges are common in numerous psychiatric disorders and medical conditions, and there is a need for better treatments than the currently available drugs, counseling, and the last resort of incarceration or institutionalization. The authors state that their findings suggest that tDCS may provide "a noninvasive adjuvant to improve anger coping capabilities in individuals with pathological manifestations of anger, and perhaps also for therapeutic inoculation for populations at risk of developing such pathology."[75] In other words, acting like a pacemaker for the brain, tDCS may hold the promise of an intervention to prevent the type of ticking time bomb and the senseless tragedy that everyone involved with Bellamy (and others like him) saw coming and tried, but failed, to prevent.

## AMPING UP BRAIN FUNCTION WITH DC CURRENT

Methods to deliver electrical stimulation through the skull to manipulate electrical circuits in the brain eliminate the risk and expense of brain surgery, and thus open the technique to wider applications. The simplest approach is to apply electrodes to the scalp and deliver either a direct current of the appropriate polarity or apply alternating currents to stimulate or inhibit neuronal excitability (depending on the polarity of the voltage) in local brain regions. This is a technique that has been used to enhance learning.

One of the most difficult tasks to teach air force pilots who guide unmanned attack drones is how to pick out targets in complex radar or video images. Pilot training is currently one of the biggest bottlenecks in deploying these new, deadly weapons. Air force researchers found that they were able to cut training time in half by delivering a mild electrical current (two milliamperes of direct current for thirty minutes) to pilots' brains during training sessions on video simulators. The current is delivered through EEG electrodes placed on the scalp. I first heard about this research at the annual Society for Neuroscience meeting in 2011.[76] Biomedical engineer Andy McKinley and colleagues at the Air Force Research Laboratory at Wright-Patterson Air Force Base contrasted the cognitive boost of tDCS with caffeine and other stimulants that have been tested as enhancements to learning. "I don't know of anything that would be comparable," he told me. The application of tDCS not only accelerated learning, it enhanced performance. Pilot accuracy was sustained in trials lasting up to forty minutes whereas accuracy in identifying threats normally declines steadily after twenty minutes. This method of boosting brain performance could also have many medical

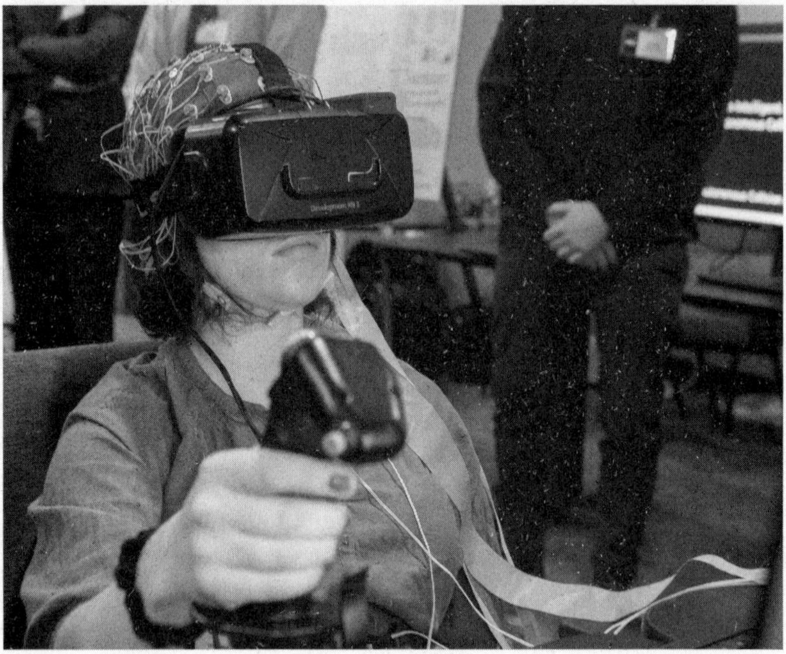

**FIGURE 39:** *The US military is using EEG and brain-computer interface methods to apply neuroscience to increase combat capabilities, as shown here where an army researcher's brainwaves are being monitored in a virtual reality computer simulation.*

applications in accelerating retraining and recovery after brain injury or loss of function due to disease.

Subjects who have experienced tDCS perceive the stimulation but state that it is not unpleasant. "It feels like a mild tickling or slight burning," undergraduate student Lauren Bullard told me, describing how the stimulation she applied to the test subjects felt to her. "Afterward I feel more alert," she says. It is not clear why. It could be a placebo effect (a positive effect that results from expecting a result) or the peculiar sensation on the scalp might heighten arousal. Increased arousal and attention are well known to enhance learning. However, control experiments were done using a sham electrical stimulus that tickled the scalp but would not be effective in stimulating the brain.

Bullard and her coauthors sought to determine if they could measure any tangible changes in the brain after tDCS, which could explain how the treatment accelerates learning. The researchers looked for both functional changes in the brain (altered brain-wave activity) and physical changes (by examining MRI brain scans) after tDCS.

McKinley, Bullard, and colleagues used magnetoencephalography (MEG) to record brainwaves that are evoked by sensory stimulation (sound, touch, and light, for example) while test subjects received tDCS. They found that tDCS gave a six-times baseline boost to the amplitude of a brainwave generated in response to stimulating a sensory nerve in the arm. This effect was not seen when mock tDCS was used, which produced a similar sensation on the scalp but did not excite brain tissue.

The effects persisted long after tDCS was stopped. Fifty minutes after tDCS, the sensory-evoked brainwave was two and a half times greater than normal. These results suggest that tDCS may increase cerebral cortex excitability, thereby heightening arousal, increasing responses to sensory input, and accelerating information processing in cortical circuits.

Remarkably, MRI brain scans revealed clear structural changes in the brain that could be detected five days after tDCS. The fiber bundles in subcortical white matter were found to be more robust and more highly organized after tDCS. This was detected by a type of MRI called diffusion tensor imaging, which showed increased anisotropy of water diffusion in the tracts (that is, water diffusion through brain tissue that was more highly restricted in one direction along the fiber tracts). Like a paintbrush wicking up paint between bristles, more highly organized axons in white matter tracts restrict water diffusion along the axons. This more ordered structure can be caused by less branching and crossing of axons and by increased myelination, which is the electrical insulation on axons that increases the speed that impulses travel from neuron to neuron. No changes were seen on the opposite side of the brain that was not stimulated by the scalp electrodes.

Discussing the results of the air force study with other scientists at the Society for Neuroscience meeting, an expert on brain imaging, Robert Turner of the Department of Neurophysics at the Max Planck Institute for Human Cognitive and Brain Sciences in Leipzig, Germany, favored the possibility that increased myelin was responsible for the structural changes detected following tDCS to accelerate learning. "In my present view . . . the previously unmyelinated axonal fibers within white matter become rapidly myelinated when they start to carry frequent action potentials," he says. By this he means that bare axons in these fiber tracts in the brain became myelinated afterward, which could contribute to improved performance in the learning task.

I also favor this view, because this has been a focus of research in my own lab. Over the last thirty years, my colleagues and I have shown that neural impulses are sensed by the cells that wrap myelin insulation around axons (a type of cell called oligodendrocytes), and we have identified many details about how action potentials are sensed by oligodendrocytes and many ways that this stimulates the formation of myelin.[77] The most recent research from my lab takes this one step further by finding that fully formed mature myelin can change its thickness to adjust the velocity at which neural impulses are transmitted. As will be explained later, the proper speed of impulse transmission can improve learning and neural circuit performance by ensuring that neural impulses arrive at critical relay points in neural networks at precise times.[78]

The air force researchers who performed this study were confident that transcranial stimulation could accelerate many kinds of learning, and this has been shown by a wide range of experiments in many different laboratories. Even at this scientific meeting in 2011, another researcher, Matthias Witkowski, at the Institute for Medicine, Psychology, and Behavioral Neurobiology at the University of Tübingen in Germany, demonstrated the potential of tDCS to accelerate learning in a very different situation. Witkowski conducts research on brain-computer interfaces to control prosthetic limbs. As was seen for the patients using BCI devices described earlier in this book (see chapter eight), learning to generate the proper brainwaves to control an artificial hand through thought alone requires considerable training. Witkowski found that if his patients received twenty minutes of tDCS stimulation once a day during five days of training, they learned to control the prosthetic hand with their thoughts much more rapidly. TDCS influences neuronal firing by driving electrical current into neurons, which will either hyperpolarize or depolarize them, and decrease or increase their ability to fire impulses, depending on the polarity of the current. When delivered to the correct brain region, tDCS is a simple method to directly manipulate electrical activity in the brain to provide a harmless and drug-free method to double the speed of learning.

Despite these findings concerning tDCS, a paper published in 2018 by György Buzsáki's laboratory at New York University and colleagues at the University of Szeged in Hungary questions whether this technique, as it is usually applied, can alter brainwave oscillations or neuronal firing at all.[79] Their studies measuring electric current flow into human cadaver heads and in rodents found that three quarters of the current applied to the scalp fails to penetrate into brain tissue because the path of the current flow is shunted through the skin and skull before reaching the brain. They find that much stronger currents than are usually used in tDCS must be delivered to influence electric fields inside brain tissue enough to alter neuronal firing.

Alternating currents (AC), in contrast to direct currents (DC), are also used for transcranial electrical stimulation, and often applied to test whether altering the frequency of brain oscillations modifies information processing and behavior. When sufficiently strong stimulation is applied through transcranial electrodes, brain oscillations and neuronal firing are clearly affected. Buzsáki's studies showed that when 6 milliamp currents are applied to humans in a slowly oscillating 1 Hz sine wave pattern while alpha-wave activity was monitored by EEG, alpha waves were clearly modified. The amplitude of the alpha waves increased and the transcranial stimulation coupled them to surge in phase with the transcranial stimulus. Alpha-wave power peaked at the crest of the wave of oscillating electricity delivered by transcranial stimulation, and was suppressed at the troughs. However, stimulation at 4.5 milliamp and higher causes a tingling and burning feeling, and the test subjects reported feeling dizzy and seeing a horizontally oscillating light in sync with the stimulation frequency, even though their eyes were closed and they were in a darkened room. The subjects also reported hearing a noise moving left to right at the same 1 Hz stimulus frequency accompanied by a metallic taste in their mouth. All of these effects stopped when the stimulus was stopped.

Using this new information, the researchers devised a new and more effective way to modulate brainwaves in humans using transcranial stimulation, by delivering higher frequency pulses from multiple electrodes arranged on the scalp so that they would intersect and generate sufficiently large voltages in localized regions of the brain, called intersectional short pulse (ISP) stimulation, and their experiments demonstrated that this stimulation had a robust effect on neuronal firing and brainwave oscillations.

Buzsáki and colleagues agree that the large body of research on tDCS shows that it is effective in modulating neuronal function and several behaviors, but many of these results may be due to indirect effects of the stimulus or possibly placebo effects. At the same time, however, they acknowledge that a neuron with a transmembrane voltage close to the threshold of firing an electrical impulse will be triggered to fire by a very small voltage, so that even weak tDCS may trigger firing of a subset of such neurons poised on the threshold of firing.

It is important to appreciate that our ability to detect electrophysiological changes in the brain, including by EEG, is limited. The brain and behavior are typically much more sensitive to stimulation than we can show with electrode recordings. In other words, the effects of tDCS in controlling psychiatric conditions and in boosting learning, alertness, and performance, as reported in the study on drone pilots and others, likely demonstrate true biological effects. At the same time, however, the method of transcranial stimulation used is very important; if ineffective methods are applied and experimental controls are not adequate, placebo effects are likely.

## TRANSCRANIAL MAGNETIC STIMULATION

Transcranial magnetic stimulation (TMS) has been discussed as a technique to increase or decrease excitability of neural circuits by beaming high-energy pulses of magnetic fields into the brain from coils placed above the head. This technique is also being used therapeutically for a number of psychological conditions. The FDA approved TMS for treatment of obsessive-compulsive disorder (OCD) in August 2018.[80] The FDA reached this decision after reviewing data from a randomized, multicenter study of 100 patients, half of which received TMS and the other half received sham TMS. The results showed that 38 percent of the patients responded positively to the treatment as assessed by psychological testing. Interestingly, 11 percent of patients responded positively to the sham device. This approval for treating OCD follows FDA approval of TMS for treating depression in adults for whom antidepressants are ineffective, granted in October 2008. Because TMS is considered safe and is approved for treatment, the method can be applied in an effort to help patients with many other neurological and psychological conditions. TMS is used in poststroke treatment; for severe depression, PTSD, and postconcussion syndrome; to improve working memory; to relieve pain; and to treat ADHD, Parkinson's disease, Alzheimer's disease, headache, addiction to methamphetamine, anorexia nervosa, schizophrenia, multiple sclerosis, and more.

TMS is considered safe, but it can induce seizures, and it frequently causes a painful or unpleasant sensation by stimulating facial nerves that are also activated by the magnetic field beamed into the brain.

The method modifies excitability in relatively large regions of brain tissue in the focus of the magnetic pulses, so it is believed that, depending on the frequency of stimulation, neurons in the cerebral cortex are either excited or inhibited from firing. The most effective area of brain to be targeted is unclear in many disorders, but the prefrontal cortex is most frequently targeted (orbitofrontal and dorsal lateral prefrontal cortex). Some, but not all, insurance companies will cover the cost of TMS treatments.

People can respond differently to medical treatments of any kind,

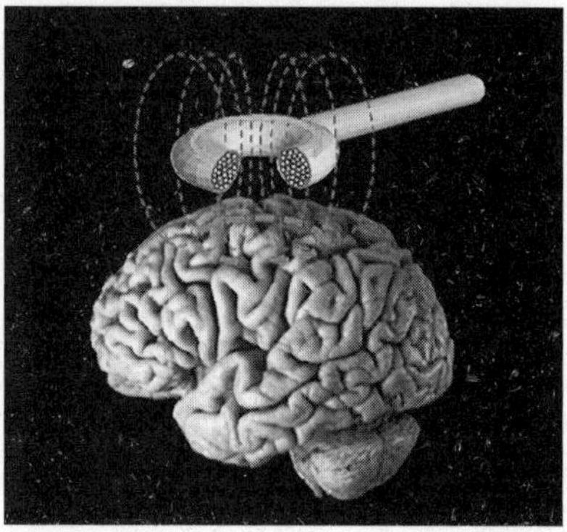

**FIGURE 40**: *Neural activity can be stimulated or inhibited in neural circuits by transcranial magnetic stimulation (TMS), in which strong magnetic fields from an electromagnet placed above the scalp are pulsed into the brain. TMS is used in experimental research and for treating several neurological and psychological disorders.*

and this is true for TMS treatment of PTSD. A recent study found that EEG analysis could predict whether or not TMS would be effective in treating this condition. Patients at the Veterans Affairs Medical Center in Providence, Rhode Island, were given a resting-state EEG before and after receiving TMS at 5 Hz to the dorsolateral prefrontal cortex for forty daily sessions to treat PTSD and major depressive disorder. The treatment was effective for about half of the twenty-nine patients, but when researchers examined the EEGs of these patients before receiving TMS, they observed that patients with higher alpha-wave coherence in the frontal brain regions predicted the amount of clinical improvement in depression and PTSD.[81]

This area of research is moving from lab to clinic. In 2019, I visited the National Intrepid Center of Excellence at Walter Reed National Military Medical Center to speak about my research on brain plasticity, and I saw how brain stimulation by TMS is being used there to help veterans with PTSD.[82]

## SIMPLE WAYS TO REGULATE BRAINWAVES

Because brainwaves are so responsive to sensory input and to mental state, it is clear that rhythmically modulating sensory experience, with such techniques as meditation, which are based on controlled breathing rhythms and thoughts, will alter brainwaves. These methods have long been used to alter cognitive function, mood, and brainwaves. This effect is most clearly experienced in rhythmic music, where everyone in a crowd will begin to sway and tap their feet rhythmically to the beat of music, which alters both brainwave activity and mood. In the extreme, rhythmic lights can synchronize and boost the activity of brainwaves in some people to the point that they will experience an epileptic seizure.

### *MEDITATION*

Written records document the practice of meditation to as far back as 1500 BCE in India, but archeological artifacts showing people seated in meditative postures date the practice to as early as 5000 BCE. Naturally, over this long time span, many variations of meditation have evolved. These different meditative techniques are accompanied by differences in brainwave activity. So you don't need any more sophisticated equipment to alter your brainwaves than prehistoric people had. Meditation or drumbeat will do it.

Meditation techniques span a continuum, from methods that tightly focus attention to those that encourage achieving an open mindful awareness state. Himalayan Yoga tradition relies on focused attention, for example, while, at the opposite extreme, the Isha Yoga tradition uses open awareness. Vipassana meditation falls midway between these two approaches to achieve a meditative state. Focused meditation employs repetition of a mantra, whereas open meditation concentrates on awareness of body sensations. The approach used in Vipassana meditation is to mentally scan each body part one by one and feel the sensations of each part. The Isha Yoga tradition focuses on breathing techniques and awareness of body sensations.

There have been a large number of EEG studies on meditation to better understand what is happening in the brain in these altered states of consciousness and mood. Most EEG studies find an increase or decrease in power of lower-frequency theta and alpha waves,[83] but more recent research has also found differences in the high-frequency gamma waves.[84] A 2017 study by Claire Braboszcz, at the Université de Toulouse, and colleagues at universities in the United States conducted an EEG analysis of practitioners using each of these three approaches to meditation, and the data showed that all of them altered brainwave activity compared to a control group of participants instructed to simply let their minds wander. Regardless of which technique the meditators used, they all had higher gamma-wave power (60 to 110 Hz) in the parietal and occipital regions than control subjects. In addition, higher alpha activity was found in the Vipassana group compared to all the other groups.

The researchers suggest that the increase in gamma power reflects a heightened level of attention and may represent top-down control of mental state (top-down control is the conscious, willful control of behavior dependent on the prefrontal cortex). Higher alpha-wave power is associated with redirecting attention to internal mental states and shutting out sensory input. Several resting-state EEG studies link the mind-wandering state to activity in the default-mode network, but meditation is associated with deactivation of this network. This is reasonable because during meditation, subjects are striving to free their thoughts from memories and autobiographical experiences and focus their awareness in some way; thus, the DMN becomes inactivated.

### SENSORY STIMULATION

The most dramatic example of how rhythmic stimulation can alter brainwaves is seen in a certain form of epilepsy. When someone has an epileptic seizure, the busily fluctuating brainwave oscillations associated with neurons processing information in the cerebral cortex simultaneously in many different circuits is suddenly disrupted by large-amplitude waves of electrical activity that are highly synchronized sweeping through the cortex. (These are called sharp waves or spikes because of the abrupt, jagged squiggling produced in the EEG record.) A seizure is excessive abnormal neural activity—not necessarily accompanied by a convulsion. Partial seizures localized to specific spots in the brain can occur, and the person may not be consciously aware of having one. The person having a partial seizure may "space out" temporarily, abruptly change behavior in various ways, or show no outward signs at all. However, disruptions in normal EEG activity are associated with disruption of normal function in the neural circuits. If all the neurons in a circuit are firing in synchrony—doing the same thing—they cannot process complex information, any more than you could carry on a conversation in a crowd while you were all chanting in unison. In a generalized seizure, the entire brain becomes involved in the highly synchronized abnormal brainwaves, globally disrupting brain function and causing unconsciousness and often convulsion. (At the opposite extreme, a temporary silencing of EEG, called spreading depression, is associated with migraine.)

About 3 percent of people with epilepsy will have seizures that are triggered by certain patterns of flashing lights, a condition known as photosensitive epilepsy. Flickering TV screens and computer monitors or rolling images will do it. Video games or TV broadcasts with rapid flashes of light or alternating patterns of different colors and intense strobe lights flashing at a frequency of five to thirty per second will trigger these seizures. What's happening in the brain is that the rhythmic sensory input is causing ongoing brainwave activity to become synchronized with the stimulus and, in people prone to seizure, the heightened amplitude and synchrony of brainwaves causes a seizure. A Maryland man was charged with aggravated assault with a deadly weapon by tweeting an animated strobe image to a journalist who has epilepsy along with the message, "You deserve a seizure for your post."[85]

Seizures induced by rhythmic input to the visual system dramatically demonstrate how sensory input can organize ongoing brainwaves into highly synchronized patterns, but synchronization of brainwaves by rhythmic stimulation happens in all of us (without inducing seizures). Rhythmic sound and lights can be used to modulate the power of specific brainwaves; there are many consumer devices marketed to use this approach to induce desired cognitive states to improve mental and even physical performance. US Olympic swimmer Michael Phelps famously wears headphones to listen to music until the last possible moment before diving into the pool. "It helps me to relax and get into my own little world," he says.[86] Of course, we all listen to music to alter our moods. It is the rhythmic aspects of music that lull us to sleep or jazz our bodies into frenzied dance.

You walk into a bar and all heads are bobbing and feet tapping in synchrony with the throbbing rhythm of music. That's because the brainwaves (alpha and beta waves) of everyone in the room are synchronized to the musical rhythm. The moment your alpha wave peaks as it oscillates happens precisely at the same time as everyone else in the room (likewise for the faster-oscillating beta waves). Moreover, the rhythmic pattern of brainwave activity gates the input of information into the brain (letting the information in) in sync with the rhythm, such that an event or information presented to a person on the beat will evoke a greater brainwave response and heightened awareness of the event than if it is presented out of rhythm.[87] So too does language, which is highly rhythmic, synchronize brain activity of speaker and listener as detected by fMRI. The synchronization between brains of people in conversation is thought to promote communication.[88]

Returning to brainwave synchronization and athletic competition, consider that whether or not a competitor stands on the podium wearing an Olympic metal can depend on a thousandth of a second difference in finishing times. Brainwaves can be what separates winners from losers when the margin is that close.

Springing from the starting blocks too late can mean failure to medal, but jumping the gun results in disqualification. What happens inside the brain in the roughly one to two tenths of a second reaction to hearing the starting pistol is an extremely complex process. Rapid sensory perception to detect the starting signal, massive information processing to launch a widespread response throughout the brain, and rapid execution of precise motor commands are all carried out unconsciously—faster than the speed of thought.

Oscillations in the electric field of brain tissue help coordinate the firing of neurons in phase with each other, just as boats are rocked in synchrony by waves in a harbor. That coordinated firing is believed to be critical for organizing and coupling information transmission across distant regions of the brain. One of the most important brainwave frequency bands for controlling movements is mu waves. These waves oscillate in the motor cortex at 7.5 to 12.5 Hz. The motor cortex is a strip of cerebral cortex that runs roughly ear to ear like a plastic headband. As discussed in association with BCI devices that operate by detecting changes in mu waves, just before a movement is initiated, mu waves are quashed (or desynchronized). This sudden calming of mu waves is believed to reflect different circuits of motor control neurons breaking pace with the larger group and starting to fire independently to execute specific motor commands to appropriate muscles.

But mu waves are themselves riding on top of slower brainwave oscillations, in the same way that windblown ripples oscillate on top of slower ocean swells. These very slow oscillations in brain activity were once dismissed as artifacts caused by fluctuations in blood pressure as the heart beats; however, blood flow affects brain function. Thus, there is an interactive mind-body loop between brain and heart, and both slowly oscillate together. Although an athlete's heart rate may be sixty-five beats per minute, when followed over several tens of minutes the heart rate will vary slightly, oscillating at about 0.1 Hz (one cycle every

10 seconds). These very slow oscillations in blood flow affect the firing of neurons in the brain.

So as Olympic swimmers are poised motionless and ready to leap into the pool, inside their brain neuronal excitability in their cerebral cortex is slowly oscillating on a 0.1 Hz cycle. By chance, some competitors will be at the peak of the neuronal excitability wave when the starting pistol fires, and others will be in the trough. "This excitability cycle modulates the central mu rhythms," says Gert Pfurtscheller, a neuroscientist who specializes in brainwaves in motor control working at the Graz University of Technology in Austria. "It can be expected, therefore, that there exists a relationship between the 0.1 Hz excitability cycle, the starting gun, and the race outcome," he says. [89]

The connection between neural oscillations and brain function provides a link that can be used to enhance cognitive and athletic performance, mood, and learning by directly influencing brainwaves through many different means, including electrical stimulation, drugs, neurofeedback, meditation, and sensory stimulation.

## CHAPTER 12

# Into the Future

From cognitive enhancement to control of prosthetics, from self-driving cars to surveillance, the electric brain is on a new frontier. The technology to use brainwaves to interface mind with machine is beginning to leave the scientific laboratory and become available in the marketplace. There are a number of new companies offering headsets of various designs to consumers that claim to enable users to dispense with keyboard and mouse to control their laptops and other electronic devices through a simple EEG recording device.

I interviewed high school students in Washington, DC, who connected one such consumer device to a motorized wheelchair to drive it around campus with "no hands," using only their brainwaves. I also tested an EEG device to control a video game, to make an object move on the screen by mental exertion. In both cases, however, I am not convinced that the operations were actually responding to my EEG, because the companies making the brain interface devices guard the proprietary information about how the device works and what it actually detects. The devices might be sensing muscle tension (EMG) as I furrowed my brow, willing the interface to follow my commands, rather than reading brainwaves.

Transcranial direct current stimulation (tDCS) to modulate brainwaves by applying DC voltage to the scalp is a simple method that can directly influence brain function. As described in chapter eleven, tDCS is being used by the United States Air Force to accelerate training of drone pilots and to improve their performance on the job by heightening and sustaining mental focus while scanning the stream of video feeds from the battlefield. No need to join the air force if you are interested in amping up your brainpower this way; you can pick up what you need on the internet or at a local hardware store.

## BRAINWAVE GADGETS

In contrast to TMS, which requires sophisticated instrumentation to deliver high-energy magnetic pulses to the brain from electrical coils positioned over the head, tDCS is a perfect device for the DIY hobbyist. All it takes to make a DIY model is four 9-volt batteries, some adhesive electrodes, and wire. Tech-savvy college students are already using homemade tDCS devices to improve mood and concentration for study and to boost their brainpower before exams. College student Matt Herich, for example, uses a DYI brain stimulator made by another college student at Northern Arizona University to put himself into an intense state of mental focus to help him learn math much faster in online college courses.[1] For those who are less handy with a soldering iron, there is a flurry of new companies selling low-cost tDCS headsets online for home use that sell for about $300. Of course, there are risks of using DYI medical devices that are not approved by the FDA and not manufactured to exacting standards, and detrimental neurological consequences of delivering electrical current into the brain, such as seizure, are a possibility.

Just before Christmas 2016, Professor Christopher James, director of Engineering in Biomedicine at the University of Warwick in England, announced that his team developed improved headsets that enable children to operate toy robots and remote-controlled cars and helicopters by mental concentration alone. "The exciting bit is what comes next—how long before we start unlocking the front door or answering the phone through brain-computer interfaces?" he asks.[2]

New brainwave devices are under development that could transform daily life. Devices using brainwaves as a biometric ID can replace passwords. Going beyond lie detector tests, which are based on reading emotions influencing heart rate, respiration, and perspiration, brainwaves can reveal true or false thoughts. Reading a witness's brainwaves that are evoked by a photograph of a crime scene or other evidence could reveal whether or not the person has seen the material before. The P300 "oddball" visually evoked brainwave potential (see page 317) that erupts when we see or hear something novel will not appear if the object or information presented to the witness is not new to them.

Japanese automaker Nissan Motor Co. announced at the Consumer Electronics Show (CES) in Las Vegas in January 2018 that they are developing a self-driving car that uses the driver's brainwaves to steer the vehicle. Drivers of the future, Nissan says, will don a fashionable lightweight helmet, resembling an open mesh bike helmet, that picks up brainwaves. The brainwaves will be fed into an onboard computer and the car will be autonomously guided without any conscious effort on the part of the driver.

Researchers at Nissan claim that they can read the driver's preconscious intentions of whether to turn left or right or slam on the brakes. Senior Research Innovator Dr. Lucian Gheorghe at Nissan in Japan says, "Let's say you are changing lanes. Before you turn the steering wheel to change the direction of your car, you should have made a decision in your mind, and before that, your brain must have sensed some circumstance that made you think about changing lanes."[3] The ability to drive an automobile by tapping into a person's brainwaves would greatly enhance the pleasure of driving, in Gheorghe's view, by relieving the driver from being consciously aware of the harrying and niggling details that make driving tedious. Imagine riding in a car motoring among other vehicles on a crowded Los Angeles freeway, swerving in and out of traffic or inching spasmodically in stop-and-go traffic jams while you freely carry on conversations, text or talk on the phone, or turn your conscious mind to more pressing thoughts. No worries about being crushed in a multicar pileup; preconscious brainwaves would transport you and those around you effortlessly through traffic as if being chauffeured in a limo.

It must be acknowledged that much of the appeal of the CES and auto shows is in showcasing futuristic electronic devices and concept vehicles, many of which will look ridiculously fanciful in retrospect.[4] For example, the development of rockets and the jet engine in the 1940s launched a squadron of concept vehicles through the late 1950s and early 1960s that incorporated design features of jets and rockets. Bizarrely innovative vehicles hinged at the midpoint to pivot around corners; amphibious and even flying vehicles have dazzled the public in past shows. However, some of the incredible innovations displayed at auto shows in the past eventually evolved into features in cars that are on the road today: for example, automatic brake systems, turbine-powered engines, and airbags automatically deployed in the blink of an eye to protect passengers in a crash in the milliseconds before their body would impact the dashboard or fly through the windshield. In theory, the idea of a vehicle guided by brainwaves may not be as absurd as it sounds, because your brain normally drives your car largely without your conscious awareness. The hurdles faced for this technology are primarily technological, but they are also biological. The neuroscience of brainwave research does not currently provide the necessary capability or level of understanding to do much more than demonstrate rudimentary feasibility of the concept at present, but the science of brainwave analysis and development of technology to record them are rapidly advancing.

If, as it appears, cars and highways of the future will evolve into a system of transportation that uses self-driving vehicles, there is still the need for a "kill switch" to wrest control away from the computer in the car and allow the driver to act, as Neil Armstrong did in wresting control from the computer-guided landing of *Apollo 11* seconds before reaching the surface of the moon and narrowly avoiding a potentially tragic impact.

"Our [onboard] systems will be able to tell an autonomous vehicle, 'the driver will be steering in the next 300 milliseconds,'" Gheorghe says, by detecting the driver's preconscious intentions reflected in brainwaves. "Then we can use this window in time to enhance the execution synchronizing the support of the AV without your own actions."[5]

## NEW APPROACHES TO EAVESDROPPING ON BRAIN ACTIVITY

Simpler and far less invasive methods of monitoring brain activity are necessary if brainwaves are to be used for consumer devices. A novel approach for a brain-computer interface is being developed at a start-up company, CTRL-Labs, launched by the inventor of Microsoft Internet Explorer Thomas Reardon and his partners.[6] To get around the problem of implanting electrodes into the brain or wearing an unsightly EEG cap for brain-computer interface, Reardon and colleagues are developing a BCI device that taps into your brain through your arm.

Nine stories up in an office building in midtown Manhattan, the elevator doors open to the CTRL-Labs nerve center of lab benches and computer screens, with energetic people fiddling with electronic components and manipulating robotic arms. Others are conferring inside a sleek glass-enclosed conference room. Off-road bikes lean against the wall, and on the table directly in front of the elevator is a copy of the book by Eric Kandel, James Schwartz, and Thomas Jessell, *Principles of Neuroscience.*

Thomas Reardon, wearing a navy-blue hooded sweatshirt with Columbia (the university) stenciled in white letters across its chest, glances up as I walk in. Chief Strategy Officer Josh Duyan greets me and makes introductions. With Reardon, Patrick Kaifosh cofounded the company, which grew out of their work together as graduate students at Columbia University, where they were pursuing PhD degrees in neuroscience in the laboratories of Thomas Jessell and Attila Losonczy, two premier researchers in cellular neuroscience.

Duyan slips something onto his wrist that resembles a spiked choker that a goth punker might favor, but the golden square studs are on the inside of the band, pressed against his skin. A ribbon cable drapes from the wristband to a laptop, where on the screen a white mannequin-like hand spreads her fingers, waves, bends her wrist, and makes a fist, all mimicking Duyan's own hand movements, picked up electronically through this wristband.

The wristband is not sensing neural impulses from the brain; it is sensing voltage bursts produced when a muscle cell contracts. The computer analyzes the electrical discharges from arm muscles and calculates what hand motions would be created by that electronic signature, as well as the speed of motion and force of grip. Then the computer drives the same motions in the virtual hand, but the limb on the screen could just as easily be one of the real black-and-chrome robotic arms I see lying about the office like pieces of Iron Man's suit. In a video demonstration,

Reardon taps his fingers on a blank tabletop, and letters string out on a computer screen at twenty words a minute. The system is recognizing specific muscle contractions encoded to a particular letter—no keyboard needed!

This method of typing by encoding letters to individual muscles twitching is a major accomplishment because the electrical signals picked up on the skin are a cacophony of thousands of individual muscle fibers contracting in his forearm. Streaking across the computer screen, the raw recording of voltage fluctuations picked up by the armband looks like the erratic trace of a seismometer when an earthquake hits. In recent years, neurophysiologists have developed mathematical methods to isolate the firing of individual muscle fibers among the thousands that are simultaneously active when a muscle contracts. This process can be likened to sophisticated acoustical analysis and filtering to isolate the sound of one person talking in a noisy room. With this wristband intercepting neuromuscular signals, you can dance your fingers in the air or twitch them in your pockets and type away (in principle). Forget about pecking at a cell phone touch screen or keyboard; specific muscle twitches can be assigned to specific letters. The applications go beyond typing. Imagine, a robotic arm that could go places and do things that your own arm could not—snake through the aorta and do surgery on a faulty heart valve beating inside a patient's chest, perhaps. But there is far greater potential.

The human hand is a marvel, and with it our brain interacts with the world to transmit our thoughts to others. Consider the richness of American Sign Language, for example, but clay tablets, pencils, and QWERTY keyboards were the primitive technologies that *Homo sapiens* first developed to transmit information via motor control. In the twenty-first century, we have much more powerful capabilities than these simple tools.

"Your kids will not type," Reardon says, meaning that they will have no use for typing when commands can be transmitted to computers through muscle twitches in our forearms. Not only will this electronic interface between mind, muscle, and machine be faster and more capable than the old tools, Reardon claims the new technology will enable the human brain to transcend the anatomical limits imposed by evolution. "The input of the human brain is enormous," Reardon says, considering the phenomenal capacity of our five senses to take in streams of information about the world, "but its output is poor." (The brain's only means of output is through muscles—written or spoken language, and both of these are very slow relative to the speed and capacity of information processing in the brain.) The goal at CTRL-Labs is to overcome this keyhole effect limiting the human brain's output.

A pianist communicates the music in his mind to a listener's mind by tapping keys on a piano. Duyan could do the same with only a picture of a piano keyboard, imagining himself playing *Chopsticks* or Chopin. It is not necessary for finger muscle to move bone and forcefully depress a wooden lever (piano key); imperceptible twitching of a microscopic muscle fiber is enough to create an electrical discharge that is easily detected by this brain-computer interface. Rather than being limited to the five digits on one hand that evolution of mammals has given us, what if Duyan imagines having twelve fingers, one for each note on the chromatic scale (all the white and black keys in an octave)? Now the frenzied chords and blizzard of arpeggios in Chopin's *Polonaise* are a piece of cake. Of course, the pianist would have to learn to activate the muscle fibers in his hand in a different way than he is accustomed to when moving only the five digits we have, but why not? Learning to control twelve digits is not that different from learning to control five. Spiders control eight legs with ease, and you don't see millipedes tripping over themselves. Those are bug brains. Imagine what a human brain could do.

Neurofeedback is used to train the brain and computer to interface with each other. The process is as effortless as playing a video game. For example, Duyan simply imagines trying to point the on-screen mannequin's index finger and it responds. What is happening inside his motor cortex is beyond his perception, but when the mannequin points her finger, his brain learned. "You are training the computer too," Reardon says. Eventually skill in moving the hand on the screen develops unconsciously and automatically.

But what is the limit? The goal these researchers are working toward is to tap into electrical activity from every single motor neuron operating your arm. For example, we have 774 motor axons that control all the muscle fibers in our biceps. Each motor axon typically forms connections to several individual microscopic muscle fibers, called a motor unit, that contract together when action potentials transmitted through the axon stimulate them. The simultaneous contraction of several motor units is required to make the entire muscle contract. Reardon claims that CTRL-Labs' hardware and software advances have enabled them to achieve "single unit detection." This capability would mean that rather than the 774 motor units in our biceps doing only one interaction with the world—pulling back the forearm—they could in principle do 774 different types of interactions with a computer. For that to work, the human brain would have to have the ability to control its single motor units individually. The brain's capability to do so was in question, but their research has shown that it is able to have that level of control. When you think about it, single-unit control of motor units is what our brain is doing all the time to give us that fine, fluid control of our limbs. Training to achieve single-unit control of a computer interface is simply doing what the brain is already doing but taking it to a brain-computer interface.

I was not permitted to give the wristband a test drive, which suggests to me that the device is still at an early stage of development. Instead, Patrick Kaifosh, who is a skilled expert, sits down, straps on the device, and begins playing a game of Asteroids on his cell phone wired up to the wristband, flying his spaceship around the screen, shooting down invaders with his lasers, and dodging incoming missiles from aliens, all while his palm rests motionless on the table, with only the slightest twitching of skin here and there reflecting individual motor units being activated, as he carries on a conversation with me.

"What does it feel like?" I ask.

Perplexed by the question, he asks, "What does it feel like for you to think about moving your fingers?" he asks.

Indeed, I am effortlessly interfacing now with my computer through my fingertips, automatically pressing twenty-six keys of the alphabet as I transmit my thoughts to you. What if instead of only two dozen keys, I had hundreds of them, all automatically controlled by my brain?

## FANTASTIC VOYAGE

Another futuristic method of tapping into the electric brain evokes, in my mind at least, scenes of Raquel Welch, the bombshell American beauty and actress of the 1960s and '70s, clad in her glistening wet, skin-tight white rubber wet suit in the 1966 sci-fi movie *Fantastic Voyage*. In the movie, based on Isaac Asimov's tale, rescuers in a miniature submarine are injected into a man's carotid artery and swept through the vascular system on a wild adventure to destroy a blood clot.

Neurosurgery hasn't quite reached that point yet, but when the neuroscientist conducting groundbreaking research emailed me about the technique he and his colleagues were developing to monitor brainwave activity, I wanted to see it in action. I caught the train to New York City, rode to the Bronx on the New York subway, where water drips from the tunnel walls in dim light. Fluorescent light fixtures flicker or hang from

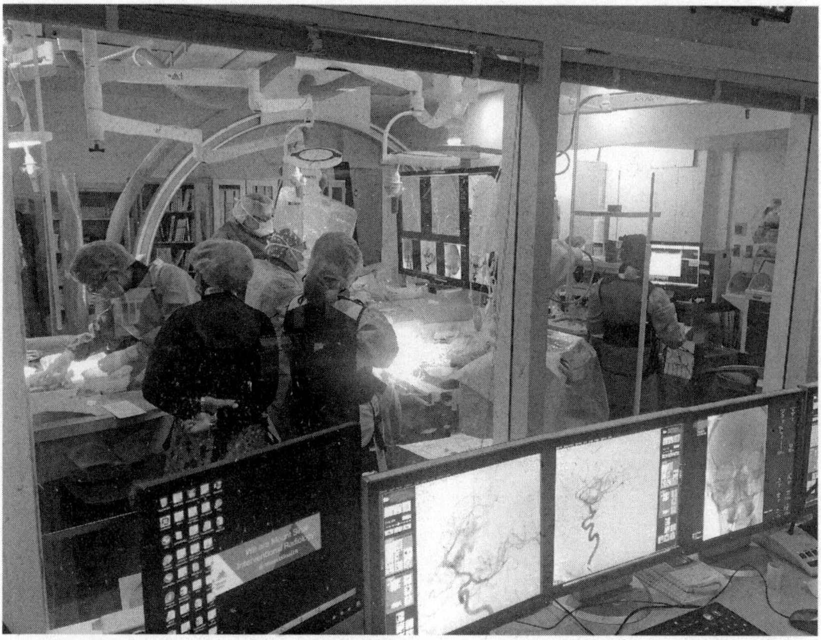

**FIGURE 41:** *Noninvasive methods of monitoring electrical activity in the brain and stimulating neurons are being developed, such as the Stentrode device, which is inserted into the brain through the vascular system. The patient here is having a vascular malformation in his brain treated by neurosurgeons who are inserting a slender cannula into the femoral artery of the leg and the jugular vein in the neck to meet at the point where blood vessels in the brain are constricted, with the potential of a life-threatening rupture.*

the ceiling, cold and broken. Paint peels in thick flakes like bark from eucalyptus trees, and the rusty steel girders shot through with holes are crumbling. The racket is deafening, and the stench of rot invades the nostrils. After emerging from the subway station, a short walk through a cold light drizzle takes me to the bustling Mount Sinai Hospital where this research is being conducted.

Dressed in a sterile white jumpsuit in a futuristic operating room resembling a scene from *Fantastic Voyage*, I watch in amazement as a team of doctors snake a slender wire through a man's jugular vein, threading it through a maze of blood vessels into his brain. At my side in the operating room is Mount Sinai Medical Center neurologist Dr. Thomas Oxley, who explains how his new invention, called a Stentrode, uses the bloodstream as a pathway into the brain to implant microelectrodes. His goal is to couple mind and machine, but in probing the brain with their new electrodes, the researchers are also probing the limits of knowledge about how the brain operates.

Only the eyes of the seven doctors, nurses, and technicians are visible as, cloaked in blue surgical gowns and masks, they fixate like NASA engineers on eight flat-screen monitors, tracking the advance of the probe wriggling like a parasite inside the patient's brain. The man, buried beneath surgical drapes, is anesthetized with his head at the hub of an enormous gyroscopic array of steel arches that suspend mobile X-ray machines beaming to the monitors' real-time 3-D displays of the probe's progress through the brain's blood vessels. The surgeons push one probe (a cannula) through the man's femoral artery in his groin and thread another through his jugular vein, to bring them together at a dangerous spot where blood vessels in his brain have ballooned under pressure, threatening to burst and kill him.

The patient is being treated for a life-threatening vascular malformation in his brain, but Oxley conceived that if microelectrodes were engineered into a vascular shunt, a device neurosurgeons routinely insert into the brain's blood vessels to treat strokes, he could also record the brain's electrical activity with high fidelity from within the blood vessel without the need to drill holes through the skull to pierce the brain with electrodes.

"How do they navigate the cannula through the tangled network of vessels?" I ask. Oxley, who is from Melbourne and speaks with a "Good on You!" Aussie accent, offers to show me. We strip off our sterile garb and dash to a nearby room, where Holly Oemke, manager of the Neurosurgery Simulation Core, and Carter Chatillon, program lead at Surgical Theater, are operating computers displaying computer-generated colorful 3-D renderings of the patient's brain, compiled from MRI imaging conducted before the patient underwent the endovascular shunt procedure. Everyone's brain and vascular anatomy are somewhat different, so before operating on a living person's brain the neurosurgeon rehearses his or her planned medical mission like a jet fighter pilot in a flight simulator. Dragging a mouse, the technicians rotate, slice, and zoom in on the 3-D image of the man's brain to reveal anatomical details from any angle.

Then, it's my turn. They hand me pistol-grip controls for each of my hands and fit virtual-reality goggles over my eyes. Now in 3-D stereo

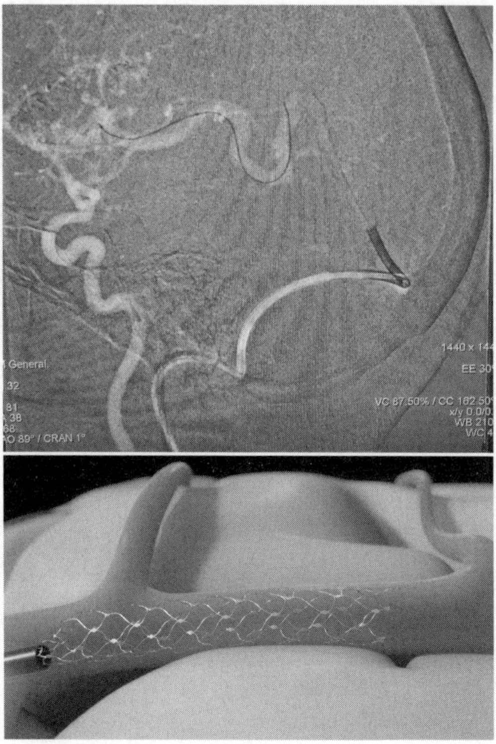

**FIGURE 42:** *Top: Surgeons watch the progress of the cannula as they thread it into the brain's vascular system on a live X-ray. Bottom: The Stentrode used for BCI is developed from a standard vascular shunt but it is equipped with electrodes to record and stimulate neurons inside the brain.*

vision, I am suspended in space above the patient's head. Flailing my hands, which guide my movement, I whorl above the scalp like a dizzy out-of-control helicopter pilot looking for an emergency landing spot. Squeezing the triggers, I flinch as a green laser beam shoots from my ray gun, pinpointing my trajectory and reporting the distance on my heads-up digital display. Following Oemke's instructions, I begin to get the hang of it. I work my hands back and forth as if crawling, drawing myself toward the scalp. Soon the scalp fills my view and the scene resembles the *Apollo 11* space capsule touching down on the moon. Suddenly I pop through the patient's skull to find myself inside his brain. Amazing! Everywhere I look I see the insides of the patient's brain in vivid detail! I look down and see the opening of his spinal cord. I look back and see the occipital lobes and their vasculature. I hover over the crimson blood vessels and fly like a drone along them, tracing their tortuous course to where the stent needs to be placed.

Oxley demonstrated the remarkably simple way they guide a cannula through the twists and turns of arteries and veins. He threaded a slender wire through a fine plastic cannula, a flexible tube that can range in diameter from that of a cooked spaghetti noodle to only half a millimeter. When the tip of the wire was thrust out the end of the cannula, he pinched a bend in it. By twisting the wire inside the cannula, the bent tip protruding from the opening swivels in any direction. If he wants to take the right fork out of the jugular, he simply twists the wire so the bend flips to the right and then progresses into that branch. He slips the cannula encasing the wire forward to advance it bit by bit, negotiating the tangle of blood vessels through hairpin turns and loops with ease. As I watched the surgeons doing this on the monitors inside the operating room, the tip of the wire writhed inside the patient's brain like a plumber's snake auguring out a clogged drain.

A typical vascular stent, which is shoved through the cannula, is a tiny tubular mesh that will expand against the walls of a blood vessel, but Oxley's modified stent—the "Stentrode"—has sixteen electrodes in it. About six days after it is inserted, the stent becomes implanted into the vessel wall permanently by the body's healing response, making good electrical contact to intercept electrical signals radiating from brain tissue through the thin blood vessel walls. Oxley and his partners are developing methods to use the Stentrode to operate prosthetic devices and to interface with computers for other purposes by using the brain's own electrical signals, as well as to activate neural circuits by stimulating electrodes in the shunt.

Many sensational articles about brain-computer interfaces (BCI) can give the impression that scientists have the ability to extract and translate electrical activity in neural networks as if the neural code was as clear to neuroscientists as computer code is to programmers. "We don't have it—not even close," says Dr. Joshua Bederson, chairman, Department of Neurosurgery at Mount Sinai, when asked if this scientific knowledge and technical capability exists.

Oxley agrees, but the key, he says, is to keep things simple. It is not currently possible to decipher the neural code of electrical impulses zipping through complex neural circuits, but the brain's overall buzz of electrical activity fluctuates and surges in waves that are associated with particular mental states and functions. By detecting critical transitions in brainwave activity that are known to change with certain motor and cognitive functions, neuroscientists can program computers to activate a motor in a prosthetic limb or to click a mouse to interface with a computer, much like how one learns to depress the clutch in an automobile when the sound of the engine reaches a specific pitch.

## THE STENTRODE

From the day brainwaves were first discovered by Richard Caton, the physician in Liverpool, England, who performed experiments on rabbits in 1875 (see pages 50–54), a surprising property was evident. The rhythmic oscillations of brainwaves are instantly disrupted at the moment a sensory or motor action commences. It is as if an orchestra suddenly stops their rhythmic coordinated playing to tune up and then resumes playing. Neuroscientists analyze the shifts in synchrony and the mixture of different frequencies of ongoing brainwaves to detect the abrupt desynchronization of brainwaves that occurs the instant a person visualizes performing a motor action.

"The brain does the work for us," Bederson says. How the human brain works is still a deep mystery, and we may never fully comprehend it, but BCI works because the brain itself will exploit the new channel of information provided to it from a computer and learn to control a prosthetic device. It learns from trial-and-error feedback to click a mouse or move a motorized limb by modulating brainwaves in the appropriate brain region. How the brain does that is obscure.

The waves of electrical activity in the cerebral cortex detected by electrodes on the scalp (EEG) are like muffled conversations inside a room heard by pressing an ear to the door to eavesdrop. Electrical signals in the brain are diminished and muffled by the loss of high frequencies dampened by the bony skull. As has been previously discussed, when the skull is opened under surgery, electrodes can be placed directly on the surface of the brain to receive crisp, strong signals. This approach, termed electrocorticography (ECoG), is like having a microphone planted inside a room for surveillance.

Both of these methods (EEG and ECoG) can be used for BCI, but they only access electrical activity in the superficial layer of brain tissue (cerebral cortex). Oxley's Stentrode can be inserted at the surface of the brain or deep inside, anywhere that can be reached through the vascular system. The Stentrode wirelessly transmits the brainwave signals it detects through a broadcast device implanted near the collarbone, similar to pacemakers used for cardiac patients.

A major practical impediment of using EEG for BCI is that it requires a person to wear a garish swimming cap of EEG electrodes sprouting from the head that feed signals into bulky electronic instruments. EEG is fine for the laboratory, but it is difficult to have normal social interactions looking like a space alien.

Better signals can be obtained by drilling holes in the skull and poking sharp electrodes or arrays of microelectrodes into the brain, like the brain implants that are helping Nathan Copeland and others with spinal cord injury overcome their disabilities (see pages 238–250). In fact, individual neural impulses can be detected with implanted electrodes rather than the waves of electrical activity generated from populations of neurons detected by EEG or ECoG, but implanting them is an invasive procedure. Over time the electrodes can move and lose contact with neurons due to scar tissue forming around them. Moreover, the prospect of brain surgery is frightening for anyone, but inserting a shunt into a blood vessel is a common procedure and much more acceptable to most people. Stentrodes provide the recording sensitivity of ECoG without the risks of open skull surgery.

Dr. Chris Kellner, a vascular neurosurgeon at Mount Sinai who routinely places shunts in patients' brains, says that the risks of inserting shunts into the brain's blood vessels is extremely low. Complications arise in "less than one in a thousand [cases]," he says. Placing shunts in arteries is riskier than placing them in veins, because if a blood clot forms in an artery it could travel to a constriction, block blood supply to part of the brain, and cause a stroke; but if there is a blockage in veins, collateral vessels will drain the oxygen-depleted blood back to the heart.

Building on successful experiments on sheep that had Stentrodes implanted in their brains for up to six months,[7] Oxley and his colleagues have applied for FDA approval to begin clinical trials in humans. In the meantime, they launched a start-up company, Synchron, to perfect the device and develop better methods to decode the neural signals the Stentrode detects to control prosthetic devices. Because the Stentrode can be used to both detect and stimulate neural signals, it has the potential to provide sensory feedback from a prosthetic limb by activating the brain's sensory cortex to give a patient the sense of touch, which is essential to handle delicate objects. In addition to prosthetic limbs, they are looking to develop computer interfaces to enable people who suffer strokes or motor neuron diseases, like ALS, to communicate by a computer interface that would type or speak words the person selects by controlling a mouse.

This new technology of tapping into the brain through its blood vessels provides fresh capabilities for basic research and medicine. Dr. Bederson believes that the diagnostic applications of the Stentrode are even more important than the BCI applications. "I would use it now, if I had it," he says. Some of his patients with traumatic brain injury are unconscious or in a coma. It is difficult to gauge the level of consciousness in these situations. "A large number of patients can exhibit profoundly altered states of consciousness but have a normal EEG," he says. Neurologists could use a Stentrode in comatose patients to easily explore the state of electrical activity throughout the brain, examining regions that are not detectable by EEG and using this information to guide appropriate therapies.

The combined force of basic science and medical research on brain-computer interface technology is propelling what was once an obscure and specialized area of neuroscience into the mainstream, but now BCI technology is being further boosted by the industrial power of visionary businessmen like Elon Musk, Mark Zuckerberg, and many other biotech start-up companies working to develop noninvasive ways to tap into electrical activity in the brain and to manipulate it. This exciting race to develop new technology to interface with the brain electronically is certain to result in more powerful and much less invasive techniques to monitor and manipulate neural activity, but unification of electronics with neurons will also bring new technological risks.

## INTERFERENCE WITH THE ELECTRIC BRAIN IN THE ELECTRONIC AGE

While monitoring and harnessing the electrical impulses in the brain can lead to innovation in therapies such as BCI, there is the possibility that the ubiquitous electronics we encounter in daily life may be having an effect on our electronic brain.

Every electronic device comes with a UL sticker of approval assuring that the device will not emit an electrical disturbance that will interfere with other electronic devices in our home, or, if it will, there is a warning of the potential disruption that an internet router, for example, may cause a cell phone or stereo receiver. Prior to the recent breakthroughs in understanding the importance of brainwaves in cognition, there was no reason to suspect that electronic devices could disrupt our brain function, but there is concern that silent and imperceptible interactions between electronics in our modern world and the waves of bioelectrical activity in our brain can occur. As more wireless electronic devices, such as cell phones, radiate electromagnetic waves through our environment in the "electronic age," the potential for their electromagnetic emissions to perturb brainwaves is becoming clear. For example, research shows that cell phone broadcasts can disrupt brainwaves in a way that prevents sleep.

According to a 2008 study, cell phone signals can interfere with a person's brainwaves.[8] Neuroscientist Rodney Croft and his colleagues at Swinburne University of Technology in Australia found that a Nokia 6110 cell phone, operating in transmission mode in contact with the heads of 120 men and women, altered the power of their alpha waves. They compared the EEG recordings when the phone was transmitting and not transmitting, under the control of the experimenters, without subjects' knowledge. The EEG recordings showed a sudden power boost in the volunteers' alpha brainwaves, only on the side of the brain in contact with the cell phone when it was transmitting. Croft speculates that the heightened alpha waves reflect the mind concentrating to overcome the electrical interference in brain circuits caused by the pulsed electromagnetic radiation from cell phones.

This alpha boost may be consistent with findings of sleep researchers at Loughborough University, United Kingdom, who found that when people tried to sleep right after a cell phone is switched off, they take nearly twice as long compared to when the phone had previously been off or in "standby" mode. James Horne, one of the study's authors, reassures that the effects are harmless and less disruptive to sleep than half a cup of coffee. Still, Horne wonders, "If with different doses, durations, or other devices, would there be greater effects?"[9]

This research also raises questions of whether electromagnetic radiation from other sources that flood our modern-day environment might interfere with electromagnetic activity in the brain. High-voltage electromagnetic fields (EMF), such as those radiating from high-voltage power lines, can have a wide range of biological effects, but studies of the possible interactions with brainwaves are few. Substantial harmful effects can be ruled out by the very common usage of these devices without obvious deleterious impacts, which diminishes incentive for further study. However, an article published in 2017 argues that more research is needed into the possible effects of cell phones and other radio frequency, electromagnetic field–emitting electronic devices on human cognition, because a review of the literature supports the conclusion that there is an increase in cortical excitability during EMF exposure, and that such hyperexcitability may disrupt sleep or have other subtle effects, but there is not a general consensus on this.[10]

For people with brain implants, however, there is no question that electromagnetic energy generated by electronics in our environment poses a safety concern.[11] Radio and TV signals, EM radiation from AC circuits, overhead power lines, security systems, citizen band and ham radio, and appliances like vacuum cleaners, microwaves, and so on are all sources of EM interferences that have a real potential to cause harm by interfering with DBS or other brain implants used in BCI, similar to risks posed for people with pacemakers.

## DON'T FORGET GLIA: THE NONELECTRIC BRAIN

From the information presented so far, one might assume that all the cells in the brain are neurons. In fact, although you rarely hear about them, by far most of the cells in brain tissue (up to 85 percent) are not neurons. These non-neuronal cells are called glia, and the name derives from the word "glue." Until very recently, glia have been regarded simply as support cells for neurons. Glia were considered to be a type of connective tissue in the brain, which explains why there is no singular noun to describe them. Like "neuron," the singular noun for neurons, the singular for glia should have been "glion," perhaps, but that word does not exist because neuroscientists regarded these non-neuronal cells as "stuff" packed between neurons. That narrow-minded perspective, betrayed by the language we use in speaking about them, was a subtle but powerful constraint on scientists' thinking about what glia are and what they might do. The majority of neuroscientists ignored them. (Most still do.) It was understood that the "stuff between neurons" (glia) provided physical and nutritional support for neurons and responded to injury and disease. But the dazzle of electric neurons attracted all the attention and constrained thinking about how learning and memory, and all the other functions our brain performs, are accomplished. The results of glial research did not get published in high-profile journals, grant funding was meager for research on these nonelectric support cells, and the scientists who dared to study them suffered by devoting their career to pursuing a backwater area of neuroscience. But as my neuroscience colleague, Alfonso Araque, at the University of Minnesota, observed: From the viewpoint of a neuron looking out, what it sees surrounding it is glia.

Surprising research has revealed that while glia do not generate electrical impulses like neurons do, they communicate without using electricity. To the shock of neuroscientists, glia were found to communicate by using neurotransmitters, and could be triggered to do so by neurons firing. Neuroscientists were gobsmacked. Glia were found to release, take up, and detect the same neurotransmitters that neurons use to communicate at synapses, and glia use neurotransmitters to relay signals from cell to cell through glial networks. Moreover, because they exchange information using the common currency of neurotransmitters that were assumed (incorrectly) to be the exclusive coin of the realm of neurons, glia can control communication between neurons. By regulating the levels of neurotransmitters at synapses and by releasing neuroactive substances, glia can amplify or quash synaptic transmission between neurons. We now know that glia can sense action potentials and synaptic activity, communicate that information through a glial network by using nonelectrical means of communication (neurotransmitters), and then influence synaptic communication between neurons at a different place in the brain. All of this communication and information processing by glia is outside the neuron doctrine, which is the bedrock belief on which all neuroscience is built. The neuron doctrine, conceived by Nobel laureate Ramón y Cajal, states that all communication in the nervous system takes place by electrical signals passing in one direction across synapses between neurons. That fundamental theory is not entirely correct.

The implicit assumption that all activity in the brain was carried out by electrical signals transmitted between neurons meant that the tools scientists used to study the brain were deaf to glial communications. Electrodes were useless for tapping into glial communications that were being transmitted without electricity. A new technique for studying neural activity revealed glial communication, communication between non-neuronal cells that had been taking place undetected in nearly all previous studies of the cellular mechanisms of nervous system function.

In the late 1980s and early 1990s, neuroscientists developed new fluorescent dyes that glowed more brightly or shifted the color of their fluorescence when they bound specific types of ions inside neurons. Most important were dyes developed to study the flow of calcium ions in cells, because calcium is called the cellular "second messenger." Cells have a plethora of sensors on their cell membrane that constantly monitor the environment, looking for specific chemicals, temperature changes, ion fluxes, and transmembrane voltage shifts. These receptors are the cellular "first messengers." When the signals are detected, calcium ions are most commonly how these receptors alert the biochemistry inside the cell to the event. Calcium is ideal as a second messenger to relay information from events detected outside the cell to the interior of the cell, because cells live in a virtual sea of calcium ions. Inside the cell, however, calcium ions are scarce, because they are actively extruded or sequestered inside special intracellular compartments. Just as a dim flashlight beam can brightly signal in a dark night, so too can a small spurt of calcium ions in a cell's cytoplasm signal events that the cell detects. The intracellular calcium signals drive biochemical reactions that marshal an appropriate response to the event detected, and they even relay the message deep into the nucleus of cells to activate or suppress appropriate genes to synthesize appropriate proteins for an adaptive response. For example, calcium signaling is what strengthens synapses, according to Hebb's postulate that neurons that fire together wire together, and the calcium signals that reach the nucleus switch on genes to convert short-term memory into long-term memory.[12]

When scientists first used calcium-sensitive dyes to soak neurons in cell culture or in slices of brain tissue kept alive in a dish and looked at the living cells through their microscopes, they were delighted to see neurons flash when they spontaneously fired action potentials or when scientists stimulated them with electrodes.

The first time I saw this phenomenon was in cell cultures I made from dorsal root ganglion neurons, the cells that give us the sense of touch. Cheers of joy erupted in the middle of the night, echoing through the dark empty halls of the laboratory, when we stimulated the neurons to fire. At the time, live cell calcium imaging was something of a black art. The dyes were finicky and the specially designed computers and microscopes to study calcium signaling in live cells were pushing the edges of technology and were prone to failure. I traveled to Colorado State University in Fort Collins to learn firsthand from the masters of this new technique, Stan Kater and Peter Guthrie. After our success in watching neurons light up when they fired electrical impulses, I brought the technology back to my lab at the NIH and began to study how calcium signaling controlled the growth of axons from neurons and turned on and off specific genes according to the pattern of neural impulses they were generating. Ultimately I began to investigate how calcium signaling in neurons of the hippocampus converts short-term memory into long-term memory.[13] But what I soon saw, together with others doing calcium imaging at the time, was that calcium signaling was going on right next to the neurons—in glia. Still more surprising was that when neurons were stimulated to fire action potentials, the glia responded by flashing back, indicating that they too had generated calcium signals in response to the neuron firing.

I have described the discovery and our current understanding of glia in my previous book *The Other Brain*,[14] so there is no need to do so here, but it is critical in a book about the electric brain to recognize that electricity is only half the story. With an understanding of what brainwaves are and how they are generated, it should be obvious that cells that control neurotransmitters and ions and thus influence the voltage across a neuron's membrane must be involved in neural oscillations and brainwaves. Indeed, they are. Glial cells called astrocytes are involved in electrical storms of brain seizure, and they have recently been found to influence gamma waves and to intervene in the cellular mechanisms of learning that are dependent on gamma wave activity.[15]

My recent research on a different type of glial cell, called an oligodendrocyte, shows glial involvement in brainwave activity through a different mechanism—controlling the speed of impulse transmission through nerve fibers. These cells form the wrapping of electrical insulation on nerve fibers, called myelin, which is essential for rapid transmission of neural impulses.[16] Our research shows that oligodendrocytes sense and respond to neural impulses to regulate the formation of myelin,[17] which has been confirmed and extended by research in many other labs. Most recently our research has found that myelin that has already formed can change its structure and this transformation alters the speed of impulse transmission through axons. By affecting the spike time arrival at critical relay points in the brain and altering the function of these neural circuits, the behaviors they enable animals to perform are altered by these type of glia; specifically, in our experiments vision was affected.[18]

I view this form of plasticity as a companion to synaptic plasticity. What Hebb and others overlooked in formulating the cellular learning rule of "neurons that fire together, wire together" is the time it takes for impulses to travel between synaptic relay points in complex neural networks. If the distance between two neurons connected to a common target neuron are different, the impulse will arrive first from the neuron with the shortest distance to travel. Spike arrival time is critical in synaptic plasticity, just as arrival time is critical in any transportation or information network. Yet this concept applied to what we like to think of as the most complicated network in the universe, the human brain, eluded neuroscientists who were focused on mechanisms that alter synaptic transmission in learning and memory.[19]

These glial cells influence the speed of impulse transmission through the myelin sheath on axons. When one considers coupling neural oscillations and brainwaves of specific frequencies across large distances in the brain, obviously the speed of transmission is critical. Our mathematical modeling shows that the frequency of brainwave transmission and the phase and amplitude coupling of neuronal oscillators will be very sensitive to the speed of impulse transmission and thus to myelin.[20]

## ELECTRIC BRAIN 2.0

We are only seeing the proverbial tip of the iceberg. Our current understanding of brainwaves is superficial. Current technology to interface with neural circuits is horribly primitive and utterly inadequate to achieve the potential we imagine is possible if we could really interface seamlessly with the electric brain, and if we understood in detail how the brain codes, processes, and retrieves information. But we are at a turning point in human history. For the first time, we are beginning to analyze and manipulate the human brain through electricity—the electricity that infuses this bodily organ with life and the electricity that we deliver into its cellular circuitry. This is significant, because in using this force of nature we are dealing directly with the fundamental mechanism by which this astonishing organ operates.

In the past we diagnosed brain and mental function and treated dysfunctions by talking, drugs, and surgery. All of these approaches are lifesaving, and in many cases miraculous, but they are also feeble, indirect approaches. We are crossing a threshold, much as we did by monitoring the heart's electrical activity with an EKG and controlling its rhythms with pacemakers to diagnose and treat cardiac function. We are moving toward a future where electricity will be how we interact with the supreme bodily organ inside our skull. It is impossible to predict when we will fully achieve this capability. It may be decades or a century away before we achieve anything like the current capability that we have in using electricity in cardiology to being able to use electricity to equal effect in brain biology, but, if we look at past scientific progress, that day will come.

EEG, ECoG, and fMRI are now enabling scientists to glean a person's cognitive strengths and weaknesses, to diagnose neurological and psychological abnormalities, and to access a person's thoughts. These methods will continue to develop into the future, and more sophisticated analytical methods to decode the meaning of electrical activity in neural circuits will be devised to increase capabilities to probe and diagnose brain function. This capability will leave the experimental laboratory and enter into the medical community and beyond. Astute technological diagnosis and "mind reading" will rely on deeper understanding of fundamental brain biology learned from animal experiments, together with advancements in artificial intelligence, to enable computers to analyze and comprehend the complex signals swirling through the brain that elude the primitive analytical power we have today. By analyzing electricity in a person's brain, scientists of the future will be able to know as much about brain function as we can now discern about body function from medical tests.

Brain-computer interface technology will continue to advance and build on such transformative successes as the cochlear implant that brings the delight of sound to the deaf, to achieving the dream of compensating for other sensory and motor dysfunctions and for alleviating cognitive dysfunctions such as memory loss and dementia. However, scalp electrodes for recording EEG are not practical outside the laboratory, and they do not have sufficient capability to intercept the origin and detailed behavior of electrical activity that makes our brain function. Dry electrodes, which do not require an electrical paste to increase conductivity through skin, or some other method to access electrical activity in the brain through the skull, must be developed. Even current electrodes being implanted into the brain cannot sample neural circuitry adequately. The electrode materials corrode, induce tissue scarring, break physically and fail electrically, and must be replaced after only a few years. Electrodes made of flexible and biocompatible materials that can record and stimulate thousands of neurons with precision must be developed. Although in looking at current technology we don't know how this essential capability can be achieved, it is certain that in time it will.

Currently a laboratory of bulky, costly, sophisticated electronic equipment, computers, and highly skilled technicians and scientists are required to interface between the brain and prosthetic devices. In the future, electronic chips implanted in the brain will communicate wirelessly with compact electronic devices that are as simple to use and as portable as a tablet computer. From the perspective of today's tablet computer and cell phone, ENIAC, the first electronic computer that filled an entire room in 1945, dazzled the world by performing what is considered trivial by today's standards.[21] We can be assured that the same trends in engineering more powerful and miniaturized BCI devices in the future is a certainty. The only question is when.

The surgery required to implant computer chips in the brain is highly invasive. Brain surgery is major surgery—frightening to consider, challenging to perform, and risky. Noninvasive means of recording and stimulating neurons are being developed that use ultrasound, for example, but the future could bring nanoparticle-type detectors infused into the brain that wirelessly interface with computers. Stimulating electrodes, such as those used in deep brain stimulation, will always have the problem of being incapable of exclusively stimulating the appropriate neurons. Electricity follows the path of least resistance, stimulating any neurons in the path of an electric current, and nerve fibers passing by the electrode from neurons located long distances away will be inadvertently stimulated. Delivering electrical stimulation through the scalp by tDCS and TMS is far too crude at present to control activity in a precise manner in selected neural circuits. These methods alter the general level of excitation in a mass of brain tissue beneath the electrodes, not the precision firing in specific neural circuits.

The current technology of optogenetics—using light to control electrical firing in cells—has solved this problem in research with animals. Foreign genes can be precisely introduced into the specific neurons of interest, which enable stimulating or inhibiting electrical activity in individual neurons by flashes of light beamed from lasers into the brain through a slender fiber-optic cable. To treat neurological disorders like Parkinson's and psychiatric disorders like schizophrenia or depression, delivering appropriate excitatory or inhibitory stimulation in the right pattern to the right neurons is essential. When this technology is perfected for use in people, our current approaches of shock therapy, TMS, DBS, and tDCS will look barbaric, much as prefrontal lobotomy is viewed today. However, there are serious ethical obstacles to applying optogenetics to humans. Unlike a drug, which can be discontinued if side effects develop, a foreign gene or a virus introduced to infect neurons or glia with specific genes cannot be removed. Many experts, however, firmly believe that we will have no choice: In the foreseeable future this approach must be used to relieve the considerable human suffering caused by neurological and psychiatric diseases. Developing treatments for serious psychological illnesses is especially urgent. Psychosurgery is too primitive and dangerous. Electrical stimulation is too blunt. Drug therapies can be effective, but they don't always work. They are often applied in a trial-and-error manner, and they affect the entire brain and body. As a result, people with severe mental illness are often condemned to institutionalization. In time, as optogenetic technology and neuroscience advance, ethical considerations will demand the use of this approach.

Other developments on the near horizon are devices that regulate brain stimulation automatically by closed-loop systems that sense aberrant electrical activity in the brain and instantly deliver stimulation to control it. Through appropriate electrical feedback, the brain will command devices to control itself. Electrical stimulation will be increasingly used to speed recovery from stroke and injury. Already, electrical stimulation of the spinal cord has been used to restore motor function to people who are paralyzed, giving them some ability to operate their once paralyzed limbs and to control blood pressure and bladder and bowel function. Electrical stimulation in combination with physical therapy is accelerating recovery from traumatic brain injury and disease.

Another effective route into the brain is via the vagus nerve, which has widespread connections between the brain and body. Vagus nerve stimulation is being used effectively to improve rehabilitation after stroke by driving plasticity in the spared neurons, as well as to treat many other conditions.

Advances in science and technology on the electric brain are being made at an astonishing pace. When I began writing this book a little over three years ago, the prospect of being able to decode language without speech from the brainwaves of people simply thinking seemed a distant dream or fantasy, but this capability was achieved as this book was in the final stages of production. A few months ago, I heard, with my own ears, speech generated by a computer interpreting brainwaves from people who had electrodes implanted in their brains to guide surgery on their brains for epilepsy. In a noisy poster session at a scientific meeting, the scientist who was part of a team that had accomplished this feat stretched headphones over my ears for me to listen. It sounded as clear as Alexa, the humanoid computer in the cloud that can comprehend and produce human speech, but this speech was coming directly from a person's thoughts rippling in waves of electrical activity through the skull. With such advances, new ethical challenges confront us.

## ETHICS OF BRAIN HACKING

Most people are well aware of the profound ethical dilemmas personal genomics raises. Do you want to know if you have a gene for Alzheimer's? Do you want your employer or insurance company to know? Few people realize the greater ethical dilemma we now face as a result of the profound insights into the cognitive ability of a person that can be obtained by monitoring their brainwaves for five minutes at rest. Resting-state brainwaves reveal a great deal about how an individual's brain is wired. A person's gender, IQ, propensity to learn specific types of information (foreign languages, for example), creativity, and personality characteristics are all identifiable from brainwave recordings. How will this ability to assess a child's (or adult's) cognitive and mental aptitude influence his or her educational and professional opportunities?

In addition to enhancing normal brain function, this new science and technology of reading brainwaves affords many potential therapeutic benefits, such as alleviating pain, treating brain disorders, and even erasing traumatic memories, but, considering the unique status of the human brain, it also raises issues of informed consent in cases of impaired mental function, and in developing mechanisms to assure protection from harm to a person's health and well-being. The risk/benefit of brain implants for medical purposes, for example, in treating neurological disorders such as in Parkinson's disease or in controlling prosthetic limbs, must be carefully considered, because implanting electrodes into the brain involves unavoidable and unique risks.

In addition to neurological applications, the use of this technology in treating psychiatric disorders, such as major depression, addiction, and schizophrenia, is now being debated by ethicists. It will be necessary to formulate new guidelines and regulations on experimental research and on applications to treat mental dysfunction. Along with revealing these new technologies, it is important to examine the issues that are of vital practical importance as we stand on the cusp of a new capability that enables man, for the first time in human history, to directly interrogate and manipulate the human brain.

In speaking with Thomas Oxley, who is developing the Stentrode method of inserting electrodes into the brain through blood vessels, I raise this question, because this method is much less invasive than brain surgery and could be used more easily. This new technology to intercept and manipulate the brain's electrical activity raises new ethical questions of privacy and mind control. Nothing is more private than a thought, as Marcel Just noted when I visited his lab where he was reading a person's thoughts using fMRI. Controlling a person's brain pits issues of free will and informed consent against the benefit to people with neurological or psychiatric disorders—the same concern that halted deep brain stimulation research in the 1970s. How does a person in a coma give informed consent to have his or her brain probed and manipulated electronically, for example? "There are major security issues here," Oxley says. "We learned that with Facebook [which exposed unanticipated privacy concerns]. But I am cautiously optimistic about where the technology is going. How the ethical issues surrounding genetic engineering have been handled provides a beacon for us in how to do this technology properly."

Cardiac specialist Dr. William Whang, director of Clinical Electrophysiology Research at Mount Sinai, sees parallels between electrical manipulation of the brain to the development of cardiac pacemakers. "It took seventy years for clinical ECG [electrocardiography] to reach the point where it is today," he says. But neuroscientists lack the basic scientific understanding that cardiologists have about how the heart works so that it can be manipulated therapeutically by electronic interfaces. Neurosurgeons using BCI are in effect experimenting on their patients in an attempt to help them. For example, neurosurgeons apply deep brain stimulation, which is approved for alleviating Parkinson's disease, in an effort to treat a wide range of other conditions, including OCD, Tourette's syndrome, addiction, obesity, anorexia, depression, and many others.

Testing the human brain for what *may* work can seem like it is crossing a line. Oxley says, "But that's medicine. Experimentation is how medicine works."

## NEUROWEAPONS

A twenty-six-year veteran Sheriff Deputy of Colusa County, California, had the honor of pinning a police badge on his twenty-two-year-old daughter as she followed in his footsteps and became a rookie police officer in December 2018. Only a week later, around 7 PM, Officer Corona responded to the report of a three-car accident near 5th and D streets in Davis. A 1996 Infinity, driven by Christian Pascual, a twenty-five-year-old UC Davis graduate, had been hit by another vehicle, sending his car careening into a third vehicle. The drivers were out of their dented cars exchanging information when Officer Natalie Corona rolled up in her squad car to assist. After collecting Pascual's driver's license and recording the information, she handed it back to him.

"That's when I heard the shots," Pascual said.[22] A gunman standing only fifteen feet behind him unleashed a burst of gunfire, sending bullets whizzing so close to his ear that he suffered hearing loss. Pascual saw Officer Corona hit repeatedly and fall to the pavement as the gunman emptied all the rounds from his handgun and then reloaded.

This was not a case of road rage. In fact, the gunman, later identified as forty-eight-year-old Kevin Douglas Limbaugh, was not involved in the car accident at all. Spotting the policewoman, he immediately ambushed and killed her. Limbaugh was later found in a nearby home, dead from a self-inflicted gunshot. Next to his body police recovered a typed message explaining the motive—the police, Limbaugh claimed, had been bombarding him for years with ultrasonic waves.[23]

The belief that the government has implanted a computer chip in their brain or controls their thoughts and actions remotely by energy beams that scramble electrical activity in their brain are common diagnostic characteristics of many people with schizotypal personality disorders. This delusion is understandable in people suffering from a brain disorder that blurs the distinction between what is real and what is imagined. Their belief that their mind is being controlled by nefarious technological means is understandable as an attempt to find a rational explanation for what they are experiencing—the vivid perception of hearing voices in their head urging them to carry out compulsive actions.

But the fact that technology actually does exist to monitor brainwaves and alter them has caused some authorities to caution mental health professionals not to jump to hasty conclusions if their patients speak of this form of "delusion." In an article in the online newsletter *Global Research*, psychoanalyst Carole Smith argues for the necessity of reevaluating such symptoms used to diagnose schizophrenia in the era of mind control via modern technology.[24] Such people may not be suffering delusions, she argues. There is a real and imminent threat of "harnessing neuroscience to military capability, this technology is the result of decades of research and experimentation, most particularly in the Soviet Union and the United States," she writes.

Issuing a wake-up call, she warns, "We have failed to comprehend that the result of the technology that originated in the years of the arms race between the Soviet Union and the West has resulted in using satellite technology not only for surveillance and communication systems but also to lock on to human beings, manipulating brain frequencies by directing laser beams, neural-particle beams, electro-magnetic radiation, sonar waves, radiofrequency radiation (RFR), soliton waves, torsion fields and by use of these or other energy fields."[25]

Smith is not alone in her fear of mind control. Targeted Justice, an organization that consist of individuals who believe they too have been targeted by mind control technology, claims to have 1,080 members,[26] and companies marketing on the internet sell electronic devices that are supposed to block mind control carried out by electromagnetic transmissions.

Continuing her argument, Smith warns that the entire population is at risk. "Military technology could be used to broadcast global mind-control, as a system for manipulating and disturbing the human mental process using pulsed radio frequency (RFR) . . . psycho-electronic weaponry, the ultimate aim of which is to enter the brain and mind. Unannounced, undebated and largely unacknowledged by scientists or by the governments who employ them—technology to *enter and control minds from a distance* has been unleashed upon us." [Added emphasis is in the original quote.][27]

No longer immediately dismissed as psychotic delusions of the mentally ill, such fears have spread broadly as the general public becomes aware of emerging technology to interrogate and manipulate the electric brain for experimental research. For example, between 2016 and the time of this writing (June 2019), the US State Department has essentially closed the US embassy in Havana, Cuba, and instituted a travel warning advising Americans not to visit Cuba because they may be targeted and suffer traumatic brain injury from a sonic weapon or some other type of energy beam weapon operated by forces in Cuba that are hostile to the United States. This alarming international incident captivated the news after twenty-four embassy staff members were allegedly targeted by what Senator Bob Menendez (D-NJ) described as "brazen and vicious attacks" on our diplomats. Senator Menendez made these comments during a Senate investigative hearing titled "Attacks on U.S. Diplomats in Cuba: Response and Oversight," which was held on January 9, 2018, and chaired by Senator Marco Rubio (R-FL).[28] Citing anonymous medical authorities who had examined the diplomats in Miami and at the University of Pennsylvania, Senator Rubio reported that the embassy staff had suffered mild traumatic brain injury or concussion caused by nonnatural sources. Rubio described the sonic weapon as "very sophisticated technology that does not exist in the U.S. or anywhere else in the world."

Fears of mind control first erupted when José Delgado's research using deep brain stimulation reached the public and sparked alarm. These concerns, and abuses as in the case of Robert Heath using brain stimulation to "treat" homosexuality, resulted in federal agencies curtailing funding for such research for decades. But as the neuroscience of brainwave research and brain-computer interface technology has advanced, renewed fear is spreading, as the alleged sonic attack in Cuba demonstrates. The sensational news of these alleged attacks swept through major print and broadcast news outlets in a ceaseless series of reports for over two years since the alarming attacks were first reported. The political and economic consequences of this fearful energy beam weapon that the US government claims can incapacitate the brain and cause traumatic brain injury are real. Travel by Americans to Cuba has been curtailed, and Cubans who need US embassy services to travel abroad are stymied.

The profound fear of neuroweapons and mind control is amplified in the Cuban "sonic attack" case by the sheer improbability that an impoverished small island country suffering under an economic embargo for sixty years, lacking modern technology and basic necessities of life under the challenges of central control that all Communist governments face, could be threatening Americans with a neuroweapon. Fear is a powerful motivator, and fear of the unknown is natural. Few people understand brainwaves and few Americans know Cuba.

Intrigued and concerned, I investigated the Cuban incident in depth. I traveled to Havana and countryside villages to interview scientists and

Cubans from all walks of life, and I examined with my own eyes the sites of the alleged sonic attacks. What I saw and learned from speaking with Cuban scientists and the man on the street in that isolated island nation was a different perspective from that portrayed in the press and US Senate committees.

The battered 1942 Ford Fleetwood jostles its rickety way through the streets of Havana in the predawn rain, the darkness pierced by one working headlight as the taxi driver squints through a streaky windshield scraped by a spasmodic wiper blade feebly squeaking. Like the dead wiper on the passenger side, needles on all the gauges in the instrument panel are lifeless, pegged at zero. The Spanish Barreiros motor grafted under the Ford's rattling hood screams in whining waves of protest, like a marooned tractor, as the driver grinds through gears to swerve around deadly potholes. The nauseating stench of diesel twists my stomach. On the dashboard a cell phone glows with the colorful image of a two-year-old girl hugging her young father. The device communicating via satellite is a technological anachronism, as if the object were beamed in by time travelers from the future.

"Story is false," the driver answers in broken English when asked about the sonic attack on workers at the US embassy.

"Who is responsible, then?" I ask.

"Donald Trump," he replies.

This is the driver's way of transcending the language barrier to say that the US government is responsible for the sonic attack story. Cuba lacks the technology, and the facts do not support an attack of any kind, he says. When pressed for evidence, he cites the analysis by Cuban scientists who investigated the alarming incident, which has gripped US citizens in fear and essentially shut down the US embassy in Havana.

"But why?" I ask.

Twisting his two extended fists as if breaking a branch, he says, "to break the *cuerda* [rope]" tying the United States and Cuba together.[29]

I meet with some of those Cuban scientists. At the Cuban

Neuroscience Center, a towering concrete building situated near Raul Castro's heavily guarded estate well outside the tourist area, Dr. Mitchell Valdés-Sosa, a neuroscientist and director of the Center, reads through the report. Point by point, he methodically lays out the evidence. Valdés-Sosa is a member of the committee of Cuban scientists who conducted the scientific investigation into the reported attack, and he is an expert in auditory physiology, which makes him especially well qualified.

The take-home message is that the allegations that a neuroweapon was used against Americans in Cuba are not supported by the available facts, and in many cases they are inconsistent with the laws of nature. There is no precedence in the scientific literature showing that sound can cause traumatic brain injury. This sonic weapon could not be heard, except by the particular individual in the room who was "targeted" with the sound, which tracked him or her moving around the room among others who heard nothing. Ultrasound (frequencies above the audible range) are used for imaging a fetus, and they are considered harmless. The transducer must be applied directly to the mother's belly slicked with a gel to transmit the ultrasound waves, which would be shielded by a sliver of an airgap with the skin. How could ultrasonic sound waves be secretly beamed over a long distance and penetrate the thick concrete and glass walls of the US embassy? Low-frequency sound beneath the range of human hearing cannot be focused to track an individual around a room, and the equipment to generate low-frequency sound would be enormous and difficult to conceal. Inside the building even deafening sound would be dampened to levels well below those that can damage hearing. And another reality check—even deafening sound, which can damage hearing, does not cause traumatic brain injury.

When cell phone recordings of the alleged sonic attack weapon were identified by Cuban scientists as crickets chirping, the weapon was later attributed to a microwave beam of some kind that created the illusion of a sound inside the person's mind. This was pure speculation, not the deduction from any evidence from the scene. When that theory was dismissed by scientists, the attack was then attributed to an unknown type of neuroweapon energy beam that causes brain injury.

The hysteria went unabated in the media even as no rational means, motive, or culprit could be identified. Speculation shifted from the Cubans, to a rogue faction in Cuba, to the Russians. The motive became more obscure when similar attacks were reported on US embassy staff in China.

From a scientific perspective, I concluded that the available evidence did not support the claim that the Cuban embassy personnel who reported various complaints were suffering from what Dr. Charles Rosenfarb, medical director of the State Department's Bureau of Medical Services, called a new medical syndrome in a Senate hearing on the alleged attacks.[30] Their symptoms ranged widely and were easily explained by many common conditions that cause tinnitus, dizziness, and difficulties with mental focus or sleep problems. (These various symptoms were scattered among all the different patients, not shared by all who were allegedly attacked.)

Neither did the evidence support the claim that these patients had suffered traumatic injury, as the US State Department claims. Objective measures, such as MRI, showed no evidence of traumatic brain injury. All the patients had normal audiograms (hearing tests), except for three individuals who demonstrated different types of mild hearing impairments that are commonly seen in any population. Later studies showed that two of these individuals had prior hearing loss. Why would all the others report suffering hearing loss, but the objective tests of their hearing showed that it was absolutely normal?

Yet people remained afraid. And fear foments harm. The people who reported their medical complaints, only to be told by the US government that they had suffered brain damage in a targeted attack using a sophisticated neuroweapon, have had their suffering cruelly amplified.

José Delgado ignited fear of mind control the minute he switched on his radio-controlled brain stimulator in the 1950s. His rational, scientific responses failed to calm fears of mind control among the general public or the funding agencies that eventually ceased their support for such research.

"Fears have been expressed that this new technology brings with it the threat of possible unwanted and unethical remote control of the cerebral activities of man by other men," Delgado wrote, "but . . . this danger is quite improbable and is outweighed."[31]

He cited the lack of predictability seen in nearly all brain stimulation: "When a point of the brain is stimulated for the first time, we cannot predict the effects which may be evoked." Far from being able to control the thoughts in another person's mind or command them to perform a desired task or behavior, brain stimulation typically disrupts normal function. "In some cases the evoked response is directed by the animal in a purposeful way, but the movements and sequential responses are usually out of context," meaning that the evoked responses are ineffectual.

Delgado, with his advanced knowledge of neuroscience and vast experience with brain stimulation, sought to calm fears over mind control using rational argument. "It is very unlikely that we could electrically direct an animal to carry out predetermined activities such as opening a gate or performing an instrumental response. We can induce pleasure or punishment and therefore the motivation to press a lever, but we cannot control the sequence of movements necessary for this act in the absence of the animal's own desire to do so."[32] (Although Benchenane's brain stimulation of a rat during sleep to encourage it to go to a specific place in a maze after awakening approaches this ability [see pages 124–126].) )

The alarming prospect of authoritarian governments controlling the masses through radio transmission or other energy beams used to invade people's brains and overwhelm their free will has no scientific basis, he argued. "Could a ruthless dictator stand at a master radio transmitter and stimulate the depth of the brains of a mass of hopelessly enslaved people? This Orwellian possibility may provide a good plot for a novel, but fortunately it is beyond the theoretical and practical limits of electrical brain stimulation . . . we cannot substitute one personality for another, nor can we make a behaving robot of a human being."[33]

This is not to say that concerns over the means to apply the neuroscience of brainwaves to develop a neuroweapon should be dismissed. On the contrary, history shows that nearly every scientific discovery, from chipping flint rocks to forging steel, to gunpowder, to atomic physics, to satellites, to genetic engineering, has been developed into weapons. US patents have been issued, as mentioned previously, to apply neuroscience for the purposes of interrogation, to incapacitate enemy combatants on the battlefield, to use brain-computer interface to elevate the mental capability of the human brain, and to enrich war machines with human intelligence.[34] Prefrontal lobotomy was horribly abused. An FDA-approved electrical stimulator designed for use in behavioral modification therapy for children with autism and other disorders was allegedly used to torture unruly children in Judge Rotenberg Center in Canton, Massachusetts, leading to court action and calls for the FDA to recall approval of the medical device.[35] Video of the horrific application of this electrical stimulator to eighteen-year-old Andre McCollins in 2002 was widely circulated.[36] The boy was strapped down and shocked for hours at the facility as he screamed and begged for the shocks to stop. He was left in a catatonic state for days as a result.[37]

In January 1998, an annual public meeting of the French National Bioethics Committee was held in Paris. Its chairman, Jean-Pierre Changeux, a neuroscientist at the Institut Pasteur in Paris, told the meeting that "advances in cerebral imaging make the scope for invasion of privacy immense. Although the equipment needed is still highly specialized, it will become commonplace and capable of being used at a distance. That will open the way for abuses such as invasion of personal liberty, control of behavior and brainwashing. These are far from being science-fiction concerns . . . and constitute 'a serious risk to society.'"[38]

In January 1999, the European Parliament passed a resolution calling for a global ban on all development and deployment of weapons that might enable any form of manipulation of human beings.[39] In October 2001, Congressman Dennis J. Kucinich introduced a bill to the House of Representatives, H.R. 2977, to ban "the use of land-based, sea-based, or space-based systems using radiation, electromagnetic, psychotronic, sonic, laser, or other energies directed at individual persons or targeted populations for the purpose of information war, mood management, or mind control of such persons or populations," which would ban, expose, and stop psycho-electronic mind control experimentation on involuntary, nonconsensual citizens.[40]

## SEPARATING FACT FROM FICTION, SCIENCE FROM SENSATIONALISM

What fascinates me about this subject, and what drove me to write this book, is that research on the "electric brain" is science in action. There is no consensus: "Don't talk to me about oscillations," a very respected colleague of mine, who is renowned worldwide for his research on how the brain operates at the level of neural circuits, tells me, dismissing the whole idea that oscillations are anything other than brain "noise." While another equally renowned neuroscientist and colleague, György Buzsáki, argues on the flyleaf of his highly recommended scholarly book on brain oscillations, *Rhythms of the Brain*, that "far from being mere noise, [neural oscillations are] actually the source of our cognitive abilities."

While neuroscientists painstakingly struggle through careful experiments to understand brainwaves and the electric brain in general, taking care never to confuse hypothesis with conclusion, the public at large and the popular media seem to be swept up in a frenzy where the distinction between fact and fantasy is of no apparent concern in penning a sensational story. The internet is abuzz with reports of neuroscientists linking the human brain to computers to upload our mind to the internet, to eliminate the keyboard and empower us to control our personal electronic devices or manipulate robotic arms, and to fuse mind and machine (frequently using the Vulcan mind meld analogy), with bold declarations that "it's not science fiction anymore." But the science beneath the sizzle doesn't support all the sensational claims.

Unless you have a medical condition that can be helped by deep brain stimulation, or you need a prosthetic, would you want anyone to drill a hole through your skull to embed a computer chip in your brain? It is also difficult to imagine a swimming cap studded with EEG electrodes becoming the next fashion rage, but even if wearing such a bizarre device in public did not provoke 911 calls, how much thinking zipping through the trillions of neural circuits inside your cerebral cortex can be gleaned from the weak electric fields swirling over your entire scalp? Even if the dream came true of an advanced technological device that could record from thousands of neural circuits in our brain at once, current understanding of how human cognition operates at a circuit level is far too rudimentary to interpret and transmit thoughts.

To read accounts in the popular press, the future is here. Here are some examples I encountered in the course of writing this book:

> This has the potential to revolutionize prosthetic limbs for the disabled, but may also encourage people to rid themselves of their biological arms in favour of mechanical super limbs. Who knows!

(Really? How many people are willing to give their right arm for one manufactured by GE? Raise your hands.)

> In just 32 years, humans won't speak to each other, and will communicate through a worldwide-consciousness instead—using just our brains.

(Really? Not thirty or thirty-one years?)

Also from the same article:

> The technology will enable people to use brain-to-brain communication everywhere on the planet . . . to help fight crimes by looking directly into the minds of criminals and victims.

Then there's Elon Musk's and Mark Zuckerberg's bold claims in public of their plans to meld mind and machine, while remaining tight-lipped about the specifics until a publicity event was held in July 2019:

> Elon Musk announced his plan to integrate computers into the human body with a "neural lace" to merge human and machine—that would augment the human brain, adding another layer to the cortex and limbic system that is capable of communicating with a computer (essentially creating cyborgs).
>
> It [neural lace] is capable of both sending messages and creating information in the brain. The high-bandwidth interface could allow you to wirelessly transmit information

> to the cloud, to computers, or even directly to the brains of other people with a similar interface in their head. There is also the possibility of downloading content to augment your consciousness: think Neo learning kung-fu in *The Matrix*. If brains and computers speak the same language, then computers can impart information to the brain. [41]

From another article:

> As computers and brains would essentially be speaking the same language, emotions could be read as data using electrodes.

(I wonder which computer language? Windows 10, probably. Fortran, BASIC, and C are probably less compatible with most brains except for the really nerdy ones.)

> You can text your friends without using your phone's keyboard.

(No need for spell-check, but we are definitely going to need mind-check if you want to keep your friends.)

> Just imagine someone using telemetry going into a smart home and being able to operate all these devices merely by thinking about them.

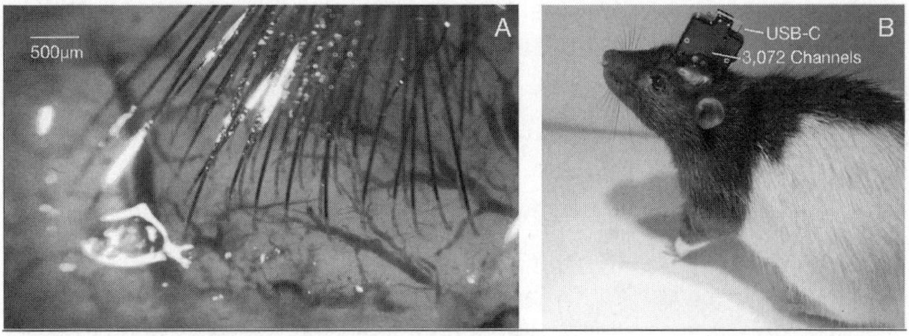


**FIGURE 43:** *Neuralink brain-computer interface by Elon Musk and Neuralink uses threadlike electrodes implanted into the cerebral cortex of mice.*

(Now there's a modern twist on the Midas touch fable.)

But scientists live in the future, and science is what gets us there. Could Orville and Wilbur Wright during their December 17, 1903, 120-foot, 12-second flight, clinging with white knuckles to a wood-and-paper airplane built of bicycle parts, have imagined Chicago O'Hare Airport with airplanes the size of office buildings taking off and landing every thirty to forty seconds at all hours with hundreds of people on board from all over the world? The idea that you could watch movies on your telephone while flying in an airplane was ludicrous in 1950, and it could be proven scientifically impossible because the cord would get hopelessly tangled.

No doubt with the wisdom of hindsight, people will look back on some of the current research and technology on the electric brain in the same way that we look on past scientists who replaced a kitten's brain with an artificial one made out of material used for metal dental fillings, or who used electricity to create life in the form of microscopic mites, or to bring corpses back to life. But while there are always missteps and misdirection along the way, it is all part of a long quest throughout history to come to grips with how and what it is to be human. The human brain is a machine, a biological one, and understanding how it works and how to manipulate it will bring benefits in health and achieve ambitions of philosophers, scholars, religious leaders, and scientists over eons. But can the human mind be reduced to a machine, and if we succeed in throwing open the curtain on the mysterious operation behind it all, which is precisely what drove Hans Berger in his experiments on his mental patients in Jena, will we lose our humanity?

I asked Thomas Reardon, who leads the start-up company CTRL-Labs in Manhattan working on BCI technology with the goal of directly interfacing machines with the human brain.[42] Given his unique background, I thought his views would be especially insightful. Reardon was taking advanced mathematics courses at MIT while he was in high school. After graduating he started a software company, which led to meeting Bill Gates. Gates hired him, and Reardon, in return, invented Internet Explorer, the highly successful internet browser that dominated all others in the 1990s and early 2000s. He left the company and began thinking about college when he met the famous physicist Freeman

Dyson. Dyson advised Reardon to broaden his mind. So the software engineer did so, by majoring in classics and literature and graduating from Columbia University in 2008 with high honors, magna cum laude. Moving from the scholarship of arts and letters, where the quest to understand human nature is at the heart of intellectual inquiry, Reardon decided to learn how the human brain works from a scientific perspective. In 2010 he earned an MS degree in neurobiology from Duke University and then went on to receive a PhD in Neuroscience and Behavior from Columbia University in 2016, working under the well-known neuroscientists Thomas Jessell and Attila Losonczy. On my visit to his company's new headquarters in Manhattan (see pages 370–374), I asked him if his work had diminished his appreciation of the human mind and of our humanity.

"No," he replied without hesitation. "It increases my appreciation of humanity."

And so it is with every scientist I know who has been fortunate to have a career devoted to marveling at nature, and trying through their own work and in collaboration with others around the world to understand a bit more of it, even as that new knowledge always magnifies what we do not know.

Science is a jigsaw puzzle, but you can never be confident that you have all the pieces, no matter how lovely the picture pieced together may appear. The breakthrough that gives us our current understanding of the brain was the startling discovery of a mysterious new force of nature—electricity. Do you suppose nature has no more secrets? What mysteries of natural forces and mechanisms that are beyond our imagination today might transform our understanding of the brain in the future?

"Do you think that we could be missing something fundamental about how the brain works?" I asked my dinner guest, neuroscientist and expert on brain oscillations György Buzsáki, at the end of a long day of presentations at a scientific meeting on brain plasticity in Tel Aviv. "Will we ever understand the human brain?"

"Not in my lifetime," he said.

Forty thousand years ago, in a cave at the edge of the Cantabrian Sea in Spain, the tomb-like silence and darkness were broken by flickering, crackling flame, as someone placed their warm palm on the wall of wet, cold rock and marveled at the contrast between flesh and stone. In her right hand she held the hollow leg bone of a deer packed with red clay ochre, and she aimed it carefully at her hand. Inhaling deeply, she put her lips to the bone and blew; the blast of breath expelled from her lungs sent a cloud of rust-red pigment spraying over her hand and stone. She withdrew her hand and gazed at the pale silhouette of four radiating fingers, laterally jutting thumb, and palm, a figure unique in nature—the human hand. This same scene was repeated around the world, leaving human hand prints on the walls of caves over tens of thousands of years wherever humans walked, from Indonesia, through Europe, to the southern tip of Argentina.[43]

From the age when tools were chipped from stone and the materials were bone and clay, through the age of bronze, and into the period of recorded human history when Greeks chiseled and polished marble into lifelike statues of men and women, mankind has sought to understand and re-create human life. Today we have television and movies that re-create human likeness and human experience in the most vivid manner our technology will enable, but it is not enough. We dream of holographic 3-D television and humanlike robots, endowed with intelligence and consciousness.

As mankind is beginning to understand the organ that makes us human, we endeavor to manipulate and re-create the human mind by linking it with computers or by making a silhouette of it using electrons, which we call "artificial intelligence." Karl August Weinhold, striving to re-create a kitten's brain with an amalgam of metal; Hans Berger, trying to capture the essence of the human psyche in the material world with his scalp electrodes; and now neuroscientists, probing the electric brain and making artificial brains from cells or electric circuits, are all striving for the same thing: to comprehend the same mystery burning in the mind of man since the Stone Age—human life, what it is to be human, and how our life emerges from matter transiently and then returns to dust.

The writer Joseph Conrad put it in a nutshell:

*The mind of man is capable of anything—because everything is in it, all the past as well as all the future.*

—*Joseph Conrad,* Heart of Darkness, *1899*

# Acknowledgments

There would be no *Electric Brain* if not for the talented expertise of my superb literary agent, Andrew Stuart, of the Stuart Agency in New York. I am grateful for his guidance in helping me formulate my ideas for this book from its earliest stages, for his acumen in finding a proper home for it at BenBella, and for his consistent support throughout the process of writing and producing this book. I thank Glenn Yeffeth and all at BenBella for their dedicated work to bring this endeavor from manuscript to book. I especially thank Alexa Stevenson and Sheila Curry Oakes for all their work as my editors; Elizabeth Degenhard for her diligent copyediting; Leah Wilson, editor-in-chief at BenBella Books, for top editing and adroitly managing all the myriad tasks involved in producing a book; and Jessika Rieck for her work as production editor. What you see in print is hardly the sole effort of author alone. It is a creative collaboration.

I am indebted to *Outside* magazine and *Scientific American* for permission to adapt portions of articles that I published in those magazines for this book, but my gratitude to these publications goes far deeper than the business of copyright permission. The opportunity to write and report on science for these magazines as a freelance writer opened new opportunities for me as a working neuroscientist. Science writing is an adventure that broadens my experience and deepens my love for science beyond the narrow confines of my own scientific research. The ideas for this book were seeded by some of these articles, and I am indebted to my editors Mariette DiChristina, Gary Stix, John Rennie, Ricki Rusting, Claudia Wallace, Michael Lemonick, Dean Visser, Robin Lloyd, Ingrid

Wickelgren, Karen Schrock, Tanya Lewis, and many others at *Scientific American* who provided me these opportunities; to Nicholas Spitzer at BrainFacts.org; and to Mary Turner, Justin Nyberg, and colleagues at *Outside* magazine.

I am extremely grateful to every one of the scientists who allowed me to interview them about their research for this book. I'll not list them here, but every quote that animates this book is a gift from them, motivated by their desire to share their passion and contribute to the process of exploring and explaining brain science. Two individuals that I am most grateful to have met, and who have contributed substantially to this book, are Jessica Eure and Robin Bernhard, who shared their work on brainwaves and neurofeedback with me. My sincere thanks also to their colleague Jay Gunkelman, master of qEEG, who provided me the unique experience of having my own brainwaves analyzed. It was a chance email—coming out of the blue—from Jessica that led me to learn about her work with Robin in EEG-based neurofeedback. That was an essential experience for me in writing this book, and I greatly admire their dedication to helping people who are suffering. The inspiration for this book sprang from my visit to Jena, Germany, hosted by my colleague Prof. Christoph Redies, a neuroanatomist at Freidrich-Schiller Universität. He and his colleague Prof. Susanne Zimmermann, medical historian, shared with me Hans Berger's notebooks and other historical information during my visit, which deeply intrigued me. I thank my colleague Dr. Davide Lecca, neuroscientist at the University of Milan, for tracking down the obscure information about Ferdinando Cezzamalli, all of which is in Italian, one of the many languages that I do not comprehend.

It is my great pleasure to dedicate this book to my dad, Richard L. Fields. When my preschool friends were beginning to learn their alphabet, Dad, an electrical engineer, was teaching me about electrons. Sketching with pencil and paper, he explained how tiny particles called electrons are buzzing like a swarm of bees around the nucleus of an atom. I took in his explanation with the same sense of wonder as someone showing me a new butterfly. The concept of electromagnetism may seem an abstraction well beyond the grasp of a preschooler's mind, but from the perspective of a naive child opening his eyes to a very strange world, subatomic physics was no less abstract than the alphabet. While Dad was teaching me about electricity and showing me how to fuse wires together with a smoking soldering gun, flicking silver splatters of molten metal on the ground, Mom was teaching me my ABCs, an equally odd but practical abstraction. In time, these experiences blossomed into a career as a scientist and a writer.

Dad taught my brother and me to build electromagnets, putting that magic, mystical force of nature at the control of a child's fingers. We wired light circuits in the house, made electric motors, built crystal radios for fun, and played with his ham radio—radio waves, yet another dazzling dimension of electricity. When I became a biologist with an interest in the nervous system, Dad's lessons and expertise in electricity and electronics continued to serve me. So it is most fitting that this book about the discovery of electricity in the brain and the transformative new research on brainwaves should be dedicated to my father.

Most of all, my sincere thanks to my wife, Melanie, and my children, Dylan, Morgan, and Kelly, who encouraged and supported my efforts to write this book. Their eyes were often the first to see what I had written, and to them I owe everything.

# BIBLIOGRAPHY FOR FURTHER READING

The following books are recommended for further reading.

## HISTORY OF STUDIES ON ELECTRICAL ACTIVITY IN THE BRAIN

Brazier, Mary Agnes Burniston. *A History of Neurophysiology in the 19th Century.* New York, NY: Raven Press, 1988.

Morus, Iwan Rhys. *Frankenstein's Children: Electricity, Exhibition, and Experiment in Early-Nineteenth-Century London.* Princeton, New Jersey: Princeton University Press, 1998.

Dougan, Andy. *Raising the Dead: The Men Who Created Frankenstein.* Edinburgh, United Kingdom: Birlinn Limited, 2008.

## ACCESSIBLE ACADEMIC BOOKS ON BRAINWAVES AND BRAIN STIMULATION

Delgado, José M. R. *Physical Control of the Mind: Toward a Psychocivilized Society.* New York, NY: Harper and Row, 1969.

Buzsáki, György. *Rhythms of the Brain.* New York, NY: Oxford University Press, 2006.

Nunez, Paul L. *The New Science of Consciousness: Exploring the Complexity of Brain, Mind, and Self.* New York, NY: Prometheus Books, 2016.

# References and Notes

## CHAPTER 1

1. Berger, H. (1904) Über die körperlichen äusserungen psychischer zustände: Weiteree xperimentelle beiträge zur lehre von der blutzirkulation in der schädelhöhle des menschen. Jena, Germany: Gustav Fischer.
2. Berger, H. (1904 and 1907) Über die körperlichen äusserungen psychischer zustände: Weiteree xperimentelle beiträge zur lehre von der blutzirkulation in der schädelhöhle des menschen. Jena, Germany: Gustav Fischer.
3. Millett, D. (2001) Hans Berger: From Psychic Energy to the EEG. *Perspectives in Biology and Medicine* 44: 522–542. https://pdfs.semanticscholar.org/75ca/e9df1a8ddb0724b53e778cacbf4be8cae983.pdf
4. Berger, H. (1910) Untersuchungen über die Temperatur des Gehirns. Jena, Germany: Gustav Fischer.
5. Berger, H. (1940) *Psyche*. Jena, Germany: Gustav Fischer.
6. Radin, D. (2006) *Entangled Minds: Extrasensory Experiences in a Quantum Reality*. New York, NY: Paraview Pocket Books.
7. La Vaque, T. J. (1999) The History of EEG Hans Berger: Psychophysiologist. A Historical Vignette. *The Journal of Neurotherapy* 3: 1–9.
8. Berger, H. (1940) *Psyche*, 6. Jena, Germany: Gustav Fischer.
9. Berger, H. (1929) Über das Elektrenkephalogramm des Menschen. *Archiv für Psychiatrie* 87: 527–570.
10. Lemke, R. (1956) Personal Reminiscences of Hans Berger. *Electroencephalography and Clinical Neurophysiology* 8: 708.
11. Ibid.
12. *Baltimore Sun*, January 4, 1931: 33. https://www.newspapers.com/newspage/214988483/
13. Burtt [*sic*], C. Psychology and Parapsychology. In Smythies, J. R., ed. (1967) *Science and ESP*. London: Routledge, Taylor and Francis Group, 61–142. See also Mundle, C. W. K. The Explanation of ESP. Chapter 8 in Smythies, J. R., ed. (1967) *Science and ESP*. London: Routledge, Taylor and Francis Group, 201.
14. Caldwell, W. E. (1959) Some Historical Notes and a Brief Summary of the Experimental Method and Findings of Ferninando Cazzamalli. *The Journal of General Psychology* 60: 121–129.
15. Cazzamalli, F. (1925) Fenomeni Telepsichice E Radioonde Cerebrali. *Neurologica* 42: 193–218; Cazzamalli, F. (1929–1930) Esperienze, Argomenti E Problemi Di Biofisca Cerebrale. *Quad. Di Psichiat.*: 81–105; Cazzamalli, F. (1935) Di un

Fenomeno radiante cerebropsichico (reflesso cerebropsicoradiante) come mezzo di esplorazione psicobiofisca. *Giornale Di Psichiat. E. Di Neuropathol.* 63: 45–56.

16. Cazzamalli, F. (1935) Phenomenes electromagnetiques du cerveau humain en activite psychosensorielle intense et leur demonstration par des complexes oscillateurs-revelateurs a triodes pour ondes ultracourtes. *Arch. Internat. Dc Neurol.* 64: 113–142.
17. Association of Italian Science of Metaphysics. http://www.metapsichica.net/home/home.asp?sid={A92545DD-98F2-4C3F-8D4F-DE98369D46D3}&idca=3&idpa=1
18. *Metapsychic, The Italian Journal of Parapsychology.* A.I.S.M. (Associazione Italiana Scientifica di Metapsichica.) http://www.metapsichica.net/home/home.asp?sid={68F482E7-66FE-4EFE-BEEE-A768C23A76C9}&idca=2&idpa=1. Accessed September 18, 2019.
19. Association of Italian Science of Metaphysics. http://www.metapsichica.net/home/home.asp?sid={A92545DD-98F2-4C3F-8D4F-DE98369D46D3}&idca=6&idpa=1. Accessed 2018.
20. Finkler, W. (1930) Die elektrische Schrift des Gehirns [The Electric Writing of the Brain]. *Neues Wiener Journal* 38: 7. See also Borck, C. (2005) Writing Brains: Tracing the Psyche with the Graphical Method. *History of Psychology* 8:79–94.
21. Moses, D. A., et al. (2019) Real-time Decoding of Question-and-Answer Speech Dialogue Using Human Cortical Activity. *Nature Communications* 10, no. 3096. See also Musk, E., and Neuralink (2019). An Integrated Brain-Machine Interface Platform with Thousands of Channels. BioRxiv. https://www.biorxiv.org/content/10.1101/703801v4; and Hanson, T. L., et al. (2019) The "Sewing Machine" for Minimally Invasive Neural Recording. BioRxiv. https://www.biorxiv.org/content/10.1101/578542v1
22. Adrian, E. D.; and Matthews, B. H. C. (1934) The Berger Rhythm: Potential Changes from the Occipital Lobes in Man. *Brain* 57: 355–385.
23. Ibid., Figure 3.
24. Lemke, R. (1956) Personal Reminiscences of Hans Berger. *Electroencephalography and Clinical Neurophysiology* 8: 708.
25. Brazier, M. (1961) *A History of the Electrical Activity of the Brain: The First Half-Century.* London: Pitman Medical Publishing Co., 114.
26. Borck, C. (2005) Writing Brains: Tracing the Psyche with the Graphical Method. *History of Psychology* 9, 79–94.
27. Berger, H. (1901) Zur Lehre von der Blutzirkulation in der Schädelhöle des menschen namentlich unter dem Einfluss von Medikamenten (Experimentelle Untersuchungen) [On the theory of the blood circulation in the human skull, especially under pharmaceutical influences (experimental investigations)]. Jena, Germany: Gustav Fischer.
28. Fayed, N.; Modrego, P. J.; and Morales, H. (2006) Evidence of Brain Damage after High-Altitude Climbing by Means of Magnetic Resonance Imaging. *The American Journal of Medicine* 119: 168. E1–6.
29. Fields, R. D. (2009, September 30) Are the Mountains Killing Your Brain? *Outside.* https://www.outsideonline.com/1884846/are-mountains-killing-your-brain

30. Lemke, R. (1956) Personal Reminiscences of Hans Berger. *Electroencephalography and Clinical Neurophysiology* 8: 708.
31. Goldensohn, E. D. (2001) Cellular Electrical Phenomena in Focal Epilepsy. In Lüders, H. O.; and Comair, Y.; eds. *Epilepsy Surgery*, 2nd ed. Philadelphia, PA: Lippincott and Williams and Wilkins, 1–18.
32. Niedermeyer, E. (2005) Historical Aspects. In *Electroencephalography: Basic Principles, Clinical Applications, and Related Fields*, 5th ed. Philadelphia, PA: Lippincott Williams and Wilkins, 1–16.
33. Das Gesetz zur Verhütung erbkranken Nachwuchses. Reichsgesetzblatt, Part I, July 14, 1933: 529; reprinted in Meier-Benneckenstein, P.; ed. Dokumente der deutschen Politik, Volume 1: Die Nationalsozialistische Revolution 1933, edited by Axel Friedrichs. Berlin, 1935: 194–95.
34. Forced Sterilization, United States Holocaust Memorial Museum. https://www.ushmm.org/learn/students/learning-materials-and-resources/mentally-and-physically-handicapped-victims-of-the-nazi-era/forced-sterilization. Accessed September 15, 2019.
35. Redies, C.; Viebig, M.; Zimmermann, S.; and Fröber, R. (2005) Origin of Corpses Received by the Anatomical Institute at the University of Jena During the Nazi Regime. *The Anatomical Record Part B: The New Anatomist* 285B (1): 6–10. Redies, C.; Fröber, R.; Viebig, M.; and Zimmermann, S. (2012) Dead Bodies for the Anatomical Institute in the Third Reich: An Investigation at the University of Jena. *Annals of Anatomy* 194 (3): 298–303. doi: 10.1016/j.aanat.2011.12.004
36. Pierre Gloor states that at this time, before the existence of brain imaging technology, surgeons would sometimes stick probes in the brain in an attempt to localize tumors. Gloor, P. (1969) *Hans Berger on the Electroencephalogram of Man.* New York, NY: Elsevier Publishing Company, 5.
37. Fields, R. D. (2011) Imaging Learning: The Search for a Memory Trace. *The Neuroscientist* 17: 185–196.
38. Chapman, H. C.; and Brubaker, A. P. (1888) Researches upon the General Physiology of Nerve and Muscle No. 2. *Proceedings of the Academy of Natural Sciences of Philadelphia* 40: 155–161. http://www.jstor.org/stable/pdf/4061244.pdf
39. Brazier, Mary A. B. (1961) *A History of the Electrical Activity of the Brain: The First Half-Century.* London: Pitman Medical Publishing Co.: 30.
40. Beck, A. (1891) The Determination of Localization in the Brain and Spinal Cord by Means of Electrical Phenomena. Doctoral thesis published in *Polska Akademija Umiejetności* 2: 187–232.
41. Brazier, Mary A. B. (1961) *A History of the Electrical Activity of the Brain: The First Half-Century.* London: Pitman Medical Publishing Co., 33. Beck, A. (1891) The Determination of Localization in the Brain and Spinal Cord by Means of Electrical Phenomena. Doctoral thesis published in *Polska Akademija Umiejetnosci* 2: 212.
42. Brazier, Mary A. B. (1961) *A History of the Electrical Activity of the Brain: The First Half-Century.* London: Pitman Medical Publishing Co., 33. Beck, A. (1891) The Determination of Localization in the Brain and Spinal Cord by Means

of Electrical Phenomena. Doctoral thesis published in *Polska Akademija Umiejetności* 2: 187–232.

43. Hitler's Black Book—List of Persons Wanted. Forces War Records. https://www.forces-war-records.co.uk/hitlers-black-book. Accessed September, 15, 2019.
44. The Polish Ministry of Information (1943) *The German New Order in Poland.* London: Hutchinson and Co., 432, quoted in Wróbel, P. (2000) *The Devil's Playground: Poland in World War II.* Montreal: The Canadian Foundation for Polish Studies of the Polish Institute of Arts & Sciences. http://www.warsawuprising.com/paper/wrobel1.htm. Accessed September 15, 2019.
45. Krętosz, J. Likwidacja kadry naukowej Lwowa w lipcu 1941 roku. In Heska-Kwaśniewicz, K.; Ratuszna, A.; and Żurawska, E., eds. (2012) *Niezwykła więź Kresów Wschodnich i Zachodnich.* Uniwersytet Śląski: 13–21. Retrieved December 12, 2014. http://wydawnictwo.us.edu.pl/node/2822. See also https://en.wikipedia.org/wiki/Massacre_of_Lw%C3%B3w_professors. Accessed September 18, 2019.
46. The Yivo Encyclopedia of Jews in Eastern Europe. L'viv. http://www.yivo encyclopedia.org/article.aspx/Lviv. Accessed September 15, 2019.
47. Nagorski, A. (2015). *The Greatest Battle.* New York, NY: Simon and Schuster, 83.
48. Coenen, A.; and Zayachkivskar, O. (2013) Adolf Beck: A Pioneer in Electroencephalography in Between Richard Caton and Hans Berger. *Advances in Cognitive Psychology* 9: 216–221. https://www.ncbi.nlm.nih.gov/pmc/articles/PMC3902832/pdf/acp-09-216.pdf
49. Brazier, Mary A. B. (1961) *A History of the Electrical Activity of the Brain: The First Half-Century.* London: Pitman Medical Publishing Co., 48. And https://en.wikipedia.org/wiki/Janowska_concentration_camp. Accessed September 18, 2019.
50. U-S-History.com. History of Catonsville, Maryland. https://www.u-s-history.com/pages/h2789.html. Accessed September 15, 2019.
51. Caton, R. (1887) Researches on Electrical Phenomena of Cerebral Grey Matter. *Ninth International Medical Congress* 3: 246–249.
52. Brazier, Mary A. B. (1961) *A History of the Electrical Activity of the Brain: The First Half-Century.* London: Pitman Medical Publishing Co., 16.
53. Ibid., 7.
54. Ibid. Also Caton, R. (1877) Interim Report on Investigation of the Electric Currents of the Brain. *British Medical Journal* 1: 62–65.
55. Brazier, Mary A. B. (1961) *A History of the Electrical Activity of the Brain: The First Half-Century.* London: Pitman Medical Publishing Co., 62.
56. Caton, R. (1891) Die Strome der Centralnervensystems. *Centralblatte der Physiology* 4 (1890–1891): 785-786, translated by Brazier, ibid., 62, in reference to his prior paper, Caton, R. (1875) The Electric Currents of the Brain. *British Medical Journal* 2: 278.
57. Brazier, Mary A. B. (1961) *A History of the Electrical Activity of the Brain: The First Half-Century.* London: Pitman Medical Publishing Co., 20–21.
58. Berger, H. (1929) Uber das Elektrenkephalogramm des Menschen. *Archiv für Psychiatrie* 87: 527–570.

## CHAPTER 2

1. Morus, I. R. (1998) *Frankenstein's Children*. Princeton, NJ: Princeton University Press, 109.
2. BritainExpress.com. Fyne Court. https://www.britainexpress.com/attractions.htm?attraction=3623. Accessed September 15, 2019.
3. Crosse, C. A. H. (1857) *Memorials Scientific and Literary of Andrew Cross, the Electrician*. London: Longman, Brown, Green, Longmans and Robert.
4. Adams, S. Letters of Scientist who Inspired Frankenstein Go on Public Display. *Telegraph*, December 8, 2008: https://www.telegraph.co.uk/technology/3684107/Letters-of-scientist-who-inspired-Frankenstein-go-on-public-display.html. Retrieved August 23, 2016.
5. Ibid.
6. Wright, B. (2015) *Andrew Crosse and the Mite That Shook the World*. Leicester, England: Troubador Publishing.
7. Finger, S.; and Law, M. B. (1998) Karl August Weinhold and His "Science" in the Era of Mary Shelly's Frankenstein: Experiments on Electricity and the Restoration of Life. *Journal of the History of Medicine* 53: 161–180.
8. Morus, I. R. (1998) *Frankenstein's Children*. Princeton, New Jersey: Princeton University Press, reference 63.
9. Crosse, A. (1836–7) On the Production of Insects by Voltaic Electricity. *Annals of Electricity* 1: 242–244, quote on p 242–243. Also quoted in Cross, A. (1838) Description of Some Experiments Made with the Voltaic Battery, by Andrew Crosse, Esq. of Broomfield, near Taunton, for the Purpose of Producing Crystals; in the Process of which Experiments Certain Insects Constantly Appeared. Communicated in a letter dated December 27, 1847, addressed to the Secretary of the London Electrical Society and read on January 20, 1838, 246–257; and in Morus, I. R. (1998) *Frankenstein's Children*. Princeton, New Jersey: Princeton University Press, 139–140.
10. Dougan, A. (2008) *Raising the Dead: The Men Who Created Frankenstein*. Edinburgh, Scotland: Birlinn Limited, eBook edition, location 1764 of 2384.
11. Wright, B. (2015) *Andrew Crosse and the Mite That Shocked the World*. Leicester, England: Troubador Publishing, 117.
12. Stallybrass, O. (1967) How Faraday "Produced Living Animalculae": Andrew Crosse and the Story of a Myth. *Proceedings of the Royal Institution* 41: 597–619, quoted in Morus, I. R. (1998) *Frankenstein's Children*. Princeton, New Jersey: Princeton University Press, 142.
13. Turpin, M. (1838) Note on the Kind of Acarus presented to the Academy at the Session of the 30th of October, by M. Roberton to whom Mr. Crosse had communicated it; by M. Turpin. *The Annals of Electricity, Magnetism, and Chemistry and Guardian of Experimental Science* 2 (1838): 355–360.
14. Morus, I. R. (1998) *Frankenstein's Children*. Princeton, New Jersey: Princeton University Press, 142.
15. Ibid.

16. Ibid., 138, referencing Mackintosh's electrical theory of the universe, *New Moral World* 3 (1836–7): 239.
17. Ibid., 138, referencing Mackintosh's lectures in *Mechanics' Magazine* 27 (1837): 9–10.
18. Ibid., 138, quoting Thomas S. Mackintosh, The Electrical Theory of the Universe, or the Elements of Physical and Moral Philosophy, Boston 1838, 360–361.
19. Bada, J. L.; and Lazcano, A. (2000) *Origins of Life and Evolution of the Biosphere* 30: 107–112. https://doi.org/10.1023/A:1006746205180
20. Berhanu, S., et al. (2019) Artificial Photosynthetic Cell Producing Energy for Protein Synthesis. *Nature Communications* 10 (1325).
21. Devlin, H. Scientists Grow "Mini-Brain on the Move" That Can Contract Muscle. *The Guardian*, March 18, 2019. https://www.theguardian.com/science/2019/mar/18/scientists-grow-mini-brain-on-the-move-that-can-contract-muscle
22. Also spelled as Carl August Weinhold.
23. Weinhold's experiments published in Weinhold, K. A. (1817) Versuche über das Leben und seine, Grundkräfte, auf dem Wege der experimental-Physiologie [Experiments on Life and Its Primary Forces through the Use of Experimental Physiology]. Magdeburg: Creutz. https://books.google.com/books/about/Versuche_%C3%BCber_das_Leben_und_seine_Grund.html?id=viU_AAAAcAAJ. From Finger, S.; and Law, M. B. (1998) Karl August Weinhold and His "Science" in the Era of Mary Shelly's Frankenstein: Experiments on Electricity and the Restoration of Life. *Journal of the History of Medicine* 53: 169.
24. Ibid.
25. Morus, I. R. (1998) *Frankenstein's Children*. Princeton, New Jersey: Princeton University Press, eBook edition, location 1667 of 2384.
26. Finger, S.; and Law, M. B. (1998) Karl August Weinhold and His "Science" in the Era of Mary Shelly's Frankenstein: Experiments on Electricity and the Restoration of Life. *Journal of the History of Medicine* 53: 169.
27. Ibid., 179.
28. Eberle, H.; and Weinhold, K. A. Martin-Luther-Universität Halle-Wittenberg (in German). https://www.catalogus-professorum-halensis.de/weinholdkarlaugust.html. Accessed September 15, 2019, citing Hans-Theodor Koch, Karl August Weinhold (1782–1829) and his suggestion for infusion (1827), in Piechocki/Koch, contributions (1965): 182–188.
29. Mörgeli, C. (1993) Chirurgischer Eingriff gegen die Übervölkerung: Professor Weinholds Verhaut-Infibulation. *Gesnerus* 50: 264–273.
30. Weinhold, K. A. (1929) Das Gleichgewicht der Bevölkerung, als Grundlage der Wohlfahrt der Gesellschaft und der Familien. Verlag von Carl Focke, Leipzig, 67. See also Lorenz, M., Menschen-Zucht Frühe Ideen und Strategien 1500–1870, Wallstein Verlag, Göttingen, 2018, 137-138.
31. HistoryExtra.com. A Brief History of Capital Punishment in Britain. https://www.historyextra.com/period/modern/a-brief-history-of-capital-punishment-in-britain/. Accessed September 15, 2019.

32. Hurren, E. T. (2016) *Dissecting the Criminal Corpse: Staging Post-Execution Punishment in Early Modern England.* Basingstoke, UK: Palgrave Macmilan. https://www.ncbi.nlm.nih.gov/books/NBK384636/
33. Ibid.
34. Dougan, A. (2008) *Raising the Dead: The Men Who Created Frankenstein.* Edinburgh, Scotland: Birlinn Limited, eBook edition, location 2012 of 2384, eBook location 1532 of 2384.
35. Ibid., eBook location 1534 of 2384.
36. Ibid., eBook location 1547 of 2384, citing Ure, A. (1819) *The Journal of Science and the Arts* VI (John Murray, 1819): 283–293.
37. Ibid., eBook location1561 of 2384
38. Ibid., eBook location 1205 of 2384.
39. Ibid., eBook location 1605 of 2384.
40. Wright, B. (2015) *Andrew Crosse and the Mite That Shook the World.* Leicester, England: Troubador Publishing.
41. Shelley, M. W. (1869) *Frankenstein The Modern Prometheus,* Sever, Francis, and Company, Boston and Cambridge, 30.
42. Ibid., 37.
43. Dougan, A. (2008) *Raising the Dead: The Men who Created Frankenstein.* Edinburgh, Scotland: Birlinn Limited, eBook edition, eBook location 1195 of 2384.
44. Hurren, E. T. (2016) *Dissecting the Criminal Corpse: Staging Post-Execution Punishment in Early Modern England.* Basingstoke, UK: Palgrave Macmilan. See also Aldini, J. (1803) *An Account of the Late Improvements in Galvanism, with a Series of Curious and Interesting Experiments.* London: Wilks and Taylor, 221 pages. http://www.gutenberg.org/files/57267/57267-h/57267-h.htm
45. King, G. Edison vs. Westinghouse: A Shocking Rivalry. *Smithsonian.com*, October 11, 2011. http://www.smithsonianmag.com/history/edison-vs-westinghouse-a-shocking-rivalry-102146036/
46. Death Penalty Information Center. 125 Years Ago, First Execution Using Electric Chair Was Botched. DeathPenaltyInfo.org, August 7, 2015. http://www.deathpenaltyinfo.org/node/6216. Also Death Penalty Information Center. Methods of Execution. https://deathpenaltyinfo.org/methods-execution. Accessed September 15, 2019.

## CHAPTER 3

1. Center for Disease Control and Prevention. Autism Spectrum Disorder (ASD). https://www.cdc.gov/ncbddd/autism/index.html. Accessed September 15, 2019.
2. Ibid.
3. Reported to me by Jennifer Eure in July 2017, quoting from her medical notes about one of her patients with the consent of "Joey's" mother who wished to share her experience with readers of this book. The names and other details have been changed to respect privacy.

4. Fink, M. (2009) *Electroconvulsive Therapy, a Guide for Professionals and Their Patients.* Oxford, UK: Oxford University Press.
5. Fields, R. D. (2009) *The Other Brain.* New York, NY: Simon and Schuster.
6. Wright, B. A. (1990) An Historical Review of Electroconvulsive Therapy. *Jefferson Journal of Psychiatry* 8 (2): 68–74. https://jdc.jefferson.edu/cgi/viewcontent.cgi?referer=https://www.google.com/&httpsredir=1&article=1256&context=jeffjpsychiatry
7. Shorter, E.; and Healy, D. (2007) *Shock Therapy, A History of Electroconvulsive Treatment in Mental Illness.* New Brunswick, NJ: Rutgers University Press, 27.
8. Fields, R. D. (2015) A New Mechanism of Nervous System Plasticity: Activity-Dependent Myelination. *Nature Reviews Neuroscience* 16: 756–767.
9. Guarino, B. Ancient DNA Solves Mystery of Canaanites, Reveals the Biblical People's Fate. *The Washington Post,* July 27, 2017. https://www.washingtonpost.com/news/speaking-of-science/wp/2017/07/27/ancient-dna-solves-mystery-of-the-canaanites-reveals-the-biblical-peoples-fate/?utm_term=.0862ebe404a3
10. Fields, R. D. Lightning in your Brain. *BrainFacts.org,* May 23, 2014. https://www.brainfacts.org/brain-anatomy-and-function/cells-and-circuits/2014/lightning-in-your-brain
11. Liu, Z.; Ding, L.; and He, B. (2006) Integration of EEG/MEG with MRI and fMRI in Functional Neuroimaging. *IEEE Engineering in Medicine and Biology Magazine* 25: 46–53.

## CHAPTER 4

1. Snider, J.; Plank, M.; Lynch, G.; Halgren, E.; and Poizner, H.; et al. (2013) Human Cortical Theta During Free Exploration Encodes Space and Predicts Subsequent Memory. *Journal of Neuroscience* 33 (38): 15056–15068.
2. Fields, R. D. (2007) The Shark's Electric Sense. *Scientific American* 297 (2): 74–81.
3. Fields, R. D.; Bullock, T. H.; Lange, G. D. (1993) Ampullary Sense Organs, Peripheral, Central and Behavioral Eectroreception in Chimaeras (Hydrolagus, Holocephali, Chondrichthyes). *Brain, Behavior and Evolution* 41: 269–289.
4. Montgomery, J.; et al.,. (2013) *Novel Neurotechnologies: Intervening in the Brain* Nuffield Council on Bioethics, London, page xxiv.
5. Karalis, N.; et al. (2016) 4 Hz Oscillations Synchronize Prefrontal-Amygdala Circuits During Fear Behavior. *Nature Neuroscience* 19 (4): 605–612. https://www.ncbi.nlm.nih.gov/pmc/articles/PMC4843971/
6. Fields, R. D.; and Lange, G. D. (1980) Electroreception in the Ratfish (Hydrolagus colliei). *Science* 207 (4430): 547–548. https://www.ncbi.nlm.nih.gov/pubmed/7352266
7. Fields, R. D.; and Ellisman, M. H. (1985) Synaptic Morphology and Differences in Sensitivity. *Science* 12: 197–199.
8. Buzsáki, G. (1989) Two-Stage Model of Memory Trace Formation: A Role for "Noisy" Brain States. *Neuroscience* 31 (3): 551–570.
9. Girardeau, G.; et al. (2017) Reactivations of Emotional Memory in the

Hippocampus-Amygdala System During Sleep. *Nature Neuroscience* 20 (11): 1634–1642. doi:10.1038/nn.4637

10. De Laviléon, G.; et al. (2015) Explicit Memory Creation During Sleep Demonstrates a Causal Role of Place Cells in Navigation. *Nature Neuroscience* 18: 493–495. https://www.nature.com/articles/nn.3970

## CHAPTER 5

1. Saberi, K.; and Perrott, D. R. (1999) Cognitive Restoration of Reversed Speech. *Nature* 398: 760.
2. Luo, H.; and Poeppel, D. (2007) Phase Patterns of Neuronal Responses Reliably Discriminate Speech in Human Auditory Cortex. *Neuron* 54: 1001–1010; and Giraud, A. L.; and Poeppel, D. (2012) Cortical Oscillations and Speech Processing: Emerging Computational Principles and Operations. *Nature Neuroscience* 15 (4): 511–517. doi:10.1038/nn.3063
3. Wolchover, N. Breaking the Code: Why Yuor Barin Can Raed Tihs. *LiveScience.com*, February 9, 2012. https://www.livescience.com/18392-reading-jumbled-words.html
4. Fields, D. Guitar Hero. *Washington Post Magazine*, September 9, 2007. http://www.washingtonpost.com/wp-dyn/content/article/2007/09/04/AR2007090401785.html
5. Meichle, A. Albert Einstein in Princeton. *Albert Einstein in the World Wide Web.* http://www.einstein-website.de/z_biography/princeton-e.html. Accessed September 15, 2019.
6. Buschman, T. J.; et al. (2012) Synchronous Oscillatory Neural Ensembles for Rules in the Prefrontal Cortex. *Neuron* 76: 838–846.
7. Cardin, J. A. (2016) Snapshots of the Brain in Action: Local Circuit Operations Through the Lens of Gamma Oscillations. *Journal of Neuroscience* 36: 10496–10504; and Sohal, V. S. (2016) How Close Are We to Understanding What (if Anything) Gamma Oscillations Do in Cortical Circuits? *Journal of Neuroscience* 36: 10489–10495.

## CHAPTER 6

1. Celizic, M. Doctors Pull Plug, Comatose Woman Wakes Up. TODAY.com, September 10, 2007. https://www.today.com/news/doctors-pull-plug-comatose-woman-wakes-wbna20689992
2. Edlow, B. L.; et al. (2017) Early Detection of Consciousness in Patients with Acute Severe Traumatic Brain Injury. *Brain* 140 (9): 2399–2414. https://doi.org/10.1093/brain/awx176
3. Carey, B. "It's Gigantic": A New Way to Gauge the Chances for Unresponsive Patients. *New York Times*, June 26, 2019. https://www.nytimes.com/2019/06/26/health/brain-injury-eeg-consciousness.html
4. Edlow, B. L.; et al. (2017) Early Detection of Consciousness in Patients with Acute Severe Traumatic Brain Injury. *Brain* 140 (9): 2399–2414. https://doi.org/10.1093/brain/

5. Lewis, L. D.; et al. (2012) Rapid Fragmentation of Neuronal Networks at the Onset of Propofol-Induced Unconsciousness. Proceedings of the National Academy of Sciences of the United States of America 109 (49): E3377–86. doi:10.1073/pnas. And Purdon, P. L.; et al. (2013) Electroencephalogram Signatures of Loss and Recovery of Consciousness from Propofol. *Proceedings of the National Academy of Sciences of the United States of America* 110: E1142–E1151.
6. Guohua, L.; et al. (2009) Epidemiology of Anesthesia-Related Mortality in the United States, 1999–2005. *Anesthesiology* 110 (4): 759–765. https://www.ncbi.nlm.nih.gov/pmc/articles/PMC2697561/
7. Parker-Pope, T. The Pain of Being a Redhead. *New York Times*, August 6, 2009. https://well.blogs.nytimes.com/2009/08/06/the-pain-of-being-a-redhead/


## CHAPTER 7


1. Gauchat, A.; Seguin, J. R.; McSween-Cadieux, E.; and Zadra, A. (2015) The Content of Recurrent Dreams in Young Adolescents. *Consciousness and Cognition* 37: 103–111.
2. Zadra, A. Recurrent Dreams: Their Relation to Life Events. In Barrett, D., ed. (1996) *Trauma and Dreams*. Cambridge, Massachusetts: Harvard University Press, 231–247.
3. Sharpless, B. A. (2016). A Clinician's Guide to Recurrent Isolated Sleep Paralysis. *Neuropsychiatric Disease and Treatment* 12: 1761–1767.
4. Vasas, A. Amanda's Story. Narcolepsy Network. http://narcolepsynetwork.org/amandas-story/. Accessed September 15, 2019.
5. Chen, L.; et al. (2009) Animal Models of Narcolepsy. *CNS & Neurological Disorders Drug Targets* 8 (4): 296–308.
6. Robson, D. The Tragic Fate of the People Who Stop Sleeping. BBC *Future*, January 19, 2016. http://www.bbc.com/future/story/20160118-the-tragic-fate-of-the-people-who-stop-sleeping
7. Dean, R. C.; and Lue, T. F. (2005) Physiology of Penile Erection and Pathophysiology of Erectile Dysfunction. *Urologic Clinics of North America* 32 (4): 379–395, v.
8. Moruzzi, G.; and Magoun, H. W. (1949) Brain Stem Reticular Formation and Activation of the EEG. *Electroenchalography and Clinical Neurophysiology* 1: 455–473.
9. Sterman, M. B.; and Clemente, C. D. (1962) Forebrain Inhibitiory Mechanisms: Sleep Patterns Induced by Basal Forebrain Stimulation in the Behaving Cat. *Experimental Neurology* 6: 103–117.
10. Brunner, D. P.; et al. (1990) Effect of Partial Sleep Deprivation on Sleep Stages and EEG Power Spectra: Evidence for Non-REM and REM Sleep Homeostasis. *Electroencephalography and Clinical Neurophysiology* 75 (6): 492–499.
11. De Noon, D. J. Ambien Linked to "Sleep Eating": Rare Cases of Unconscious Eating and Cooking Seen in Patients Using Sleeping Pill. WebMD.com, March 15, 2006. https://www.webmd.com/sleep-disorders/news/20060315/ambien-linked-to-sleep-eating#1

12. James, S. M.; et al. (2017) Shift Work: Disrupted Circadian Rhythms and Sleep—Implications for Health and Well-Being. *Current Sleep Medicine Reports* 3: 104–112.
13. Takeuchi, T.; et al. (1992) Isolated Sleep Paralysis Elicited by Sleep Interruption. *Sleep* 15 (3): 217–225.
14. Derry, C. (2012) Nocturnal Frontal Lobe Epilepsy vs Parasomnias. *Current Treatment Options in Neurology* 14 (5): 451–463.
15. Gemignani, A.; et al. (2012) Thalamic Contribution to Sleep Slow Oscillation Features in Humans: A Single Case Cross Sectional EEG Study in Fatal Familial Insomnia. *Sleep Medicine* 13: 946–952.
16. Ujma, P. P.; et al. (2017) The Sleep EEG Spectrum Is a Sexually Dimorphic Marker of General Intelligence *Scientific Reports* 7 (1): 18070. doi:10.1038/s41598-017-18124-0
17. De Gennaro, L.; and Ferrara, M. (2003) Sleep Spindles: An Overview. *Sleep Medicine Reviews* 7 (5): 423–440.
18. Stix, G. (2013) Sleep Hits the Reset Button for Individual Neurons. *Scientific American,* March 22, 2013. https://blogs.scientificamerican.com/talking-back/sleep-hits-the-reset-button-for-individual-neurons/
19. Bukalo, O.; et al. (2013) Synaptic Plasticity by Antidromic Firing During Hippocampal Network Oscillations. *Proceedings of the National Academy of Sciences of the United States of America* 110 (13): 5175–5180.
20. Larson, J.; and Munkácsy, E. (2015) Theta-Burst LTP. *Brain Research* 1621: 38–50.
21. Dudek, S. M.; and Fields, R. D. (2002) Somatic Action Potentials Are Sufficient for Late-Phase LTP-Related Cell Signaling. *Proceedings of the National Academy of Sciences of the United States of America* 99: 3962–3967.
22. Fields, R. D. (2004) Making Memories Stick. *Scientific American* 290: 74-81.
23. Tononi, G.; and Cirelli, C. (2006) Sleep Function and Synaptic Homeostasis. *Sleep Medicine Reviews* 10 (1): 49–62.
24. Krack, P.; Hariz, M. I.; Baunez, C.; Guridi, J.; and Obeso, J. A. (2010) Deep Brain Stimulation: From Neurology to Psychiatry? *Trends in Neurosciences* 33 (10): 474–484.

## CHAPTER 8

1. The Spanish Civil War. *Encyclopedia Britannica.* https://britannica.com/event/Spanish-Civil-War. Accessed October 3, 2019.
2. The Spanish Civil War. *Holocaust Encyclopedia.* https://www.ushmm.org/wlc/en/article.php?ModuleId=10008214. Accessed October 3, 2019.
3. The Spanish Civil War. *Encyclopedia Britannica.* https://www.britannica.com/event/Spanish-Civil-War. Accessed October 3, 2019.
4. Blackwell, B. (2012) José Manuel Rodriguez Delgado. *Neuropsychopharmacology* 37 (13): 2883–2884. https://www.ncbi.nlm.nih.gov/pmc/articles/PMC3499727/
5. The Spanish Civil War. *Holocaust Encyclopedia.* https://www.ushmm.org/wlc/en/article.php?ModuleId=10008214. Accessed October 3, 2019.

6. About Simon Wiesenthal. Simon Wiesenthal Center. http://www.wiesenthal.com/about/about-simon-wiesenthal/. Accessed September 9, 2019.
7. Delgado, J. M. R. (1969) *Physical Control of The Mind: Toward a Psychocivilized Society*. New York, NY: Harper and Row.
8. Ibid., 232.
9. http://www.cajal.csic.es/departamentos/herreras-espinosa/datos.html
10. Charles, A.; and Brennan, K. C. (2009) Cortical Spreading Depression—New Insights and Persistent Questions. *Cephalalgia* 29 (10): 1115–1124.
11. Murugan, V. (2009) Embryonic Stem Cell Research: A Decade of Debate from Bush to Obama. *Yale Journal of Biology and Medicine* 82 (3): 101–103. https://www.ncbi.nlm.nih.gov/pmc/articles/PMC2744932/#s1title
12. Delgado, J. M. R. (1969) *Physical Control of The Mind: Toward a Psychocivilized Society*. New York, NY: Harper and Row, 71.
13. Ibid., 68.
14. Delgado, J. M. R. (1952) Permanent Implantation of Multilead Electrodes in the Brain. *Yale Journal of Biology and Medicine* 24: 351–358; Delgado, J. M. R. Chronic Implantation of Intracerebral Electrodes in Animals. In Sheer, D. E., ed. (1961) *Electrical Stimulation of the Brain*. Austin, TX: University of Texas Press, 25–36; Delgado, J. M. R. (1960) Emotional Behavior in Animals and Humans. *Psychiatry Research Reports* 12: 259–271; Delgado, J. M. R. (1963) Cerebral Heterostimulation in a Monkey Colony. *Science* 141: 161–163; Delgado, J. M. R. (1965) Sequential Behavior Repeatedly Induced by Red Nucleus Stimulation in Free Monkeys. *Science* 148: 1361–1363; Delgado, J.M. R. Aggression and Defense Under Cerebral Radio Control. In Clemente, C.; and Lindsley, D. (1965) Aggression and Defense. Neural Mechanisms and Social Patterns. *UCLA Forum in Medical Sciences* 7 (5): 171–193; Delgado, J. M. R.; Hamlin, H.; and Chapman, W. P. (1952) Technique of Intracranial Electrode Implacement for Recording and Stimulation and Its Possible Therapeutic Value in Psychotic Patients. *Confinia Neurologica* 12: 315–319; Delgado, J. M. R.; et al. (1968) Intracerebral Radio Stimulation and Recording in Completely Free Patients. *Journal of Nervous and Mental Disease* 147: 329–334; Delgado, J. M. R.; Roberts, W. W.; and Miller, N. E. (1954) Learning Motivated by Electrical Stimulation of the Brain. *American Journal of Physiology* 179: 587–593; Mahl, G. F.; Rothenberg, A.; Delgado, J. M. R.; and Hamlin, H. (1964) Psychological Responses in the Human to Intracerebral Electric Stimulation. *Psychosomatic Medicine* 26: 337–336.
15. Hess, W. R. (1927) Stammganglien-Reizversuche. (*Verh. Dtsch. Physiol.* Ges., Sept. 1927). *Ber. Ges. Physiol.* 42: 554–555, 1928. See also Fields, R. D. (2015) *Why We Snap: Understanding the Rage Circuit in Our Brain*. New York, NY: Dutton Press; and Fields, R. D. (2019) The Roots of Human Aggression. *Scientific American* 320: 64–71 for a further description of the discovery and operation of the brain's hypothalamic attack area.
16. Penfield, W.; and Jasper, H. (1954) *Epilepsy and the Functional Anatomy of the Human Brain*. Boston: Little, Brown.

17. Delgado, J. M. R. (1969) *Physical Control of The Mind: Toward a Psychocivilized Society.* New York, NY: Harper and Row, 75.
18. Heath, R. G.; Monroe, R. R.; and Mickle, W. (1955) Stimulation of the Amygdaloid Nucleus in a Schizophrenic Patient. *American Journal of Psychiatry* 111: 862–863.
19. Delgado, J. M. R. (1969) *Physical Control of The Mind: Toward a Psychocivilized Society.* New York, NY: Harper and Row, 133.
20. Ibid., 135.
21. Ibid. 136–137.
22. Ibid., 137.
23. Ibid., 137.
24. King, H. E. Psychological Effects of Excitation in the Limbic System. In Sheer, D. E., ed. (1961) *Electrical Stimulation of the Brain.* Austin, TX: University of Texas Press, 477–486.
25. Ibid., 485.
26. Delgado, J. M. R. (1969) *Physical Control of The Mind: Toward a Psychocivilized Society.* New York, NY: Harper and Row, e-book location 2353 of 4043.
27. Ibid., e-book location 2059 of 4043.
28. Kaiser, J. NIH to End All Support for Chimpanzee Research. *Science,* November 18, 2015. http://www.sciencemag.org/news/2015/11/nih-end-all-support-chimpanzee-research
29. Fields, R. D. (2016) Wireless Brain Implant Allows "Locked-In" Woman to Commmunicate. *Scientific American News,* November 17, 2016. https://www.scientificamerican.com/article/wireless-brain-implant-allows-ldquo-locked-in-rdquo-woman-to-communicate/
30. Engber, D. The Neurologist Who Hacked His Brain—and Almost Lost His Mind. *Wired,* January 26, 2016. https://www.wired.com/2016/01/phil-kennedy-mind-control-computer/
31. Pandarinath, C.; et al. (2017) High Performance Communication by People with Paralysis Using an Intracortical Brain-Computer Interface. *eLIFE* 6: e18554.
32. Velliste, M.; Perel, S.; Spalding, M. C.; Whitford, A. S.; and Schwartz, A. B. (2008) Cortical Control of a Prosthetic Arm for Self-Feeding. *Nature* 453: 1098–1101.
33. Paraplegic Wearing Robotic Suit Kicks Off World Cup in Brazil. YouTube, uploaded by SuperRoboHead, June 13, 2014. https://www.youtube.com/watch?v=inCvbDLfXBo
34. de Hond, M. What Went Wrong when the Paraplegic Man in Robotic Suit Kicked Off the World Cup 2014? YouTube, uploaded by Marc de Hond Producties, June 13, 2014. https://www.youtube.com/watch?v=WaQcC8yJmMU
35. Sutherland, S. (2016) Melding Mind and Machine. *BrainFacts.org.* https://www.brainfacts.org/Diseases-and-Disorders/Therapies/2016/Melding-Mind-and-Machine-080216. Accessed September 30, 2019.
36. LaPook, J. President Shares Monumental Handshake with Paralyzed Man. *CBS News,* October, 16, 2016. https://www.cbsnews.com/videos/president-shares-monumental-handshake-with-paralyzed-man/

37. Flesher, S. N.; et al. (2016) Intracortical Microstimulation of Human Somatosensory Cortex. *Science Translational Medicine* 8 (361): 361ra141. http://stm.sciencemag.org/content/early/2016/10/12/scitranslmed.aaf8083
38. Paralyzed Man Regains Sense of Touch. Neurological Surgery, University of Pittsburgh, October 13, 2016. http://www.neurosurgery.pitt.edu/news/paralyzed-man-regains-sense-touch
39. Utah Array. Blackrock Microsystems. http://blackrockmicro.com/electrode-types/utah-array/. Accessed October 4, 2019.
40. Geddes, L. (2016) First Paralysed Person to Be "Reanimated" Offers Neuroscience Insights. *Nature*, April 13, 2016. https://www.nature.com/news/first-paralysed-person-to-be-reanimated-offers-neuroscience-insights-1.19749
41. Wagner, F. B.; et al. (2018) Targeted Neurotechnology Restores Walking in Humans with Spinal Cord Injury. *Nature* 536: 65–71; and (2018) Paralysed People Walk Again After Spinal-Cord Stimulation. *Nature* 563: 6. https://www.nature.com/articles/d41586-018-07237-9
42. Vidal, J. J. (1977) Real-Time Detection of Brain Events in EEG. *Proceedings of the IEEE* 65: 633–642.
43. Farwell, L. A.; and Donchin, E. (1988) Talking Off the Top of Your Head: Toward a Mental Prosthesis Utilizing Event-Related Brain Potentials. *Electroencephalography and Clinical Neurophysiology* 70 (6): 510–523.
44. Krusienski, D. J.; Sellers, E. W.; McFarland, D. J.; Vaughan, T. M.; Wolpaw, J. R. (2008) Toward eEnhanced P300 Speller Performance. *Journal of Neuroscience Methods*. 167(1):15-21.
45. Wolpaw, J. R.; McFarland, D. J.; Neat, G. W.; and Forneris, C. A. (1991) An EEG-Based Brain-Computer Interface for Cursor Control. *Electroencephalography and Clinical Neurophysiology* 78 (3): 252–259; Pfurtscheller,. G; Flotzinger, D.; and Kalcher, J. (1993) Brain-Computer Interface—a New Communication Device for Handicapped Persons. *Journal of Microcomputer Applications* 16: 293–299; Müller, K. R.; and Blankertz, B. (2006) Towards Noninvasive Brain-Computer Interfaces. *IEEE Signal Processing Magazine* 23: 125–128.
46. Müller, K. R.; Tangermann, M.; Dornhege, G.; Krauledat, M.; Curio, G.; and Blankertz, B. (2008) Machine Learning for Real-Time Single-Trial EEG-Analysis: From Brain-Computer Interfacing to Mental State Monitoring. *Journal of Neuroscience Methods* 167 (1): 82–90.
47. Sitaram, R.; Caria, A.; Veit, R.; Gaber, T.; Rota, G.; Kuebler, A.; and Birbaumer, N. (2007) FMRI Brain-Computer Interface: A Tool for Neuroscientific Research and Treatment. *Computational Intelligence and Neuroscience* 2007: 25487; Yoo, S. S.; Fairneny, T.; Chen, N. K.; Choo, S. E.; Panych, L. P.; et al. (2004) Brain-Computer Interface Using fMRI: Spatial Navigation by Thoughts. *NeuroReport* 15: 1591–1595.
48. Mellinger, J.; Schalk, G.; Braun, C.; Preissl, H.; Rosenstiel, W.; Birbaumer, N.; and Kübler, A. (2007) An MEG-Based Brain-Computer Interface (BCI). *NeuroImage* 36 (3): 581–593.
49. Coyle, S. M.; Ward, T. E.; and Markham, C. M. (2007) Brain-Computer Interface

Using a Simplified Functional Near-Infrared Spectroscopy System. *Journal of Neural Engineering* 4 (3): 219–226.

50. Hochberg, L. R.; Serruya, M. D.; Friehs, G. M.; Mukand, J. A.; Saleh, M.; Caplan, A. H.; Branner, A.; Chen, D.; Penn, R. D.; and Donoghue, J. P. (2006) Neuronal Ensemble Control of Prosthetic Devices by a Human with Tetraplegia. *Nature* 442 (7099): 164–171; Taylor, D. M.; Tillery, S. I.; and Schwartz, A. B. (2002) Direct Cortical Control of 3D Neuroprosthetic Devices. *Science* 296 (5574): 1829–1832.
51. Leuthardt, E. C.; Schalk, G.; Wolpaw, J. R.; Ojemann, J. G.; and Moran, D. W. (2004) A Brain-Computer Interface Using Electrocorticographic Signals in Humans. *Journal of Neural Engineering* 1 (2): 63–71.
52. ASU Researcher Creates System to Control Robots with the Brain. *ASU Now*, July 8, 2016. https://asunow.asu.edu/20160710-discoveries-asu-researcher-creates-system-control-robots-brain
53. Rao, R. P.; Stocco, A.; Bryan, M.; Sarma, D.; Youngquist, T. M.; Wu, J.; and Prat, C. S. (2014) A Direct Brain-to-Brain Interface in Humans. *PLoS One* 9: e111332.
54. Much of this research has been published since this interview. See Jiang, L.; et al. (2019) BrainNet: A Multi-Person Brain-to-Brain Interface for Direct Collaboration Between Brains. *Scientific Reports* 9: 6115.
55. O'Doherty, J. E.; Lebedev, M. A.; Ifft, P. J.; Zhuang, K. Z.; Shokur, S.; et al. (2011) Active Tactile Exploration Using a Brain-Machine-Brain Interface. *Nature* 479: 228–231.
56. Yoo, S. S.; et al. (2013) Non-Invasive Brain-to-Brain Interface (BBI): Establishing Functional Links between Two Brains. *PloS One* 8 (4): e60410.
57. Ibid. https://journals.plos.org/plosone/article?id=10.1371/journal.pone.0060410.
58. Li, G.; and Zhang, D. (2016) Brain-Computer Interface Controlled Cyborg: Establishing a Functional Information Transfer Pathway from Human Brain to Cockroach Brain. *PLoS One* 11 (3): e0150667.

## CHAPTER 9

1. Folke, T.; et al. (2016) A Bilingual Disadvantage in Metacognitive Processing. *Cognition* 150: 119–132.
2. Prat, C. S.; Yamasaki, B. L.; and Peterson, E. R. (2018) Individual Differences in Resting-State Brain Rhythms Uniquely Predict Second Language Learning Rate and Willingness to Communicate in Adults. *Journal of Cognitive Neuroscience* 21: 1–17; and Prat, C.; Yamasaki, B.; Kluender, R.; and Stocco, A. (2016). Resting State qEEG Predicts Rate of Second Language Learning in Adults. *Brain and Language* 157–158: 44–50.
3. Mason, R. A.; and Just, M. A. (2016) Neural Representations of Physics Concepts. *Psychological Science* 27: 904–913.
4. Huth, A. C.; de Heer, W. A.; Griffiths, T. L.; Theunissen, F. E.; and Gallant, J. L. (2016) Natural Speech Reveals the Semantic Maps That Tile Human Cerebral Cortex. *Nature* 532: 453–458.
5. Stack, L. (2017) Georgia Tech Student Leader Is Shot Dead by Campus Police. *New*

*York Times*, September 18, 2017. https://www.nytimes.com/2017/09/18/us/georgia-tech-killing-student.html

6. Selk, A.; Shapiro, T. R.; and Lowery, W. Call about Suspicious Man Was Made by Georgia Tech Student Killed by Police, Investigators Say. *Washington Post*, September 18, 2017. https://www.washingtonpost.com/news/grade-point/wp/2017/09/17/knife-wielding-campus-pride-leader-killed-by-police-at-georgia-tech/?utm_term=.eb0a1588fa59
7. Kassam, K. S.; et al. Identifying Emotions on the Basis of Neural Activation. *PLoS One* 19 (6): e66032.
8. Dosenbach, N. U. F.; Nardos, B.; Cohen, A. L.; Fair, D.; Power, J. D.; Church, J.; et al. (2010) Prediction of Individual Brain Maturity Using fMRI. *Science* 329: 1358–1361. doi:10.1126/science.1194144
9. Finn, E. S.; Shen, X.; Scheinost, D.; Rosenberg, M. D.; Huang, J.; Chun, M. M.; et al. (2015) Functional Connectome Fingerprinting: Identifying Individuals Using Patterns of Brain Connectivity. *Nature Neuroscience* 1–11. doi:10.1038/nn.4135
10. Fields, R. D. Why the First Drawings of Neurons Were Defaced. *Quanta Magazine*, September 28, 2017. https://www.quantamagazine.org/why-the-first-drawings-of-neurons-were-defaced-20170928/
11. Brainnetome Atlas. http://atlas.brainnetome.org/. Accessed October 3, 2019.
12. Fan, L.; et al. (2016) The Human Brainnetome Atlas: A New Brain Atlas Based on Connectional Architecture. *Cerebral Cortex* 26 (8): 3508–3526.
13. Krebs, R. M.; et al. (2009) Personality Traits Are Differentially Associated with Patterns of Reward and Novelty Processing in the Human Substantia Nigra/Ventral Tegmental Area. *Biological Psychiatry* 65: 103–110.
14. Van Schuerbeek, P.; et al. (2011) Individual Differences in Local Gray and White Matter Volumes Reflect Differences in Temperament and Character: A Voxel-Based Morphometry Study in Healthy Young Females. *Brain Research* 1371: 32–42.
15. Jiang, R.; et al. (2017) Predicting Temperament Dimension Scores Using Brainnetome-Atlas based Functional Connectivity. Poster Presentation at Organization for Human Brain Mapping Meeting, June 25–29, 2017.
16. Allen, J. J. B.; Kline, J. P. (2004) Frontal EEG Asymmetry, Emotion, and Psychopathology: The First, and the Next, 25 years. *Biological Psychology* 67 (1–2): 1–5.
17. Cartocci, G.; et al. (2016) Gender and Age Related Effects While Watching TV Advertisements: An EEG Study. *Computational Intelligence and Neuroscience* 2016: 3795325
18. Varikuti, D.; et al. (2017) Evaluation of Non-Negative Matrix Factorization of Grey Matter in Age Prediction. Poster Presentation at Organization for Human Brain Mapping Meeting, June 25–29, 2017; and Varikuti, D.P., (2018) Evaluation of Non-Negative Matrix Factorization of Grey Matter in Age Prediction. *Neuroimage*, 173: 394–410.
19. Zhang, C., and Micheal, A. (2017) The DMN Contributes Most to Gender Prediction: A Large Resting fMRI Study, Poster Presentation at Organization for Human Brain Mapping Meeting, June 25–29, 2017.

20. Van Essen, D. C. (2013) The WU-Minn Human Connectome Project: An Overview. *NeuroImage* 80: 62–79.
21. Ma, F.; Guntupalli, J.; and Haxby, J. (2017) Hyperalignment Improves Prediction of Fluid Intelligence from Functional Connectivity. Poster Presentation at Organization for Human Brain Mapping Meeting, June 25–29, 2017.
22. Jausovec, N.; and Jausovec, K. (2000) Differences in Resting EEG Related to Ability. *Brain Topography* 12 (3): 229–240.
23. Ujma, P. P.; et al. (2017) The Sleep EEG Spectrum Is a Sexually Dimorphic Marker of General Intelligence *Scientific Reports* 7: 18070.
24. Pótári, A.; et al. (2017) Age-Related Changes in Sleep EEG Are Attenuated in Highly Intelligent Individuals. *Neuroimage* 146: 554–560.
25. Tessier, S.; et al. (2015) Intelligence Measures and Stage 2 Sleep in Typically-Developing and Autistic Children. *International Journal of Psychophysiology* 97 (1): 58–65.
26. Geiger, A.; et al. (2011) The Sleep EEG as a Marker of Intellectual Ability in School Age Children. *Sleep* 34 (2): 181–189.
27. Fonseca, L. C.; et al. (2006) Quantitative EEG in Children with Learning Disabilities: Analysis of Band Power. *Arquivos de Neuro-psiquiatria* 64 (2B): 376–381.
28. Posthuma, D.; et al. (2001) Are Smarter Brains Running Faster? Heritability of Alpha Peak Frequency, IQ, and Their Interrelation. *Behavior Genetics* 31 (6): 567–579.
29. Smith, D. J.; et al. (2015) Childhood IQ and Risk of Bipolar Disorder in Adulthood: Prospective Birth Cohort Study. *BJPsych Open* 1 (1): 74–80.
30. Beaty, R. E.; et al. (2018) Robust Prediction of Individual Creative Ability from Brain Functional Connectivity. *Proceedings of the National Academy of the Sciences of the United States of America* 115 (5): 1087–1092.
31. Boot, N.; et al. (2017) Widespread Neural Oscillations in the Delta Band Dissociate Rule Convergence from Rule Divergence During Creative Idea Generation. *Neuropsychologia* 104: 8–17; and Marmpena, M.; et al. (2016) Phase to Amplitude Coupling as a Potential Biomarker for Creative Ideation: An EEG Study. *Engineering in Medicine and Biology Society, Annual International Conference of the IEEE* 2016: 383–386. doi:10.1109/EMBC.2016.7590720; and Schwab, D.; et al. (2014) The Time-Course of EEG Alpha Power Changes in Creative Ideation. *Frontiers in Human Neuroscience* 8: 310; and Fink, A.; and Benedek, M. (2014) EEG Alpha Power and Creative Ideation. *Neuroscience & Biobehavioral Reviews* 44: 111–123. doi:10.1016/j.neubiorev.2012.12.002; and Junk, E.; Benedek, M.; and Neubauer, A. C. (2012) Tackling Creativity at Its Roots: Evidence for Different Patterns of EEG α Activity Related to Convergent and Divergent Modes of Task Processing. *International Journal of Psychophysiology* 84 (2): 219–225.
32. Fink, A; and Neubauer, A. C. (2006) EEG Alpha Oscillations During the Performance of Verbal Creativity Tasks: Differential Effects of Sex and Verbal Intelligence. *International Journal of Psychophysiology* 62 (1): 46–53.
33. Lustenberger, C.; et al. (2015) Functional Role of Frontal Alpha Oscillations in Creativity. *Cortex* 67: 74-82.

34. Przysinda, E.; Zeng, T.; Maves, K.; Arkin, C.; and Loui, P. (2017) Jazz Musicians Reveal Role of Expectancy in Human Creativity. *Brain and Cognition* 119: 45–53.
35. Lopata, J. A.; Nowicki, E. A.; and Joanisse, M. F. (2017) Creativity as a Distinct Trainable Mental State: An EEG Study of Musical Improvisation. *Neuropsychologia* 99: 246–258.
36. Adhikari, B. M.; Norgaard, M.; Quinn, K. M.; Ampudia, J.; Squirek, J.; and Dhamala, M. (2016) The Brain Network Underpinning Novel Melody Creation. *Brain Connectivity* 6 (10): 772–785.
37. Habibi, A.; Wirantana, V.; and Starr, A. (2014) Cortical Activity during Perception of Musical Rhythm: Comparing Musicians and Non-musicians. *Psychomusicology* 24 (2): 125–135.
38. Park, M.; et al. (2017) Neural Connectivity in Internet Gaming Disorder and Alcohol Use Disorder: A Resting-State EEG Coherence Study. *Scientific Reports* 7: 1333.
39. Schiavone, G.; Linkenkaer-Hansen, K.; Maurits, N. M.; Plakas, A.; Maassen, B. A.; Mansvelder, H. D.; van der Leij, A.; and van Zuijen, T. L. (2014) Preliteracy Signatures of Poor-Reading Abilities in Resting-State EEG. *Frontiers in Human Neuroscience* 8: 735.
40. Rumsey, J. M.; Coppola, R.; Denckla, M. B.; Hamburger, S. D.; and Kruesi, M. J. (1989) EEG Spectra in Severely Dyslexic Men: Rest and Word and Design Recognition. *Electroencephalography and Clinical Neurophysiology* 73 (1): 30–40; and Rippon, G.; and Brunswick, N. (2000) Trait and State EEG Indices of Information Processing in Developmental Dyslexia. *International Journal of Psychophysiology* 36 (3): 251–265; and Klimesch, W.; Doppelmayr, M.; Wimmer, H.; Gruber, W.; Röhm, D.; Schwaiger, J.; and Hutzler, F. (2001) Alpha and Beta Band Power Changes in Normal and Dyslexic Children. *Clinical Neurophysiology* 112 (7): 1186–1195; and Babiloni, C.; Stella, G.; Buffo, P.; Vecchio, F.; Onorati, P.; Muratori, C.; Miano, S.; Gheller, F.; Antonaci, L.; Albertini, G.; and Rossini, P. M. (2012) Cortical Sources of Resting State EEG Rhythms Are Abnormal in Dyslexic Children. *Clinical Neurophysiology* 123 (12): 2384–2391.
41. Papagiannopoulou, E. A.; and Lagopoulos, J. (2016) Resting State EEG Hemispheric Power Asymmetry in Children with Dyslexia. *Frontiers in Pediatrics* 4: 11.
42. Whitford, T. J.; Rennie, C. J.; Grieve, S. M.; Clark, C. R.; Gordon, E.; and Williams, L. M. (2007) Brain Maturation in Adolescence: Concurrent Changes in Neuroanatomy and Neurophysiology. *Human Brain Mapping* 28 (3): 228–237. doi .org/10.1002/hbm.20273
43. Anderson, A. J.; and Perone, S. (2018) Developmental Change in the Resting State Electroencephalogram: Insights into Cognition and the Brain. *Brain and Cognition* 126: 40–52.
44. Takano, T.; and Ogawa, T. (2007) Characterization of Developmental Changes in EEG-gamma Band Activity During Childhood Using the Autoregressive Model. *Pediatrics International* 40: 446–452.
45. Fraga González, G.; Van der Molen, M. J. W.; Žarić, G.; Bonte, M.; Tijms, J.; Blomert, L.; Stam, C. J.; and Van der Molen, M. W. (2016) Graph Analysis of EEG

Resting State Functional Networks in Dyslexic Readers. *Clinical Neurophysiology* 127 (9): 3165–3175.

46. Dhar, M.; Been, P. H.; Minderaa, R. B.; and Althaus, M. (2010) Reduced Interhemispheric Coherence in Dyslexic Adults. *Cortex* 46 (6): 794–798.
47. Lohvansuu, K.; Hämäläinen, J. A.; Tanskanen, A.; Ervast, L.; Heikkinen, E.; Lyytinen, H.; and Leppänen, P. H. (2014) Enhancement of Brain Event-Related Potentials to Speech Sounds Is Associated with Compensated Reading Skills in Dyslexic Children with Familial Risk for Dyslexia. *International Journal of Psychophysiology* 94 (3): 298–310.
48. Bruni, O.; Ferri, R.; Novelli, L.; Finotti, E.; Terribili, M.; Troianiello, M.; Valente, D.; Sabatello, U.; and Curatolo, P. (2009) Slow EEG Amplitude Oscillations During NREM Sleep and Reading Disabilities in Children with Dyslexia. *Developmental Neuropsychology* 34 (5): 539–551.
49. Bruni, O.; Ferri, R.; Novelli, L.; Terribili, M.; Troianiello, M.; Finotti, E.; Leuzzi, V.; and Curatolo, P. (2009) Sleep Spindle Activity Is Correlated with Reading Abilities in Developmental Dyslexia. *Sleep* 32 (10): 1333–1340.
50. Schneps, M. H. The Advantages of Dyslexia: With Reading Difficulties Can Come Other Cognitive Strengths. *Scientific American*, August 19, 2014. https://www.scientificamerican.com/article/the-advantages-of-dyslexia/
51. Hamilton, J. Orphans' Lonely Beginnings Reveal How Parents Shape a Child's Brain. *NPR Morning Edition*, February 24, 2014. https://www.npr.org/sections/health-shots/2014/02/20/280237833/orphans-lonely-beginnings-reveal-how-parents-shape-a-childs-brain
52. Nelson, C. A.; Fox, N. A.; and Zeanah, C. H. (2013) Tragedy leads to study of severe child neglect. *Scientific American*, April 1, 2013.

## CHAPTER 10

1. Jasper, H. H.; and Carmichael, L. (1935) Electrical Potentials from the Intact Human Brain. *Science* 81: 51–53.
2. Long, X.; et al. (2017) Prediction of Alzheimer's Disease based on MRI Deformation Poster presentation 3912 at Organization for Human Brain Mapping Meeting, June 25–29, 2017; and Long, X.; et al., (2017) Prediction and Classification of Alzheimer Disease Based on Quantification of MRI Deformation. *PloS ONE* 12(3) e0173372.
3. Alzheimer's Disease Neuroimaging Initiative. http://adni.loni.usc.edu/.
4. Brenner, R. P.; Ulrich, R. F.; Spiker, D. G.; Sclabassi, R. J.; Reynolds, C. F.; and Marin R. S.; et al. (1986). Computerized EEG Spectral Analysis in Elderly Normal, Demented and Depressed Subjects. *Electroencephalography and Clinical Neurophysiology* 64: 483–492; and Dierks, T.; Ihl, R.; Frölich, L.; and Maurer, K. (1993) Dementia of the Alzheimer Type: Effects on the Spontaneous EEG Described by Dipole Sources. *Psychiatry Research* 50: 151–162; and Huang, C.; Wahlund, L. O.; Dierks, T.; Julin, P.; Winblad, B.; and Jelic, V. (2000) Discrimination of Alzheimer's Disease and Mild Cognitive Impairment by Equivalent EEG Sources: a

Cross-Sectional and Longitudinal Study. *Clinical Neurophysiology* 111: 1961–1967; and Jeong, J. (2004) EEG Dynamics in Patients with Alzheimer's Disease. *Clinical Neurophysiology* 115: 1490–1505.

5. Wang, J.; Fang, Y.; Wang, X.; Yang, H.; Yu, X.; and Wang, H. (2017) Enhanced Gamma Activity and Cross-Frequency Interaction of Resting-State Electroencephalographic Oscillations in Patients with Alzheimer's Disease. *Frontiers in Aging Neuroscience* 26 (9): 243. doi:10.3389/fnagi.2017.00243
6. Buzsáki, G.; and Watson, B. O. (2012) Brain Rhythms and Neural Syntax: Implications for Efficient Coding of Cognitive Content and Neuropsychiatric Disease. *Dialogues in Clinical Neuroscience* 14: 345–367.
7. Axmacher, N.; Henseler, M. M.; Jensen, O.; Weinreich, I.; Elger, C. E.; and Fell, J. (2010) Cross-Frequency Coupling Supports Multi-Item Working Memory in the Human Hippocampus. *Proceedings of the National Academy of the Sciences of the United States of America* 107: 3228–3233.
8. Zhang, X.; Zhong, W.; Brankack, J.; Weyer, S. W.; Müller, U. C.; and Tort, A. B.; et al. (2016) Impaired θ-γ Coupling in APP-deficient Mice. *Scientific Reports* 6: 21948.
9. Smailovic, U.; Koenig, T.; Kåreholt, I.; Andersson, T.; Kramberger, M. G.; Winblad, B.; and Jelic, V. (2017) Quantitative EEG Power and Synchronization Correlate with Alzheimer's Disease CSF Biomarkers. *Neurobiology of Aging* 63: 88–95.
10. Brueggen, K.; Fiala, C.; Berger, C.; Ochmann, S.; Babiloni, C.; and Teipel, S. J. (2017) Early Changes in Alpha Band Power and DMN BOLD Activity in Alzheimer's Disease: A Simultaneous Resting State EEG-fMRI Study. *Frontiers in Aging Neuroscience* 9: 319.
11. Mazaheri, A.; Segaert, K.; Olichney, J.; Yang, J. C.; Niu, Y. Q.; Shapiro, K.; and Bowman, H. (2017) EEG Oscillations During Word Processing Predict MCI Conversion to Alzheimer's Disease. *Neuroimage: Clinical* 17: 188–197.
12. Han, C. X.; Wang, J.; Yi, G. S.; and Che, Y. Q. (2013) Investigation of EEG Abnormalities in the Early Stage of Parkinson's Disease. *Cognitive Neurodynamics* 7: 351–359; and Chaturvedi, M.; Hatz, F.; Gschwandtner, U.; Bogaarts, J. G.; Meyer, A.; Fuhr, P.; and Roth, V. (2017) Quantitative EEG (QEEG) Measures Differentiate Parkinson's Disease (PD) Patients from Healthy Controls (HC). *Frontiers in Aging Neuroscience* 23 (9): 3.
13. Benz, N.; Hatz, F.; Bousleiman, H.; Ehrensperger, M. M.; Gschwandtner, U.; Hardmeier, M.; Ruegg, S.; Schindler, C.; Zimmermann, R.; Monsch, A. U.; and Fuhr, P. (2014) Slowing of EEG Background Activity in Parkinson's and Alzheimer's Disease with Early Cognitive Dysfunction. *Frontiers in Aging Neuroscience* 6: 314.
14. Chaturvedi, M.; et al. (2017) Quantitative EEG (QEEG) Measures Differentiate Parkinson's Disease (PD) Patients from Healthy Controls (HC). *Frontiers in Aging Neuroscience* 9: 3. doi:10.3389/fnagi.2017.00003
15. Caviness, J. N.; Beach, T. G.; Hentz, J. G.; Shill, H. A.; Driver-Dunckley, E. D.; and Adler, C. H. (2017) Association Between Pathology and Electroencephalographic Activity in Parkinson's Disease. *Clinical EEG and Neuroscience* 49(5): 321–327.

16. Annanmaki, T.; Palmu, K.; Murros, K.; and Partanen, J. (2017) Altered N100-potential Associates with Working Memory Impairment in Parkinson's Disease. *Journal of Neural Transmission* 124 (10): 1197–1203.
17. Lubar, J. E. (1991) Discourse on the Development of EEG Diagnostic and Biofeedback for Attention-Deficit/Hyperactivity Disorders. *Biofeedback and Self-Regulation* 16: 201–225.
18. Snyder, S. M.; and Hall, J. R. (2006) A Meta-Analysis of Quantitative EEG Power Associated with Attention-Deficit Hyperactivity Disorder. *Journal of Clinical Neurophysiology* 23: 440–455.
19. Arns, M.; Conners, K.; and Kraemer, H. C. (2012) A Decade of EEG Theta/Beta Ratio Research in ADHD: A Meta-Analysis. *Journal of Attention Disorders* 17: 374–383.
20. Gloss, D.; Varma, J. K.; Pringsheim, T.; and Nuwer, M. R. (2016) The Utility of EEG Theta/Beta Power Ratio in ADHD Diagnosis. Report of the Guideline Development, Dessimination, and Implementation Subcommittee of the American Academy of Neurology. *Neurology* 87: 2375–2379.
21. Li, H.; Zhao, Q.; and Huang, F. (2017) Increased Beta Activity Links to Impaired Emotional Control in ADHD Adults with High IQ. *Journal of Attention Disorders* 23: 754–764.
22. Ellingson, R. J. (1954) The Incidence of EEG Abnormality Among Patients with Mental Disorders of Apparently Nonorganic Origin: A Critical Review. *American Journal of Psychiatry* 111 (4): 263–275.
23. Hill, D. (1952) EEG in Episodic Psychotic and Psychopathic Behaviour; a Classification of Data. *Electroencephalography and Clinical Neurophysiology* 4 (4): 419–442.
24. Niv S.; Ashrafulla, S.; Tuvblad, C.; Joshi, A.; Raine, A.; Leahy, R.; and Baker, L. A. (2015) Childhood EEG Frontal Alpha Power as a Predictor of Adolescent Antisocial Behavior: A Twin Heritability Study. *Biological Psychology* 105: 72–76.
25. Ibid., 9.
26. Yasuhara, A. (2010) Correlation Between EEG Abnormalities and Symptoms of Autism Spectrum Disorder (ASD). *Brain and Development* 32: 791–798.
27. Wang, Y.; et al. (2017) Classification of Schizophrenia Using Functional Connectivity Based on fMRI Data. Poster Presentation 3913 at the Organization for Human Brain Mapping Meeting, June 25–29, 2017.
28. Gheiratmand, M.; et al. (2017) Functional Network Patterns as Multivariate Predictors of Symptom Severity in Schizophrenia. Poster presentation 3939 at the Organization for Human Brain Mapping Meeting, June 25–29, 2017; and Gheiratmand, M.; et al, (2017) Learning Stable and Predictive Network-Based Patterns of Schizophrenia and Its Clinical Symptoms. *NPG Schizophrenia*, 3, number 22.
29. Azurii, K.; Collier, D. H.; Wolf, J. N.; Valdez, B. I.; Turetsky, M. A.; Elliott, R. E.; and Gur, R. C. (2014) Comparison of Auditory and Visual Oddball fMRI in Schizophrenia: Progressive Reduction of Visual P300 Amplitude in Patients with First-Episode Schizophrenia: An ERP Study. *Schizophrenia Research* 158: 183–188; and Oribe, N.; et al. (2015) Progressive Reduction of Visual P300 Amplitude in

Patients with First-Episode Schizophrenia: An ERP Study. *Schizophrenia Bulletin* 41: 460–470.

30. Schilling, C.; Schlipf, M.; Spietzack, S.; Rausch, F.; Eisenacher, S.; Englisch, S.; Reinhard, I.; Haller, L.; Grimm, O.; Deuschle, M.; Tost, H.; Zink, M.; Meyer-Lindenberg, A.; and Schredl, M. (2017) Fast Sleep Spindle Reduction in Schizophrenia and Healthy First-Degree Relatives: Association with Impaired Cognitive Function and Potential Intermediate Phenotype. *European Archives of Psychiatry and Clinical Neuroscience* 267 (3): 213–224.
31. Sharma, A.; Sauer, H.; Smit, D. J.; Bender, S.; and Weisbrod, M. (2011) Genetic Liability to Schizophrenia Measured by P300 in Concordant and Discordant Monozygotic Twins. *Psychopathology* 44 (6): 398–406.
32. Barr, M. S.; Rajji, T. K.; Zomorrodi, R.; Radhu, N.; George, T. P.; Blumberger, D. M.; and Daskalakis, Z. J. (2017) Impaired Theta-Gamma Coupling During Working Memory Performance in Schizophrenia. *Schizophr Research* 189: 104–110.
33. Huang, Y.; Mohan, A.; De Ridder, D.; Sunaert, S.; and Vanneste, S. (2018) The Neural Correlates of the Unified Percept of Alcohol-Related Craving: A fMRI and EEG Study. *Scientific Reports* 8 (1): 923.
34. Alcohol Use Disorder. National Institute on Alcohol Abuse and Alcoholism. https://www.niaaa.nih.gov/alcohol-health/overview-alcohol-consumption/alcohol-use-disorders. Accessed October 3, 2019.
35. Mumtaz, W.; Vuong, P. L.; Xia, L.; Malik, A. S.; and Rashid, R. B. A. (2017) An EEG-Based Machine Learning Method to Screen Alcohol Use Disorder. *Cognitive Neurodynamics* 11 (2): 161–171.
36. Hu, B.; Dong, Q.; Hao, Y.; Zhao, Q.; Shen, J.; and Zheng, F. (2017) Effective Brain Network Analysis with Resting-State EEG Data: a Comparison Between Heroin Abstinent and Non-addicted Subjects. *Journal of Neural Engineering* 14 (4): 046002.
37. Wang, G. Y.; Kydd, R. R.; and Russell, B. R. (2015) Quantitative EEG and Low-Resolution Electromagnetic Tomography (LORETA) Imaging of Patients Undergoing Methadone Treatment for Opiate Addiction. *Clinical EEG and Neuroscience* 47: 180–187.
38. Ieong, H. F.; and Yuan, Z. (2017) Resting-State Neuroimaging and Neuropsychological Findings in Opioid Use Disorder during Abstinence. *Frontiers in Human Neuroscience* 11: 169. https://www.frontiersin.org/articles/10.3389/fnhum.2017.00169/full
39. Ding, X.; et al. (2017) Predicting Smoking Cessation Treatment Outcomes Using Dynamics Between Large-Scale Brain Networks. Poster presentation 3923 at the Organization for Human Brain Mapping Meeting, June 25–29, 2017.
40. Li, X.; Ma, R.; Pang, L.; Lv, W.; Xie, Y.; Chen, Y.; Zhang, P.; Chen, J.; Wu, Q.; Cui, G.; Zhang, P.; Zhou, Y.; and Zhang, X. (2017) Delta Coherence in Resting-State EEG Predicts the Reduction in Cigarette Craving after Hypnotic Aversion Suggestions. *Scientific Reports* 7 (1): 2430.
41. Olbrich, S.; and Arns, M. (2013) EEG Biomarkers in Major Depressive Disorder:

Discriminative Power and Prediction of Treatment Response. *International Review of Psychiatry* 25 (5): 604–618.

## CHAPTER 11

1. Cortese, S.; et al. (2016) Neurofeedback for Attention-Deficit/Hyperactivity Disorder: Meta-Analysis of Clinical and Neuropsychological Outcomes from Randomized Controlled Trials. *Child and Adolescent Psychiatry* 55: 444–455.
2. Lofthouse, N.; Arnold, L. E.; Hersch, S.; Hurt, E.; and DeBeus, R. (2012) A Review of Neurofeedback Treatment for Pediatric ADHD. *Journal of Attention Disorders* 16 (5): 351–372.
3. Thibault, R. T.; and Rax, A. (2017) The Psychology of Neurofeedback: Clinical Intervention Even If Applied Placebo. *American Psychologist* 72 (7): 679–688.
4. Schönenberg, M. (2017) Neurofeedback, Sham Neurofeedback, and Cognitive-Behavioural Group Therapy in Adults with Attention-Deficit Hyperactivity Disorder: a Triple-Blind, Randomised, Controlled Trial. *Lancet Psychiatry* 4 (9): 673–684.
5. Biofeedback Certification International Alliance. https://www.bcia.org/i4a/pages/index.cfm?pageid=1. Accessed October 3, 2019.
6. Hohenfeld, C.; et al. (2013) Cognitive Improvement and Brain Changes after Real-Time Functional MRI Neurofeedback Training in Healthy Elderly and Prodromal Alzheimer's Disease. *Frontiers in Neurology* 8: 384.
7. Zhang, G.; Yao, L.; Zhang, H.; Long, Z.; and Zhao, X. (2013) Improved Working Memory Performance Through Self-Regulation of Dorsal Lateral Prefrontal Cortex Activation Using Real-Time fMRI. *PLoS One* 8 (8): e73735; and Sherwood, M. S.; Kane, J. H.; Weisend, M. P.; and Parker, J. G. (2016) Enhanced Control of Dorsolateral Prefrontal Cortex Neurophysiology with Real-Time Functional Magnetic Resonance Imaging (rt-fMRI) Neurofeedback Training and Working Memory Practice. *Neuroimage* 124 (Pt A): 214–223.
8. Ghaziri, J.; Tucholka, A.; Larue, V.; Blanchette-Sylvestre, M.; Reyburn, G.; Gilbert, G.; Lévesque, J.; and Beauregard, M. (2013) Neurofeedback Training Induces Changes in White and Gray Matter. *Clinical EEG and Neuroscience* 44 (4): 265–272.
9. Wang, J. R.; and Hsieh, S. (2013) Neurofeedback Training Improves Attention and Working Memory Performance. *Clinical Neurophysiology* 124 (12): 2406–2420.
10. Staufenbiel, S. M.; Brouwer, A. M.; Keizer, A.W.; and van Wouwe, N. C. (2014) Effect of Beta and Gamma Neurofeedback on Memory and Intelligence in the Elderly. *Biological Psychology* 95: 74–85.
11. Gruzelier, J. H.; Hirst, L.; Holmes, P.; and Leach, J. (2014) Immediate Effects of Alpha/Theta and Sensory-Motor Rhythm Feedback on Music Performance. *International Journal of Psychophysiology* 93 (1): 96–104.
12. Fetz, E. E. (1969) Operant Conditioning of Cortical Unit Activity. *Science* 163 (3870): 955–958; and Taylor, D. M.; Tillery, S. I. H.; and Schwartz, A. B. (2002) Direct Cortical Control of 3D Neuroprosthetic Devices. *Science* 296 (5574):

1829–1832; and Carmena, J. M.; Lebedev, M. A.; Crist, R. E.; O'Doherty, J. E.; Santucci, D. M.; Dimitrov, D. F.; Patil, P. G.; Henriquez, C. S.; and Nicolelis, M. A. L. (2003) Learning to Control a Brain-Machine Interface for Reaching and Grasping by Primates. *PLOS Biology* 1 (2): e42; and Schafer, R. J.; and Moore, T. (2011) Selective Attention from Voluntary Control of Neurons in Prefrontal Cortex. *Science* 332 (6037): 1568–1571.

13. Engelhard, B.; Ozeri, N.; Israel, Z.; Bergman, H.; and Vaadia, E. (2013) Inducing Gamma Oscillations and Precise Spike Synchrony by Operant Conditioning via Brain-Machine Interface. *Neuron* 77 (2): 361–375; and So, K.; Dangi, S.; Orsborn, A. L.; Gastpar, M. C.; and Carmena, J. M. (2014) Subject-Specific Modulation of Local Field Potential Spectral Power During Brain-Machine Interface Control in Primates. *Journal of Neural Engineering* 11 (2): 026002; and Khanna, P.; and Carmena, J. M. (2015) Changes in Reaching Reaction Times Due to Volitional Modulation of Beta Oscillations. *7th International IEEE/EMBS Conference on Neural Engineering (NER)* 2015: 340–343; and Schalk, G.; Miller, K. J.; Anderson, N. R.; Wilson, J. A.; Smyth, M. D.; Ojemann, J. G.; Moran, D. W.; Wolpaw, J. R.; and Leuthardt, E. C. (2008) Two-Dimensional Movement Control Using Electrocorticographic Signals in Humans. *Journal of Neural Engineering* 5 (1): 75–84; and Wang, W.; Collinger, J. L.; Degenhart, A. D.; Tyler-Kabara, E. C.; Schwartz, A. B.; Moran, D. W.; Weber, D. J.; Wodlinger, B.; Vinjamuri, R. K.; Ashmore, R. C.; Kelly, J. W.; and Boninger, M. L. (2013) An Electrocorticographic Brain Interface in an Individual with Tetraplegia. *PLoS One* 8 (2): e55344.
14. Wolpaw, J. R.; and McFarland, D. J. (2004) Control of a Two-Dimensional Movement Signal by a Noninvasive Brain-Computer Interface in Humans. *Proceedings of the National Academy of the Sciences of the United States of America* 101 (51): 17849–17854; and Millan, J. R.; Renkens, F.; Mourino, J.; and Gerstner, W. (2004) Noninvasive Brain-Actuated Control of a Mobile Robot by Human EEG. *IEEE Transactions on Biomedical Engineering* 51 (6): 1026–1033; and Gevensleben, H.; Holl, B.; Albrecht, B.; Schlamp, D.; Kratz, O.; Studer, P.; Wangler, S.; Rothenberger, A.; Moll, G. H.; and Heinrich, H. (2009) Distinct EEG Effects Related to Neurofeedback Training in Children with ADHD: A Randomized Controlled Trial. *International Journal of Psychophysiology* 74 (2): 149–157; and Tan, G.; Thornby, J.; Hammond, D. C.; Strehl, U.; Canady, B.; Arnemann, K.; and Kaiser, D. A. (2009) Meta-analysis of EEG Biofeedback in Treating Epilepsy. *Clinical EEG and Neuroscience* 40 (3): 173–179; Subramanian, L.; Hindle, J. V.; Johnston, S.; Roberts, M. V.; Husain, M.; Goebel, R.; and Linden, D. (2011) Real-Time Functional Magnetic Resonance Imaging Neurofeedback for Treatment of Parkinson's Disease. *Journal of Neuroscience* 31 (45): 16309–16317; and Jensen, M. P.; Gertz, K. J.; Kupper, A. E.; Braden, A. L.; Howe, J. D.; Hakimian, S.; and Sherlin, L. H. (2013) Steps Toward Developing an EEG Biofeedback Treatment for Chronic Pain. *Applied Psychophysiology and Biofeedback* 38 (2): 101–108.
15. *Brain Facts, A Primer on the Brain and Nervous System.* Published by the Society

for Neuroscience. 2008, Washington, DC. https://www.brainfacts.org/the-brain-facts-book. Accessed October 3, 2019.

16. Khanna, P.; Swann, N.; De Hemptinne, C.; Miocinovic, S.; Miller, A.; Starr, P. A.; and Carmena, J. M. (2016) Neurofeedback Control in Parkinsonian Patients Using Electrocortigraphy Signals Accessed Wirelessly with a Chronic, Fully Implanted Device. *IEEE Transactions on Neural Systems and Rehabilitation Engineering* 25 (10): 1715–1724. doi:10.1109/TNSRE.2016.2597243
17. Swann, K.P. (2017) Neurofeedback Control in Parkinsonian Patients Using Electrocortigraphy Signals Accessed Wirelessly With a Chronic, Fully Implanted Device. *IEEE Trans. Neural. Cyst. Rehabil. Eng.* 25 (10): 1715–1724.
18. Jacqueline, A.; et al. (2017) Sensorimotor Rhythm Neurofeedback as Adjunct Therapy for Parkinson's Disease. *Annals of Clinical and Translational Neurology* 4 (8): 585–590.
19. Philippens, I. H.; and Vanwersch, R. A. P. (2010) Neurofeedback Training on Sensorimotor Rhythm in Marmoset Monkeys. *NeuroReport* 21: 328–332.
20. Roth, S. R.; Sterman, M. B.; and Clemente, C. D. (1967) Comparison of EEG Correlates of Reinforcement, Internal Inhibition and Sleep. *Electroencephalography and Clinical Neurophysiology* 23: 509–520.
21. Sterman, M. B.; and Egner, T. (2006) Foundation and Practice of Neurofeedback for the Treatment of Epilepsy. *Applied Psychophysiology and Biofeedback* 31: 21–35.
22. Thompson, M.; and Thompson, L. (2002) Biofeedback for Movement Disorders (Dystonia with Parkinson's Disease): Theory and Preliminary Results. *Journal of Neurotherapy* 6: 51–70; and Thompson, M.; and Thompson, L. (2011) Improving Quality of Life Using Biofeedback Plus Neurofeedback. *NeuroConnections* Winter 2011: 18–21.
23. Subramanian, L.; Hindle, J. V.; Johnston, S.; Roberts, M. V.; Husain, M.; Goebel, R.; and Linden, D. (2011) Real-Time Functional Magnetic Resonance Imaging Neurofeedback for Treatment of Parkinson's Disease. *Journal of Neuroscience* 31 (45): 16309–16317.
24. Subramanian, L.; Morris, M. B.; Brosnan, M.; Turner, D. L.; Morris, H. R.; and Linden, D. E. (2016) Functional Magnetic Resonance Imaging Neurofeedback-guided Motor Imagery Training and Motor Training for Parkinson's Disease: Randomized Trial. *Frontiers in Behavioral Neuroscience* 10: 111.
25. Zich, C.; et al. (2017) High-Intensity Chronic Stroke Motor Imagery Neurofeedback Training at Home: Three Case Reports. *Clinical EEG and Neuroscience* 48 (6): 403–412. doi:10.1177/1550059417717398
26. Emmert, K.; et al. (2017) Active Pain Coping Is Associated with the Response in Real-Time fMRI Neurofeedback During Pain. *Brain Imaging Behavior* 11 (3): 712–721.
27. Haller, S.; Birbaumer, N.; and Veit, R. (2010) Real-Time fMRI Feedback Training May Improve Chronic Tinnitus. *European Radiology* 20 (3): 696–703; and Miller, R.; et al. (2016) Slow Cortical Potential Neurofeedback in Chronic Tinnitus Therapy: A Case Report. *Applied Psychophysiology and Biofeedback* 41 (2): 225–249.

28. De Charms, R. C.; Maeda, F.; Glover, G. H.; Ludlow, D.; Pauly, J. M.; and Soneji, D.; et al. (2005) Control over Brain Activation and Pain Learned by Using Real-Time Functional MRI. *Proceedings of the National Academy of Sciences of the United States of America* 102 (51): 18626–18631.
29. Ruiz, S.; Lee, S.; Soekadar, S. R.; Caria, A. R.; and Kircher, T.; et al. (2013) Acquired Self-Control of Insula Cortex Modulates Emotion Recognition and Brain Network Connectivity in Schizophrenia. *Human Brain Mapping* 34 (1): 200–212.
30. Linden, D. E.; Habes, I.; Johnston, S. J.; Linden, S.; Tatineni, R.; and Subramanian, L.; et al. (2012) Real-Time Self-Regulation of Emotion Networks in Patients with Depression. *PLoS One* 7 (6): e38115.
31. Papoutsi, M.; et al. (2017) Stimulating Neural Plasticity with Real-Time fMRI Neurofeedback in Huntington's Disease: a Proof of Concept Study. *Human Brain Mapping*, 39 (3): 1339–1353.
32. Hammond, C. D. (2005) Neurofeedback with Anxiety and Affective Disorders. *Child and Adolescent Psychiatric Clinics of North America* 14: 105–123.
33. Peniston, E. G.; and Kulkosky, P. J. (1991) Alpha-Theta Brainwave Neurofeedback Therapy for Vietnam Veterans with Combat-Related Posttraumatic Stress Disorder. *Medical Psychotherapy: An International Journal* 4: 47–60.
34. Hurt, E.; Arnold, L. E.; and Lofthouse, N. (2014) Quantitative EEG Neurofeedback for the Treatment of Pediatric Attention-Deficit/Hyperactivity Disorder, Autism Spectrum Disorders, Learning Disorders, and Epilepsy. *Journal of the American Academy of Child and Adolescent Psychiatry* 55: 1090–1091.
35. Cortese, S.; et al. (2016) Neurofeedback for Attention-Deficit/Hyperactivity Disorder: Meta-Analysis of Clinical and Neuropsychological Outcomes from Randomized Controlled Trials. *Journal of the American Academy of Child and Adolescent Psychiatry* 55: 444–455.
36. Mollohan, S. My Journey with DBS: Wired Up, Tuned In, and Turned On. The Michael J. Fox Foundation. September 8, 2016. https://www.michaeljfox.org/foundation/news-detail.php?my-journey-with-dbs-wired-up-tuned-in-and-turned-on
37. Tarsy, D.; Vitek, J. L.; Starr, P. A.; and Okun, M. S.; eds. (2008) *Deep Brain Stimulation in Neurological and Psychiatric Disorders.* Totowa, NJ: Humana Press.
38. Shah, D. B.; et al. (2008) Functional Neurosurgery in the Treatment of Severe Obsessive Compulsive Disorder and Major Depression: Overview of Disease Circuits and Therapeutic Targeting for the Clinician. *Psychiatry* 5: 24–33.
39. Pool, J. L. (1954) Psychosurgery of Older People. *Journal of American Geriatric Association* 2: 456–465.
40. Heath, R. G. (1963) Electrical Self-Stimulation of the Brain in Man. *American Journal of Psychiatry* 120: 571–577.
41. Herrington, T. M.; Cheng, J. J.; and Eskandar, E. N. (2016) Mechanisms of Deep Brain Stimulation. *Journal of Neurophysiology* 115: 19–38.
42. Tröster, A. I.; McTaggart, A. B.; and Ines, A. H. Neuropsychological Issues in Deep Brain Stimulation of Neurological and Psychiatric Disorders. Chapter 21 in

Denys, D.; Feenstra, M.; and Schuurman, R.; eds. (2012) *Deep Brain Stimulation: A New Frontier in Psychiatry*. New York, NY: Springer, 399–452.

43. Hariz, M. History of "Psychiatric" Deep Brain Stimulation: A Critical Appraisal. Chapter 26 in Denys, D.; Feenstra, M.; and Schuurman, R.; eds. (2012) *Deep Brain Stimulation: A New Frontier in Psychiatry*. New York, NY: Springer.
44. Caruso, J. P.; and Sheehan, J. P. (2017) Psychosurgery, Ethics, and Media: A History of Walter Freeman and Lobotomy. *Neurosurgical Focus* 43: E6.
45. Walter Freeman: The Father of the Lobotomy. MedicalBag.com May 21, 2015. https://www.medicalbag.com/despicable-doctors/walter-freeman-the-father-of-the-lobotomy/article/472966/
46. Moan, C. E.; and Heath, R. G. (1972) Septal Stimulation for the Initiation of Heterosexual Behavior in a Homosexual Male. *Journal of Behavior Therapy and Experimental Psychiatry* 3 (1): 23–26, IN1, 27–30.
47. Baumeister, R. F. (2000) The Tulane Electrical Brain Stimulation Program: A Historical Case Study in Medical Ethics. *Journal of the History of the Neurosciences* 9: 262–278.
48. Moan, C. E.; and Heath, R.G. (1972) Septal Stimulation for the Initiation of Heterosexual Behavior in a Homosexual Male. *Journal of Behavior Therapy and Experimental Psychiatry* 3 (1):23-30, 24.
49. Ibid., 25.
50. Ibid., 25.
51. Ibid., 26.
52. Ibid., 28.
53. Baumeister, R. F. (2000) The Tulane Electrical Brain Stimulation Program: A Historical Case Study in Medical Ethics. *Journal of the History of the Neurosciences* 9: 262–278.
54. Penfield, W. (1958) Some Mechanisms of Consciousness Discovered During Electrical Stimulation of the Brain. *Proceedings of the National Academy of Sciences of the United States of America* 44: 51–66; 57; 60.
55. Ibid., 60.
56. Penfield, W. (1958) Some Mechanisms of Consciousness Discovered During Electrical Stimulation of the Brain. *Proceedings of the National Academy of Sciences of the United States of America* 44: 51–66.
57. Ibid., 53.
58. Ibid., 54
59. Ibid.
60. Ibid., 57.
61. Ibid., 57.
62. Ibid.
63. Ibid.
64. Ibid., 65.
65. Ekstrom, A. D.; Kahana, M. J.; Caplan, J. B.; Fields, T.A.; Isham, E. A.; Newman,

E. L.; and Fried, I. (2003) Cellular Networks Underlying Human Spatial Navigation. *Nature* 425: 184–188.

66. Aghajan, M.; et al. (2005) Theta Oscillations in the Human Medial Temporal Lobe during Real-World Ambulatory Movement. *Current Biology* 27 (24): 3743–3751.
67. Quiroga, R. Q.; Reddy, L.; Kreiman, G.; Koch, C.; and Fried, I. (2005) Invariant Visual Representation by Single Neurons in the Human Brain. *Nature* 435 (7045): 1102–1107.
68. Gelbard-Sagiv, H.; Mukamel, R.; Fried, I.; et al. (2008) Internally Generated Reactivation of Single Neurons in Human Hippocampus During Free Recall. *Science* 322: 96–101; and Mormann, F.; et al. (2017) Scene-Selective Coding by Single Neurons in the Human Parahippocampal Cortex. *Proceedings of the National Academy of Sciences of the United States of America* 114 (5): 1153–1158; and De Falco, E.; et al. (2016) Long-Term Coding of Personal and Universal Associations Underlying the Memory Web in the Human Brain. *Nature Communications* 7: 13408.
69. Titiz, A. S.; et al. (2017) Theta-Burst Microstimulation in the Human Entorhinal Area Improves Memory Specificity. *eLife Sciences* 6: e29515.
70. Reinhart, R. M. G.; and Nguyen, J. A. (2019) Working Memory Revived in Older Adults by Synchronizing Rhythmic Brain Circuits. *Nature Neuroscience* 22: 820–827.
71. Keith, A. Convicted of Killing Two when He Was 17, District Man Is Sentenced to 65 Years in Prison. *Washington Post*, January 4, 2019. https://www.washingtonpost.com/local/public-safety/convicted-of-killing-two-when-he-was-17-district-man-is-sentenced-to-65-years-in-prison/2019/01/04/6e20e5b2-1051-11e9-84f0c-d58c33d6c8c7_story.html?noredirect=on&utm_term=.768f402bce10
72. Fields, R. D. (2019) The Roots of Human Aggression. *Scientific American* 320: 64–71.
73. Fields, R. D. (2015) *Why We Snap: Understanding the Rage Circuit in Your Brain.* New York, NY: Dutton/Penguin.
74. Gilam, G.; et al. (2018) Attenuating Anger and Aggression with Neuromodulation of the vmPFC: A Simultaneous tDCS-fMRI Study. *Cortex* 109: 156–170b.
75. Ibid., 170.
76. Fields, R. D. Amping Up Brain Function: Transcranial Stimulation Shows Promise in Speeding Up Learning, *Scientific American News*, November 25, 2011. https://www.scientificamerican.com/article/amping-up-brain-function/
77. Fields, R. D. (2012) Regulation of Myelination by Functional Activity. Chapter 45 in Kettenmann, H.; and Ransom, B. R.; eds. (2012) *Neuroglia,* third edition. Oxford, UK: Oxford University Press; and Fields, R. D. (2015) A New Mechanism of Nervous System Plasticity: Activity-Dependent Myelination. *Nature Reviews Neuroscience* 16: 756–767.
78. Fields, R. D. (2008) White Matter in Learning, Cognition and Psychiatric Disorders. *Trends in Neurosciences* 31: 361–370.

79. Vöröslakos, M.; et al. (2018) Direct Effects of Transcranial Electric Stimulation on Brain Circuits in Rats and Humans. *Nature Communications* 9 (1): 483.
80. FDA Permits Marketing of Transcranial Magnetic Stimulation for Treatment of Obsessive Compulsive Disorder. U.S. Food and Drug Administration press announcement, August 17, 2018. https://www.fda.gov/news-events/press-announcements/fda-permits-marketing-transcranial-magnetic-stimulation-treatment-obsessive-compulsive-disorder
81. Zandvakili, A.; et al. (2019) Use of Machine Learning in Predicting Clinical Response to Transcranial Magnetic Stimulation in Comorvid Posttraumatic Stress Disorder and Major Depression: A Resting State Electroenchapaography Study. *Journal of Affective Disorders* 252: 47–54.
82. National Intrepid Center of Excellence. https://www.wrnmmc.capmed.mil/NICoE/SitePages/index.aspx. Accessed October 4, 2019.
83. Cahn, B. R.; and Polich, J. (2006) Meditation States and Traits: EEG, ERP, and Neuroimaging Studies. *Psychological Bulletin* 132: 180–211.
84. Braboszcz, C.; Cahn, B. R.; Levy, J.; Fernandez, M.; and Delorme, A. (2017) Increased Gamma Brainwave Amplitude Compared to Control in Three Different Meditation Traditions. *PLoS One* 12 (1): e0170647.
85. Park, M. (2017) Assault Charge Filed after Tweet Sent to a Journalist with Epilepsy. *CNN*, March 20, 2017. https://www.cnn.com/2017/03/17/us/twitter-journalist-strobe-epilepsy/index.html
86. Puglise, N. (2016) What Is Michael Phelps Listening to on His Trademark Olympics Headphones? *The Guardian,* August 8, 2016. https://www.theguardian.com/sport/2016/aug/08/michael-phelps-headphones-music-swimming-olympics-rio
87. Fields, R. D. The Power of Music: Mind Control by Rhythmic Sound. *Scientific American*, October 19, 2012. https://blogs.scientificamerican.com/guest-blog/the-power-of-music-mind-control-by-rhythmic-sound/
88. Fields, R. D. Of Two Minds: Listener Brain Patterns Mirror Those of the Speaker. *Scientific American*, July 27, 2010. https://blogs.scientificamerican.com/guest-blog/of-two-minds-listener-brain-patterns-mirror-those-of-the-speaker/
89. Fields, R. D. (2016) Olympic Gold for Brainwave Performance. BrainFacts.org, August 21, 2016. http://www.brainfacts.org/Brain-Anatomy-and-Function/Cells-and-Circuits/2016/Olympic-Gold-for-Brainwave-Performance


## CHAPTER 12


1. Beggin, Riley. Students Zap Their Brains for a Boost, for Better Or Worse. NPR *All Things Considered*, January 7, 2017. http://www.npr.org/sections/alltechconsidered/2017/01/07/507133313/students-zap-their-brains-for-a-boost-for-better-or-worse. Accessed October 17, 2019.
2. Mind-Controlled Toys: The Next Generation of Christmas Presents? Press release

from Warwick News and Events. December 16, 2016. https://warwick.ac.uk/newsandevents/pressreleases/mind-controlled_toys_the/
3. Finding the Seeds Lying Beyond Our Imagined Futures. Nissan Motor Corporation. May 2017. https://www.nissan-global.com/EN/NRC/FRONTLINES/LUCIAN_GHEORGHE/
4. http://oldconceptcars.com. Accessed October 4, 2019.
5. Lucchesi, N. Nissan Reveals It Is Working on "Brain-to-Vehicle" Technology. *Inverse*, January 3, 2018. https://www.inverse.com/article/39894-nissan-brain-to-vehicle-technology
6. https://ctrl-labs.com. Accessed October 3, 2019.
7. Oxley, T.; et al. (2008) Minimally Invasive Endovascular Stent-Electrode Array for High-Fidelity, Chronic Recordings of Cortical Neural Activity. *Nature Biotechnology* 34: 320–327.
8. Croft, R. J.; et al. (2008) The Effect of Mobile Phone Electromagnetic Fields on the Alpha Rhythm of Human Electroencephalogram. *Bioelectromagnetics* 29: 1–10.
9. Fields, R. D. Mind Control by Cell Phone. *Scientific American News*, May 7, 2008. https://www.scientificamerican.com/article/mind-control-by-cell/
10. Zhang, J.; Sumich, A.; and Wang, G. Y. (2017) Acute Effects of Radiofrequency Electromagnetic Field Emitted by Mobile Phone on Brain Function. *Bioelectromagnetics* 38 (5): 329–338.
11. Baker, K. B.; and Phillips, M. D. (2008) Deep Brain Stimulation Safety: *MRI and Other Electromagnetic Interactions*. Chapter 26 in Tarsy, D.; Vitek, J. L.; Starr, P. A.; and Okun, M. S.; eds. (2008) *Deep Brain Stimulation in Neurological and Psychiatric Disorders*. Totowa, NJ: Humana Press.
12. Fields, R. D. (2004) Making Memories Stick. *Scientific American* 290: 54–61.
13. Fields, R. D.; and Dudek, S. M. (2002) Somatic Action Potentials are Sufficient for Late-Phase LTP-Related Cell Signaling. *Proceedings of the National Academy of Sciences of the United States of America* 99: 3962–3967.
14. Fields, R. D. (2009) *The Other Brain*. New York, NY: Simon and Schuster.
15. Lee, H. S.; et al. (2014) Astrocytes Contribute to Gamma Oscillations and Recognition Memory. *Proceedings of the National Academy of Sciences of the United States of America* 111: E3343–3352.
16. Fields, R. D. (2008) White Matter Matters. *Scientific American* 298: 42–49.
17. Zatorre, R. J.; Fields, R. D.; and Johansen-Berg, H. (2012) Plasticity in Gray and White: Neuroimaging Changes in Brain Structure During Learning. *Nature Neuroscience* 15: 528–536.
18. Dutta, D. J.; et al. (2018) Regulation of Myelin Structure and Conduction Velocity by Perinodal Astrocytes. *Proceedings of the National Academy of Sciences of the United States of America* 115: 11832–11837; and Fields, R. D.; and Dutta, D. J. (2019) Treadmilling Model for Plasticity of the Myelin Sheath. *Trends in Neurosciences* 42: 443–447.
19. Fields, R. D. (2006) Myelination: An Overlooked Mechanism of Synaptic Plasticity? *The Neuroscientist* 11: 528–531; and Fields, R. D. (2015) A New Mechanism

of Nervous System Plasticity: Activity-Dependent Myelination. *Nature Reviews Neuroscience* 16: 756–767.

20. Pajevic, S.; Basser, P. J.; and Fields, R. D. (2015) Role of Myelin Plasticity in Oscillations and Synchrony of Neuronal Activity. *Neuroscience* 276: 135–147.
21. ENIAC. Computer History Museum. https://www.computerhistory.org/revolution/birth-of-the-computer/4/78. Accessed October 4, 2019.
22. Egel, B.; Moleski, V.; and Stanton, S. (2019) Witness: Davis Police Officer was Ambushed by Bystander Who Then Targeted "Random People." *Sacramento Bee*, January 11, 2019.
23. Helsel, P. Gunman Who Killed California Officer Left Note Saying He Was Hit by "Sonic Waves." *NBC U.S. News*, January 12, 2019. https://www.nbcnews.com/news/us-news/gunman-who-killed-california-officer-left-note-saying-he-was-n958071
24. Smith, C. On the Need for New Criteria of Diagnosis of Psychosis in the Light of Mind Invasive Technology. *Global Research*, October 18, 2007. https://www.globalresearch.ca/on-the-need-for-new-criteria-of-diagnosis-of-psychosis-in-the-light-of-mind-invasive-technology/7123
25. Ibid.
26. Targeted Justice. https://www.targetedjustice.com. Accessed October 4, 2019.
27. Smith, C. (2007) On the Need for New Criteria of Diagnosis of Psychosis in the Light of Mind Invasive Technology. Global Research, October 18, 2007. https://www.globalresearch.ca/on-the-need-for-new-criteria-of-diagnosis-of-psychosis-in-the-light-of-mind-invasive-technology/7123. Accessed October 8, 2019.
28. Attacks on U.S. Diplomats in Cuba: Response and Oversight. Senate hearing. January 9, 2018. https://www.foreign.senate.gov/hearings/watch?hearingid=6204DFB9-5056-A066-6016-5EF0A1820F88
29. Fields, R. D. "Sonic Weapon Attacks" on U.S. Embassy Don't Add Up—for Anyone. *Scientific American News*, February 16, 2018. https://www.scientificamerican.com/article/ldquo-sonic-weapon-attacks-rdquo-on-u-s-embassy-don-rsquo-t-add-up-mdash-for-anyone/
30. Attacks on U.S. Diplomats in Cuba: Response and Oversight. Senate hearing. January 9, 2018. https://www.foreign.senate.gov/hearings/watch?hearingid=6204DFB9-5056-A066-6016-5EF0A1820F88
31. Delgado, J. M. R. (1969) *Physical Control of the Mind: Toward a Psychocivilized Society*. New York, NY: Harper and Row, e-book location 1506 of 4043.
32. Ibid., eBook location 1661 of 4043.
33. Ibid., eBook location 3022 of 4043.
34. Malech, R. G. Patent #3951134. Apparatus and Method for Remotely Monitoring and Altering Brainwaves. August 5, 1974. https://patents.google.com/patent/US3951134A/en
35. Pilkington, E. Human Rights Body Calls on US School to Ban Electric Shocks on Children. *The Guardian*, December 18, 2018. https://www.theguardian.com/us-news/2018/dec/18/judge-rotenberg-center-electric-shocks-ban-inter-american-commission-human-rights

36. Footage of Judge Rotenberg Center Torturing a Person with a Disability Aired in Court. YouTube, uploaded by Lydia X. Z. Brown, April 10, 2012. https://www.youtube.com/watch?v=aAj9W0ntUMI&feature=youtu.be
37. Miller, G. Judge Rotenberg Educational Center: Please Stop Painful Electric Shocks on Your Students. Petition to Speaker of the House Robert A. DeLeo and 11 others. https://www.change.org/p/judge-rotenberg-educational-center-please-stop-painful-electric-shocks-on-your-students. Accessed October 8, 2019.
38. Changeux, J. P. (1998) Advances in Neuroscience May Threaten Human Rights. *Nature* 391: 316.
39. Plenary sessions/Europarliament, 1999. http://www.europarl.europa.eu/sides/getDoc.do?pubRef=-%2F%2FEP%2F%2FTEXT+REPORT+A4-1999-0005+0+DOC+XML+V0%2F%2FEN. Accessed October 8, 2019.
40. Space Preservation Act, HR 2977. October 2, 2001. https://fas.org/sgp/congress/2001/hr2977.html
41. Neuralink Launch Event. YouTube, uploaded by Neuralink, July 16, 2019. https://www.youtube.com/watch?v=r-vbh3t7WVI&feature=youtu.be&t=5405.
42. https://en.wikipedia.org/wiki/Thomas_Reardon
43. Hughes, V. Were the First Artists Mostly Women? *National Geographic,* October 9, 2013. https://news.nationalgeographic.com/news/2013/10/131008-women-handprints-oldest-neolithic-cave-art/

## IMAGE CREDITS

Figure 1: Photo by the author, with permission of Fredrich-Schiller-Universität Jena, Germany.

Figure 2: From Berger, H. (1907) Über die körperlichen Äußerungen psychischer Zustände. Experimentelle Beiträge zur Lehre von der Blutzirkulation in der Schädelhöhle des Menschen. II. Teil. Jena: Gustav Fischer.

Figure 3: Photo by the author.

Figure 4: From Berger, H. (1910) Untersuchungen über die Temperatur des Gehirns. Jena: Gustav Fischer, 40.

Figure 5: EEG trace (*bottom*) from Berger, H. (1929) Über das Elektrenkephalogramm des Menchen. *Archives für Psychiatrie* 87: 527–570.

Figure 6: From Gloor, P. (1969) *Hans Berger on the Electroencephalogram of Man.* New York, NY: Elsevier Publishing Company.

Figure 8: Photo by the author.

Figure 9: Photo by the author.

Figure 11: Photo by Rama: https://commons.wikimedia.org/wiki/File:Instruments-de-mesure-p1010624.jpg.

Figure 12: From Beck, A. (1891) Oznaczenie lokalizacji w mózgu I rdzeniu za pomoca zjawisk elektrycznych [The determination of localization in the brain and spinal cord with the aid of electrical phenomena]. Translated from Polish by W. A. Binek and J. S. Barlow. Warsaw: Polish Scientific Publishers, 1973.

Figure 14: Photo by the author, from *The Annals of Electricity Magnetism and Chemistry and Guardian of Experimental Science* 2: 355–360.

Figure 15: From M. Turpin (1838) *The Annals of Electricity, Magnetism, and Chemistry and Guardian of Experimental Science* 2: 355–360.

Figure 16: See Seeck, M.; et al. (2017) The Standardized EEG Electrode Array of the IFCN. *Clinical Neurophysiology*, 128: 2070–2077. Upper Image from Pixabay, creative commons license. Lower image from He, Y.; et al (2018) A Mobile Brain-Body Imaging Dataset Recorded During Treadmill Walking with a Brain-Computer Interface. *Scientific Data* 5: Article Number 1800074.

Figure 17: Courtesy of Jessica Ure and Robin Bernhard of the Virginia Center for Neurofeedback, via Jay Gunkelman of Brain Science International.

Figure 18: From Xiang, Z.; Liu, J.; Lee, Chengkuo. (2016) A Flexible Three-Dimensional Electrode Mesh: An Enabling Technology for Wireless Brain-Computer Interface Prostheses. *Microsystems and Nanoengineering* volume 2, article number 16012.

Figure 19: Image by the author.

Figure 20: *Top*: Photo of the author, courtesy of Joseph Snider. *Bottom*: Photo by the author.

Figure 21: From Xi, C.; et al. (2018) *PLoS ONE* 13: e0199120.

Figure 22: From Soplata, A.; et al. (2017) Thalamocortical Control of Propofol Phase-Amplitude Coupling. *PLoS Computational Biology* 13: e1005879.

Figure 23: From Shanechi, M. M.; et al. (2013) A Brain-Machine Interface for Control of Medically-Induced Coma. *PLoS Computational Biology* 9:31003284.

Figure 24: Image by the author.

Figure 25: Image by the author.

Figure 26: With permission of Yale Events and Activities Photographs (RU 690). Manuscripts and Archives, Yale University Library.

Figure 27: From Vansteensel, M. J.; et al. Fully Implanted Brain-Computer Interface in a Locked-In Patient with ALS. *The New England Journal of Medicine*, 375:2060–2066 Copyright © (2016) Massachusetts Medical Society. Reprinted with permission from Massachusetts Medical Society.

Figure 28: Courtesy of Dr. Erik Aarnoutse and the UM Utrecht Press Office.

Figure 29: Courtesy of Blackrock Microsystems, LLC.

Figure 30: Courtesy of Ian Burkhart.

Figure 31: From Rao, R. P. N.; et al. (2014) A Direct Brain-to-Brain Interface in Humans. *PLoS One* 9: e111332. doi:10.1371/journal.pone.0111332.

Figure 32: From Li, G.; and Zhang, D. (2016) Brain-Computer Interface Controlled Cyborg: Establishing a Functional Information Transfer Pathway from Human Brain to Cockroach Brain. *PLoS One* 11 (3): e0150667.

Figure 33: Photo by Walej: https://commons.wikimedia.org/wiki/File:FNIRS_head_Hitachi_ETG4000_2.jpg.

Figure 34: Original image by Allan Aifo, aboutmodafinil.com: https://www.flickr.com/photos/125992663@N02/14601014705/

Figure 35: From Garner, J.; et al. (2013) Functional MRI in the Investigation of

Blast-Related Traumatic Brain Injury. *Frontiers in Neurology.* https://doi.org/10.3389/fneur.2013.00016.

Figure 36: From Heine, L.; et al. (2012) Resting State Networks and Consciousness. *Frontiers in Psychology.* https://doi.org/10.3389/fpsyg.2012.00295

Figure 37: Der Lange: http://commons.wikimedia.org/w/index.php?title=File:Spike-waves.png.

Figure 38: Image by Hellerhoff: https://commons.wikimedia.org/wiki/File:Tiefe_Hirnstimulation_-_Sonden_RoeSchaedel_ap.jpg.

Figure 39: US Army photo by David McNally: https://www.arl.army.mil/www/default.cfm?article=3214

Figure 40: Photo by Eric Wassermann, NINDS.

Figure 41: Courtesy of Dr. Thomas Oxley, with permission of Mt. Sinai Hospital.

Figure 42: *Top*: Courtesy of Dr. Thomas Oxley and Mt. Sinai Hospital. *Bottom*: Courtesy of Dr. Thomas Oxley and Synchron Company.

Figure 43: From Musk, E.; and Neuralink. (2019) An Integrated Brain-Machine Interface Platform with Thousands of Channels. *BioRxiv.* http://dx.doi.org/10.1101/703801

# INDEX

## A

## B

## C

## D

## E

## O

# ABOUT THE AUTHOR

R. DOUGLAS FIELDS, PHD, is a neuroscientist, an international authority on nervous system development and plasticity, and an American Association for the Advancement of Science Fellow. He received advanced degrees from UC Berkeley, San Jose State University, and UC San Diego, and he held postdoctoral fellowships at Stanford and Yale universities before joining the National Institutes of Health in Bethesda, Maryland. Fields is also an adjunct professor at the Neuroscience and Cognitive Science Program at the University of Maryland, College Park. He has published over 150 articles in scientific journals and books from his experimental research into how the brain is modified by experience and the cellular mechanisms of memory. His scientific research has been featured internationally in newspapers, magazines, radio, and television, including *National Geographic*, *ABC News Nightline*, and *NPR Morning Edition*. His research on nervous system plasticity involving non-neuronal cells (glia) in white matter regions of the brain is recognized as pioneering a new non-synaptic mechanism of nervous system plasticity. In 2004, he founded the scientific journal *Neuron Glia Biology* to advance research on interactions between neurons and glia, and he serves on the editorial boards of several neuroscience journals.

In addition to his scientific research, Fields is also the author of numerous books and magazine articles about the brain for the general reader, including *The Other Brain*, about brain cells that communicate without using electricity (glia), and *Why We Snap*, about the neuroscience of sudden aggression, as well as numerous articles in popular magazines including *Outside*, *The Washington Post Magazine*, *Scientific American* and *Scientific American Mind*, *Time*, *Undark*, *Quanta*, and online columns for *Huffington Post*, *Psychology Today*, *Scientific American*, *Society for Neuroscience*, *BrainFacts*, and others.